Advances in Intelligent Systems and Computing

Volume 235

Series Editor

Janusz Kacprzyk, Warsaw, Poland

For further volumes:
http://www.springer.com/series/11156

Sabu M. Thampi · Ajith Abraham
Sankar Kumar Pal · Juan Manuel Corchado Rodriguez
Editors

Recent Advances in Intelligent Informatics

Proceedings of the Second International Symposium on Intelligent Informatics (ISI'13), August 23–24, 2013, Mysore, India

 Springer

Editors
Sabu M. Thampi
Indian Inst. of Information Technology and
    Management - Kerala (IIITM-K)
Kerala
India

Ajith Abraham
Machine Intelligence Research Labs
    (MIR Labs)
Auburn
USA

Sankar Kumar Pal
Indian Statistical Institute
Kolkata
India

Juan Manuel Corchado Rodriguez
Department of Computer Science
School of Science
University of Salamanca
Salamanca
Spain

ISSN 2194-5357   ISSN 2194-5365   (electronic)
ISBN 978-3-319-01777-8   ISBN 978-3-319-01778-5   (eBook)
DOI 10.1007/978-3-319-01778-5
Springer Cham Heidelberg New York Dordrecht London

Library of Congress Control Number: 2013945169

Preface

This Edited Volume contains a selection of refereed and revised papers originally presented at the Second International Symposium on Intelligent Informatics (ISI-2013), August 23–24, 2013, Mysore, India. ISI-2013 provided an international forum for sharing original research results and practical development experiences among experts in the emerging areas of Intelligent Informatics. This edition was co-located with Second International Conference on Advances in Computing, Communications and Informatics (ICACCI-2013).

Credit for the quality of the conference proceedings goes first and foremost to the authors. They contributed a great deal of effort and creativity to produce this work, and we are very thankful that they chose ISI-2013 as the place to present it. All the authors who submitted papers, both accepted and rejected, are responsible for keeping the ISI program vital. The program committee received 126 full submissions from 12 countries: India, Egypt, Malaysia, Iran, Saudi Arabia, USA, Albania, Algeria, Bangladesh, Kuwait, Australia and P.R. China and selected 47 of them for presentation. Among 47 presentations, 39 were regular talks and the remaining 8 were short talks. Each submission was reviewed by at least three expert referees.

The success of such an event is mainly due to the hard work and dedication of a number of people and the collaboration of several institutions. We are grateful to the members of the program committee for reviewing and selecting papers in a very short period of time. Their comments helped the authors improve the final version of their papers. Many thanks to all the Chairs and their involvement and support have added greatly to the quality of the symposium. We wish to thank also all the members of the Steering Committee, whose work and commitment were invaluable. The EDAS conference system proved very helpful during the submission, review, and editing phases.

We thank the Management of Sri Jayachamarajendra College of Engineering (SJCE) for hosting the conference. Sincere thanks to B.G. Sangameshwara, Principal, SJCE and Manjunath Aradhya, Local Arrangements Chair for their valuable suggestions and encouragement.

We wish to express our sincere thanks to Thomas Ditzinger, Senior Editor, Engineering/Applied Sciences Springer-Verlag for his help and cooperation.

August 2013   Sabu M. Thampi
                                                                                   Ajith Abraham
                                                                              Sankar Kumar Pal
                                                         Juan Manuel Corchado Rodriguez

ISI-2013 Conference Committee

ICACCI Steering Committee

John F. Buford	Avaya Labs Research, USA (Chair)
Xavier Fernando	Ryerson University, Canada
Rama Govindaraju	Google, USA
Peter Mueller	IBM Zurich Research Laboratory, Switzerland
Shambhu Upadhyaya	University at Buffalo, The State University of New York, USA
Raj Kumar Buyya	University of Melbourne, Australia
Chandrasekaran K.	NITK, India
Jaime Lloret Mauri	Polytechnic University of Valencia, Spain
Dapeng Oliver Wu	University of Florida, USA
David Meyer	Cisco Systems, USA
Ajith Abraham	MIR Labs, USA
Sattar B. Sadkhan AL Maliky (Chairman, IEEE Iraq Section)	University of Babylon, Iraq
Andreas Riener	University of Linz, Austria
Axel Sikora	University of Applied Sciences Offenburg, Germany
Bharat Jayaraman	University at Buffalo, The State University of New York, USA
Deepak Garg, Chair	IEEE Computer Society Chapter, IEEE India Council & Thapar University, India
Dilip Krishnaswamy	Qualcomm Research Center, San Diego CA, USA
Narayan C. Debnath	Winona State University, Winona, USA
Selwyn Piramuthu	University of Florida, USA
Raghuram Krishnapuram	IBM Research - India

Theodore Stergiou	Intracom Telecom, Greece
V.N. Venkatakrishnan	University of Illinois at Chicago USA
Ananthram Swami	Army Research Laboratory, USA
Venugopal K.R.	Bangalore University, India
Adel M. Alimi	University of Sfax, Tunisia
Madhukar Pitke	Nicheken Technologies, India
Tamer ElBatt	Cairo University and Nile University, Egypt
Francesco Masulli	University of Genoa, Italy
Manish Parashar	Rutgers, The State University of New Jersey, USA
Sabu M. Thampi	IIITM-K, India
Anand R. Prasad	NEC, Japan
P. Nagabhushan	University of Mysore, India
Suash Deb	President, Intl. Neural Network Society (INNS), India Regional Chapter

Organising Committee (SJCE)

His Holiness Jagadguru Sri Shivarathri Deshikendra Mahaswamiji
 (*Chief Patron*)
Sri B.N. Betkerur - Executive Secretary, JSS Mahavidyapeetha, Mysore
 (*Patron*)
M.H. Dhananjaya- Director-Technical Education Division, JSS
 Mahavidyapeetha, Mysore (*Patron*)
B. G. Sangameshwara - Principal, SJCE, Mysore (*Organizing Chair*)
V.N. Manjunath Aradhya, Dept. of MCA, SJCE, Mysore
 (*Organizing Secretary*)

Advisory Committee

K. Chidananda Gowda (Former Vice Chancellor)	Kuvempu University, Shimoga
Y.N. Srikant	Dept. of CSA, IISc, Bangalore
Mahadev Prasanna	Dept. of EE, IIT-Guwahati, Guwahati
N.P. Gopalan	Dept. of Comp. Applications, NIT-Trichy
G. Hemantha Kumar	DoS in Computer Science, University of Mysore, Mysore
T.N. Nagabhushan	Principal, JSSATE, Noida
Mritunjaya V. Latte	Principal JSSATE, Bangalore
C.R. Venugopal	Head – ECE, SJCE, Mysore
C.N. Ravikumar	Head- CSE, SJCE, Mysore
S.K. Padma	Head – ISE, SJCE, Mysore
S.K. Niranjan	Dept. of MCA, SJCE, Mysore
V. Vijaya Kumar	Dean, GIET, Rajahmundry
Sudarshan Iyengar	IIT Ropar, Punjab

Technical Program Committee

Honorary Chair

Sankar Kumar Pal   Indian Statistical Institute, Kolkata, India

General Chairs

Ajith Abraham   MIR Labs, USA
Janusz Kacprzyk   Polish Academy of Sciences, Poland
Adel M. Alimi   University of Sfax, Tunisia

Program Chairs

Juan Manuel Corchado
   Rodriguez   University of Salamanca, Spain
Alexander Gelbukh   Mexican Academy of Science, Mexico
Kuan-Ching Li   Providence University, Taiwan

Publication Chair

Sabu M. Thampi   IIITM-K, India

Technical Programme Committee Members

Aboul Ella Hassanien	University of Cairo, Egypt
Ajay Singh	Multimedia University, Malaysia
Akihiro Fujihara	Kwansei Gakuin University, Japan
Akimitsu Kanzaki	Osaka University, Japan
Alex James	Nazarbayev University, Kazakhstan
Amudha J.	Anrita Vishwa Vidyapeetham, India
Anca Daniela Ionita	University Politehnica of Bucharest, Romania
Angelos Michalas	Tecnological Educational Institute of Western Macedonia, Greece
Antonio LaTorre	Universidad Politécnica de Madrid, Spain
Arpan Kar	Indian Institute of Management, Rohtak, India
Atsushi Takeda	Tohoku Gakuin University, Japan
B.H. Shekar	Mangalore University, India
Bhushan Trivedi	GLS Institute of Computer Technology, India
Chien-Fu Cheng	Tamkang University, Taiwan
Ciprian Dobre	University Politehnica of Bucharest, Romania
Clive King	Oracle - Penglais, USA

Additional Reviewers

Aroua Hedhili	SOIE, National School of Computer Studies (ENSI), Tunisia
Fernando de la Prieta	University of Salamanca, Spain
Gaurav Raj	Punjab Technical University, Kapurthala, India
Hanumantha Raju	Shavige Malleshwara Hills, Kumaraswamy Layout, Bangalore, India
Imtiez Fliss	ENSI, Tunisia
Kamyar Mehranzamir	UTM, Malaysia
Mohammad Hasanzadeh	Amirkabir University of Technology, Iran
Muhammad Rafi	FAST-NU, Pakistan
Praveen Kumar	Kuvempu University, India
Roozbeh Zarei	Victoria University, Australia
Shanmugapriya D.	Avinashilingam Institute for Home Sc. and Higher Education, India
Sowmya Kamath	National Institute of Technology, Surathkal, India
Vaidehi Nedu	Dayananda Sagar College of Engineering, India
Yogita Thakran	Indian Institute of Technology Roorkee, India

Contents

Pattern Recognition, Signal and Image Processing

Data Mining, Clustering and Intelligent Information Systems

Multi Agent Systems

Computer Networks and Distributed Systems

Work-in-Progress

A Comparative Study on Feature Selection for Retinal Vessel Segmentation Using Ant Colony System

Ahmed H. Asad, Ahmad Taher Azar, and Aboul Ella Otifey Hassaanien

Abstract. The diabetic retinopathy disease spreads diabetes on the retina vessels thus they lose blood supply that causes blindness in short time, so early detection of diabetes prevents blindness in more than 50% of cases. The early detection can be achieved by automatic segmentation of retinal blood vessels which is two-class classification problem. Features selection is an essential step in successful data classification since it reduces the data dimensionality by removing redundant features, thus minimizing the classification complexity, time and maximizes its accuracy. In this paper, comparative study on four features selection heuristics is performed to select the best relevant features set from features vector consists of fourteen features that are computed for each pixel in the field of view of retinal image in the DRIVE database. The comparison is assessed in terms of sensitivity, specificity and accuracy of the recommended features set by each heuristic when used with the ant colony system algorithm. The results indicated that the recommended features set by the relief heuristic gives the best performance with sensitivity of 75.84%, specificity of 93.88% and accuracy of 91.55%.

Keywords: Retinal Blood Vessels, Feature selection, Segmentation, Features Extraction, Ant Colony System, computer aided diagnosis (CAD).

Ahmed H. Asad
Institute of Statistical Studies and Researches, CS Department, Cairo University
e-mail: ah_assad@hotmail.com

Ahmad Taher Azar
Faculty of computers and information, Benha University, Egypt;
Scientific Research Group in Egypt (SRGE)
e-mail: ahmad_T_azar@ieee.org

Aboul Ella Otifey Hassaanien
Faculty of Computer and Information, IT Department, Cairo University
e-mail: aboitcairo@gmail.com

S.M. Thampi et al. (eds.), *Recent Advances in Intelligent Informatics*,
Advances in Intelligent Systems and Computing 235,
DOI: 10.1007/978-3-319-01778-5_1, © Springer International Publishing Switzerland 2014

1   Introduction

Retinal blood vessels are important structures in many ophthalmological images. The automated extraction of blood vessels in retinal images is an important step in computer aided diagnosis (CAD) and treatment of diabetic retinopathy (Asaad et al. 2012; Vijayakumari and Suriyanarayanan 2012; Serrarbassa et al. 2008; Soares et al. 2006, Mendonca and Campilho 2006, Staal et al. 2004), hypertension (Leung et al. 2004), glaucoma (Wang et al. 2006), obesity (Mitchell et al. 2005), arteriosclerosis and retinal artery occlusion, etc. These diseases often result in changes on reflectivity, bifurcations, tortuosity as well as other patterns of blood vessels. Hence, analyzing vessel features gives new insights to diagnose the corresponding disease early. For example, vessel tortuosity characterizes hypertension retinopathy (Foracchia et al. 2001) and diabetic retinopathy usually leads to neovascularization. The latter is one of the most common causes of vision defects or even blindness worldwide (Morello 2007). If abnormal signs of diabetic retinopathy could be detected early, effective treatment before their initial onset can be performed. For the last two decades, retinal blood vessels segmentation attracts a lot of research in the medical image processing area since it is the critical components of circulatory blood vessel analysis systems (Fraz et al. 2012). Also, retinal blood vessels segmentation is the core stage in automated registration of two retinal blood vessels images of a certain patient to follow and diagnose his disease progress at different periods of time (Khan et al. 2011). Screening is vital to preventing visual loss from diabetes because retinopathy is often asymptomatic early in the course of the disease (Goatman et al. 2011; Verma et al. 2011; Jones and Edwards 2010; Rodgers et al. 2009; Farley et al. 2008; Bloomgarden 2007; Chew 2006; Sinclair 2006). If the retinopathy is detected in its early stages, blindness can be prevented in more than 50% of the cases (Xu et al. 2006; Jin et al. 2005). Manual segmentation of retinal blood vessels is a long and tedious task which also requires training and skill. It is commonly accepted by the medical community that automatic quantification of retinal vessels is the first step in the development of a computer-assisted diagnostic system for ophthalmic disorders (Fraz et al. 2012). Computer-aided analysis of retinal image should play a central role in diagnostic procedures, and extensive research efforts have been devoted to automating this process. Nevertheless, reliable vessel extraction encounters several challenges (You et al. 2001): ''(1) The blood vessels have a wide range of widths and diffident tortuosity. (2) Various structures appear in retinal image, including the optic disc, fovea, exudates and pigment epithelium changes, which severely disrupt the automatic vessel extraction. (3) The narrow vessels with various local surroundings may appear as some elongated and disjoint spots, which are usually lost''.

In this paper, an improvement of previous work (Asaad et al. 2012) is done by comparative study on four feature selection heuristics. The aim is to select the best relevant features set from features vector consists of fourteen features that are computed for each pixel in the field of view (FOV) of retinal image in the DRIVE database (Staal et al. 2004). The four feature selection algorithms are correlation-based features selection (CFS), fisher score, gini index and relief. They are compared in terms of sensitivity, specificity and accuracy of their recommended features sets when used with colony system (ACS) algorithm (Dorigo and Gambardella 1997). The rest of this paper is organized as follows: Section 2 describes in more details the used database and features. Section 3 presents the samples selection while section 4 presents the four features selection heuristics in more details. Section 5 reports the results and experimental evaluations of proposed approach. Finally in Section 6, conclusion and directions for future research are presented.

2 Methods

2.1 Retinal Image Database

Digital Retinal Images for Vessel Extraction (DRIVE) database (Staal et al. 2004) is used. It is available for free to scientific research on retinal vasculature consisting of a total of 40 color fundus photographs; seven of them are abnormal pathology cases. The photographs were obtained from a diabetic retinopathy screening program in the Netherlands. The screening population consisted of 453 subjects between 31 and 86 years of age. Each image has been JPEG compressed, which is common practice in screening programs. Of the 40 images in the database, 7 contain pathology, namely exudates, hemorrhages and pigment epithelium changes. The images were acquired using a Canon CR5 non-mydriatic 3CCD camera with FOV equal to forty-five degree. Each image resolution is 584*565 pixels with eight bits per color channel. The set of 40 images was divided into training set of twenty images and testing set of the other half. Inside each set, for each image there is circular FOV mask of diameter that is approximately 540 pixels. Inside training set, for each image one manual segmentation by an ophthalmological expert. Inside testing set, for each image two manual segmentations by two different observers where the first observer segmentation is accepted as the ground-truth for performance evaluation.

2.2 Features

The selected features are the gray-level of green channel of RGB retinal image, group of five features based on gray-level $(f_1, f_2, f_3, f_4, f_5)$ and group of eight features based on Hu moment-invariants $(Hu_1, Hu_2, Hu_3, Hu_4, Hu_5, Hu_6, Hu_7, Hu_8)$ (Asaad et al. 2012). Most of the vessels segmentation approaches extract and use the green color image of RGB retinal image for further processing since it has the best contrast between vessels and background. The five gray-level based features group is presented by Marin et al. (2011) and its features describe the gray-level variation between vessel pixel and its surrounding. The Hu moment-invariants (Hu 1962) are best shape descriptors which are invariant to translation, scale and rotation change. So they are used by the second group of eight features to describe vessels have variant widths and angles. These features are simple, better discriminate between vessel and non-vessel classes and needn't be computed at multiple scales or orientations. The features computation is more detailed in Asaad et al. (2012).

2.3   Samples Selection

The DRIVE training set contains 569415 vessel pixels and 3971591 non-vessel pixels needed to select samples from them for features computation. Since the ratio of vessel pixels to non-vessel pixels in each image and overall images is 1:7 on average, the samples set consists of 1000 vessel pixels and 7000 non-vessel pixels from each image; there are 20 images give 160000 total samples. In the previous work (Asaad et al. 2012), the samples set consists of 100000 vessel and 100000 non-vessels pixels selected randomly from the whole training set; so different features set was selected by CFS heuristic.

3   Features Selection Heuristics

Feature selection plays an important role in building classification systems (Hua et al. 2009). It can not only reduce the dimension of data, but also lower the computation consumption, and so that it can gain good classification performance. In general, the feature selection algorithms designed with different evaluation criteria contain two categories: the filter methods and wrappers methods (Blum and Langley, 1997; Talavera 2005). The filter methods mainly identify a feature subset from the original feature set with a given evaluation criterions which are independent of learning algorithms. The wrapper methods choose those features with high prediction performance estimated by specified learning algorithms (Kohavi and John 1997). This paper presents four heuristics for selecting the best relevant features set that gives the best classification performance with ACS in terms of sensitivity, specificity and accuracy.

3.1   Correlation-Based Feature Selection (CFS)

It is a heuristic approach for evaluating the worth or merit of a subset of features (Hall 2000; Hall and Smith 1999). The main premise behind this selection method is that the features that are most effective for classification are those that are most highly correlated with the classes (intensifiers and dissipaters), and at the same time are least correlated with other features. The method is therefore used to choose a subset of features that best represent these qualities. The best individual feature based on the following merit metric:

$$M_s = \frac{k\,\bar{r}_{cf}}{\sqrt{k + k\,(k-1)\,\bar{r}_{ff}}} \tag{1}$$

Where M_s is the heuristic merit of a features subset S containing k features, $\overline{r}_{cf}$ is the average feature-class correlation, and $\overline{r}_{ff}$ is the average feature-feature inter-correlation. The numerator gives an indication of how predictive a group of features are; the denominator of how much redundancy there is among them".

3.2 Fisher Score

Fisher Score (Bishop 1996) is a type of supervised feature selection methods consists in examining the correlation between projected data samples and their class labels on each feature axis. It looks for features on which the classes are compact and far from each others. By considering the sample coordinates on the feature f_r, each class ω, $\omega = 1,. . ., c$, populated with n_ω labeled samples is characterized by its mean $\mu_{\omega r}$ and its variance $\sigma^2_{\omega r}$. Moreover, let us denote μ_r the mean of all data samples on the feature f_r. The Fisher score F_r used to evaluate the relevance of the feature f_r is defined by:

$$F_r = \frac{\sum_{\omega=1}^{c} n_\omega (\mu_{\omega r} - \mu_r)^2}{\sum_{\omega=1}^{c} n_\omega \sigma^2_{\omega r}} \qquad (2)$$

In order to select the most relevant features, they are sorted according to the decreasing order of their Fisher score F_r.

3.3 Gini Index

Gini index is a non-purity split method. It fits sorting, binary systems, continuous numerical values, etc. It was put forward by Breiman et al. (1984) and is used to select the feature at each internal node of the decision tree. The main idea of Gini index algorithm is as follows: Suppose S is the set of s samples. These samples have m different classes (C_i, $i = 1,. . .,m$). According to the differences of classes, S can be divided into m subset (S_i, $i = 1,. . .,m$). Suppose S_i is the sample set which belongs to class C_i, s_i is the sample number of set S_i, then the Gini index of set S is:

$$Gini(S) = 1 - \sum_{i=1}^{m} P_i^2 \qquad (3)$$

where P_i is the probability that any sample belongs to C_i and estimating with s_i / s. Gini(S)'s minimum is 0, that is, all the members in the set belong to the same class; this denotes it can get the maximum useful information. When all the samples in the set distribute equably for the classified, Gini(S) is maximum; this denotes it can get the minimum useful information. If the set is divided into n subset, then the Gini after splitting is:

$$Gini_{split}(S) = \sum_{j=1}^{m} \frac{S_j}{S} Gini(S_j) \qquad (4)$$

The minimum $Gini_{split}$ is selected for splitting attribute. The main idea of Gini index is: for every attribute, after it traverses all possible segmentation methods, if it can provide the minimum Gini index then it is selected as the divisive criterion of this node no matter it is the root node or a sub node.

3.4   Relief

Relief is a classical instance-based attribute selection scheme introduced by Kira and Rendell (1992), and enhanced by Kononenko (1994). The key idea of Relief is to estimate attributes according to how well their values distinguish among in-stances of different classes that are near each other (Kononenko 1994). For that purpose, Relief for a given instance searches for its two nearest neighbours: one from the same class (called nearest hit H) and the other from a different class (called nearest miss M). For each attribute, it calculates the relevance scores and updates its value according to Eq. (5).

$$W(A) = W(A) - diff(A, X, H) / m + diff(A, X, M) / m \qquad (5)$$

where $W(A)$ represents the relevance scores for any attribute A, diff(A, X, H) is the difference between the values of attribute A for the two instances X and H, and m is the number of instances sampled. This process is repeated for a user-specified number of instances m. The rationale is that a useful attribute should differentiate between instances from different classes and have the same value for instances from the same class (Hall and Holmes 2003).

4   Experimental Results and Analysis

Three measures are calculated for evaluating the classification performance of selected features set with ACS. The first measure is the sensitivity (*SN*) which is the ratio of well-classified vessel pixels. The second measure is the specificity (*SP*) which is the ratio of well-classified non-vessel pixels. The third measure is the accuracy (*ACC*) which is the ratio of well-classified vessel and non-vessel pixels. The resulting features sets by the four heuristics are shown in Table 1. Since CFS as selection heuristic gives features set of six features, only six features of highest weighting were taken in the features sets of the other three weighting heuristics to compare between unified heuristics. Some features were selected by all heuristics as f_2 and Hu_1, other features were selected by majority of heuristics as f_3, f_4, f_5 and Hu_4 while others were selected only one time by some heuristics as f_1, Hu_2, Hu_3, and Hu_5 or weren't selected by any heuristic as *green gray*, Hu_6, Hu_7 and Hu_8.

Table 1 Features sets recommended by each heuristic

Feature	No of selected features	Selected Features
CFS	6	$\{f_2, f_3, f_4, f_5, \mathrm{Hu}_1, \mathrm{Hu}_4\}$
Fisher Score	6	$\{f_2, f_3, \mathrm{Hu}_1, \mathrm{Hu}_2, \mathrm{Hu}_3, \mathrm{Hu}_4\}$
Gini Index	6	$\{f_2, f_4, f_5, \mathrm{Hu}_1, \mathrm{Hu}_4, \mathrm{Hu}_5\}$
Relief	6	$\{f_1, f_2, f_3, f_4, f_5, \mathrm{Hu}_1\}$

As shown from Table 1, f_2 and Hu_1 are the cores of any features set by any heuristic because they give the best separation between vessels and non-vessels classes in the feature space. To emphasis this result, two elementary features selection methods were applied; the sequential forward selection (SFS) and the sequential backward selection (SBS). Both methods select the same features set that consisted of $f2$ and $Hu1$. Table 2 shows in descending order the classification performance values for each heuristic when its recommended features set was used with ACS for retinal vessels segmentation. The results are also represented graphically in Fig. 1.

Table 2 Classification performance values of ACS with features set recommended by each heuristic

Features Selection Heuristic	SN (%)	SP (%)	ACC (%)
Relief	75.84	93.88	91.55
CFS	75.41	93.81	91.43
Fisher Score	73.88	93.49	90.94
Gini Index	73.50	93.36	90.78

As shown from Table 2, there's no significant difference between relief and CFS which is expected since they intersect in five features according to table 1. Also more difference between all heuristics is in sensitivity than specificity which means that all resulting features sets are more capable of discriminating vessels than non-vessels. The current CFS of six features overcomes the previous CFS of eight features in the previous work (Asaad et al. 2012) (SN=73.88%, SP=92.66% and ACC=90.25%) due to different sampling technique. Table 3 shows the computation time of each features set recommended by each heuristic. As shown, there's no significant difference between them. The features computation takes time about one minute and half on PC with Intel Core-i3 CPU at 2.53 GHz and 3 GB of RAM.

Since all features sets by all heuristics take about the same computation time, so it's clear that these features are simple and easy to compute beside their discrimination power. There is a recent work by Lupas et al. (2009) compared to the current study where the feature vector consists of forty-one feature that were filtered to sixteen by FS heuristic (MIT-based CFS) but the features computation time after FS takes about two minutes on PC with Intel Core 2 Duo CPU at 3.16 GHz and 3326 MB RAM (Lupas et al. 2010). They have used Gaussian and its derivatives up to order 2 at multiple scales, the maximum response of a multi-scale matched filter using a Gaussian vessel profile, Frangi's vesselness measure, Lindeberg's ridge strengths, two-dimensional Gabor wavelet transform response taken at multiple scales and the feature containing information about Staal's ridges.

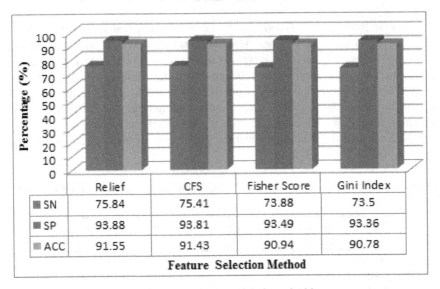

Fig. 1 Fig. 4 Performance Comparison of FS Methods for retinal images

Table 3 Computation Time of each features set recommended by each heuristic

Features Selection Heuristic	CFS	Fisher Score	Gini Index	Relief
Computation Time in Seconds	91	90	90	90

5   Conclusion and Future Work

With the increasing demands of automatic processing systems, human identification for data mining and processing has recently begun to gain more interest. On the other hand, retinal imaging offers rich potential for biometric recognition. The automatic analysis of blood vessels is a very important task in many clinical investigations and scientific research related to vascular features. The early diagnosis of several pathologies, such as arterial hypertension, arteriosclerosis or diabetic retinopathy could be achieved by analyzing the vascular structures. Moreover for many clinical investigations, vessel segmentation is becoming a prerequisite for the analysis of vessel parameters such as tortuosity and variation of the vessel width along the vessel and the ratio between the venous and arterial vessel width. The automatic retinal blood vessels segmentation is a classification problem where each pixel in the field of view (FOV) of the retinal image is classified as vessel-like or non-vessel. The pixel classification is based on computation of many features from the retinal image itself. The large number of computed features increases the classification complexity and time and reduces its accuracy, so features selection is an essential step for successful classification since it removes irrelevant features and achieves less complex, more accurate and faster classification. There are large amount of features selection heuristics serve different purposes and all have their own advantages and disadvantages. This paper has described four features selection heuristics to select the best relevant features set from features vector consists of fourteen features that are computed for each pixel in the field of view of retinal image in the DRIVE database. The comparison is assessed in terms of sensitivity, specificity and accuracy of the recommended features set by each heuristic when used with the ant colony system algorithm. The results indicated that the recommended features set by the relief heuristic gives the best performance with sensitivity of 75.84% , specificity of 93.88% and accuracy of 91.55.

As a future work, other FS heuristics can be used to improve the overall accuracy of the system. The promise of greater image resolution (modern digital cameras exceed 16 megapixels) and computer processing power in the future may allow more sensitive detection of retinal microvascular changes and lead to automated diagnosis from fundal images being a practical and efficient adjunct to ophthalmic diagnostics. Digital retinal vascular image analysis may also permit an assessment of a particular individual's specific risk stratification for a variety of cardiovascular conditions, and may have particular relevance to cerebrovascular risk.

References

Assad, A., Azar, A.T., Hassaanien, A.E.: Ant Colony-based System for Retinal Blood Vessels Segmentation. In: Seventh International Conference on Bio-Inspired Computing: Theories and Application, 2012 (BIC-TA 2012), Gwalior, India, December 14 - 16 (2012)

Bloomgarden, Z.T.: Screening for and managing diabetic retinopathy: current approaches. Am. J. Health Syst. Pharm. 64 (17 suppl. 12), S8–S14 (2007)

Blum, A., Langley, P.: Selection of relevant features and examples in machine learning. Artificial Intelligence 97(1-2), 245–271 (1997)

Breiman, L., Friedman, J., Olshen, R., Stone, C.: Classification and Regression Trees. Wadsworth & Brooks/Cole Advanced Books & Software, CA, USA (1984)

Chew, E.Y.: Screening options for diabetic retinopathy. Curr. Opin. Ophthalmol 17(6), 519–522 (2006)

Dorigo, M., Gambardella, L.M.: Ant colony system: a cooperative learning ap-proach to the traveling salesman problem. IEEE Trans. Evol. Comput. 1(1), 53–66 (1997)

Fraz, M.M., Remagnino, P., Hoppe, A., Uyyanonvara, B., Rudnicka, A.R., Owen, C.G., Barman, S.A.: Blood vessel segmentation methodologies in retinal images–a survey. Comput. Methods Programs Biomed. 108(1), 407–433 (2012), doi:10.1016/j.cmpb.2012.03.009.

Farley, T.F., Mandava, N., Prall, F.R., Carsky, C.: Accuracy of primary care clinicians in screening for diabetic retinopathy using single-image retinal photography. Ann. Fam Med. 6(5), 428–434 (2008)

Foracchia, M., Grisan, E., Ruggeri, A.: Extraction and quantitative description of vessel features in hypertensive retinopathy fundus images. In: Book Abstracts 2nd International Workshop on Computer Assisted Fundus Image Analysis (2001)

Goatman, K., Charnley, A., Webster, L., Nussey, S.: Assessment of auto-mated disease detection in diabetic retinopathy screening using two-field photography. PLoS One 6(12), e27524 (2011)

Hall, M.A., Smith, L.A.: Feature Selection for Machine Learning: Comparing a Correlation-Based Filter Approach to the Wrapper. In: FLAIRS Conference, pp. 235–239 (1999)

Hall, M.A.: Correlation-based Feature Selection for Discrete and Numeric Class Machine Learning. In: ICML, pp. 359–366 (2000)

Hall, M., Holmes, G.: Benchmarking attribute selection techniques for discrete class data mining. IEEE Transact. Knowl. Data Eng. 15(6), 1437–1447 (2003)

Hua, J.P., Tembe, W.D., Dougherty, E.R.: Performance of feature-selection methods in the classification of high-dimension data. Pattern Recognition 42(3), 409–424 (2009)

Hu, M.K.: Visual Pattern Recognition by Moment Invariants. IRE Trans. Inform. Theory. 8(2), 179–187 (1962)

Jin, X., Guangshu, H., Tianna, H., Houbin, H., Bin, C.: The Multifocal ERG in Early Detection of Diabetic Retinopathy. Conf. Proc. IEEE Eng. Med. Biol. Soc. 7, 7762–7765 (2005)

Jones, S., Edwards, R.T.: Diabetic retinopathy screening: a systematic review of the economic evidence. Diabet. Med. 27(3), 249–256 (2010)

Khan, M.I., Shaikh, H., Mansuri, A.M.: A Review of Retinal Vessel Segmentation Techniques and Algorithms. Int. J. Comp. Tech. Appl. 2(5), 1140–1144 (2011)

Kira, K., Rendell, L.A.: A practical approach to feature selection. In: The Proceedings of Ninth International Conference on Machine Learning, Aberdeen, Scotland, pp. 249–256. Morgan Kaufmann, Los Altos (1992)

Kohavi, R., John, G.H.: Wrappers for feature subset selection. Artificial Intelligence 97(1-2), 273–324 (1997)

Kononenko, I.: Estimating attributes: analysis and extensions of relief. In: Bergadano, F. (ed.) Proceedings of the Seventh European Conference on Machine Learning, vol. 784, pp. 171–182. Springer, Berlin (1994)

Leung, H., Wang, J.J., Rochtchina, E., Wong, T.Y., Klein, R., Mitchell, P.: Impact of current and past blood pressure on retinal arteriolar diameter in an older population. J. Hypertens 22(8), 1543–1549 (2004)

Marin, D., Aquino, A., Gegundez-Arias, M.E., Bravo, J.M.: A New Supervised Method for Blood Vessel Segmentation in Retinal Images by Using Grey-Level and Moment Invariants-Based Features. IEEE Trans. Med. Imaging 30(1), 146–158 (2011)

Mendonca, A.M., Campilho, A.: Segmentation of retinal blood vessels by combining the detection of centerlines and morphological reconstruction. IEEE Trans. Med. Imaging 25(9), 1200–1213 (2006)

Mitchell, P., Leung, H., Wang, J.J., Rochtchina, E., Lee, A.J., Wong, T.Y., Klein, R.: Retinal vessel diameter and open-angle glaucoma: the Blue Mountains Eye Study. Ophthalmology 112(2), 245–250 (2005)

Morello, C.M.: Etiology and natural history of diabetic retinopathy: an overview. Am. J. Health Syst. Pharm. 64 (17 suppl. 12), S3–S7 (2007)

Rodgers, M., Hodges, R., Hawkins, J., Hollingworth, W., Duffy, S., McKib-bin, M., Mansfield, M., Harbord, R., Sterne, J., Glasziou, P., Whiting, P., Westwood, M.: Colour vision testing for diabetic retinopathy: a systematic review of diagnostic accuracy and economic evaluation. Health Technol. Assess. 13(60), 1–160 (2009)

Serrarbassa, P.D., Dias, A.F., Vieira, M.F.: New concepts on diabetic retinopathy: neural versus vascular damage. Arq Bras Oftalmol. 71(3), 459–463 (2008)

Sinclair, S.H.: Diabetic retinopathy: the unmet needs for screening and a review of potential solutions. Expert Rev. Med. Devices 3(3), 301–313 (2006)

Soares, J.V., Leandro, J.J., Cesar Júnior, R.M., Jelinek, H.F., Cree, M.: Retinal vessel segmen-tation using the 2-D Gabor wavelet and supervised classification. IEEE Trans. Med. Imaging 25(9), 1214–1222 (2006)

Staal, J.J., Abramoff, M.D., Niemeijer, M., Viergever, M.A., van Ginneken, B.: Ridge based vessel segmentation in color images of the retina. IEEE Transactions on Medical Imaging 23(4), 501–509 (2004)

Talavera, L.: An evaluation of filter and wrapper methods for feature selection in categorical clustering. In: Proceeding of 6th International Symposium on Intelligent Data Analysis, Madrid, Spain, pp. 440–451 (2005)

Verma, K., Deep, P., Ramakrishnan, A.G.: Detection and classification of diabetic retinopathy using retinal images. In: Annual IEEE India Conference (INDICON), pp. 1–6 (2011), doi:10.1109/INDCON.2011.6139346

Vijayakumari, V., Suriyanarayanan, N.: Survey on the Detection Methods of Blood Vessel in Retinal Images. Eur. J. Sci. Res. 68(1), 83–92 (2012)

Wang, J.J., Taylor, B., Wong, T.Y., Chua, B., Rochtchina, E., Klein, R., Mitchell, P.: Retinal vessel diameters and obesity: a population-based study in older persons. Obesity (Silver Spring) 14(2), 206–214 (2006)

Xu, J., Hu, G., Huang, T., Huang, H., Chen, B.: Using multifocal ERG re-sponses to discriminate diabetic retinopathy. Doc. Ophthalmol. 112(3), 201–207 (2006)

You, X., Peng, Q., Yuan, Y., Cheung, Y., Lei, J.: Segmentation of retinal blood ves-sels using the radial projection and semi-supervised approach. Pattern Recognition 44(10-11), 2314–2324 (2011)

Lupascu, C.A., Tegolo, D., Trucco, E.: A comparative study on feature selec-tion for retinal vessel segmentation using FABC. In: Proc 13th International Conference on Computer Analysis of Images and Patterns (CAIP), pp. 655–662 (September 2009)

Lupascu, C.A., Tegolo, D., Trucco, E.: FABC: Retinal vessel segmentation using adaboost. IEEE Trans. Inf. Technol. Biomed. 14(5), 1267–1274 (2010)

Human Skin Segmentation in Color Images Using Gaussian Color Model

Ravi Subban and Richa Mishra

Abstract. Use of color spaces for human skin detection has been efficiently used for the several decades. It also plays an efficient role in face detection, tracking and recognition. This paper presents a comparative evaluation on the performance of skin color pixel classification methods using four rarely used color spaces and three commonly used skin detection methods. The first two skin detection methods used are piecewise linear decision boundary classifier algorithm and Gaussian color model which produce better results for color images with a wide variety of human skin tones. The third method deals with the Gaussian model with combination of two color spaces. All the results are experimentally evaluated with the help of few commonly used face databases.

Keywords: Color space, Combination of two color spaces, Gaussian model, Linear decision boundary classifier, Skin detection.

1 Introduction

The process of detecting human skin regions in color images is a two class problem which works linearly and sequentially analyzing each image pixel to see if the pixel belongs to skin region or non-skin region. The skin detection methods depend on the selection of color space and skin classification algorithm. Depending upon the nature of the skin detection methods, it can also be used for other related biometric techniques like face recognition and facial expression extraction. This step should be done with utmost accuracy to yield good results that will be useful in the application of image processing applications. It has been proved that the skin color is very useful and robust cue in the detection of human faces in color images. It is used as a feature in the detection and tracking of human faces containing skin that can be computationally inexpensive. Thus, it is well suited for

Ravi Subban · Richa Mishra
Department of Computer Science, Pondicherry University, Puducherry, India

S.M. Thampi et al. (eds.), *Recent Advances in Intelligent Informatics*,
Advances in Intelligent Systems and Computing 235,
DOI: 10.1007/978-3-319-01778-5_2, © Springer International Publishing Switzerland 2014

real-time applications [3]. As the extraction of skin region is mainly based on the constancy of colors in the color images, the color based approach is considered in addition to the skin detection methods. The intensity of segmentation of skin regions of each color spaces is different as it is based on the luminance component of the color spaces.

Various color spaces have been used for the skin detection applications in the past. The commonly used color spaces used for the skin detection are RGB, Normalized rgb, YCbCr, YUV, YIQ, HSV, HSI, etc. Some of the color spaces that are rarely used for skin detection are, YPbPr, YDbDr, XYEz, and Photo YCC (YCC). These rarely used color spaces are used with the three different approaches of skin detection method. The classification of skin and non-skin pixels on the basis of the adaptive threshold is the first skin detection method that is taken into account. This method classifies each pixel of the input color images individually and independently from its neighbors. Based on the collection of large skin color samples, some inequalities are developed to enclose the skin color distribution in adopted color spaces for skin color classification. Though these inequalities can cover too large or wrong range of skin color cluster, they are still used because of its simplicity to implement and under normal situations it covers only skin colors [2]. The effect of all the color spaces are observed by implementing them with the Gaussian model, the probability is evaluated for each pixel in the color images. Finally, the concept of using the combination of two color spaces is considered.

The rest of the paper is organized as follows: Section II focuses on the related work done by some of the researchers in the past. The advantages and the conversion formulas of the color spaces are given in the Section III. The techniques proposed and used in this paper are briefly described in the next section. Section IV describes the experimental results obtained by using the color spaces with the skin detection methods. Finally Section VI concludes the paper.

2   Related Work

The researchers have done a considerable works in the field of skin detection for the past several decades. They also used this concept in various application of image processing in several ways. There are some researchers who have used traditionally used color spaces for skin segmentation whereas some of them have used rarely used color spaces [5] in the applications related to skin segmentation. The color spaces considered in this paper come under the category of rarely used color spaces for skin detection as only few researches have made use of them. Wang et al. [6] has presented a new face detection method using the combination of YCbCr and KL skin color space. Francois et al. [10] has presented a method to improve color decorrelation for lossless color image compression using the LAR codec using YDbDr color space. But the combined study and analysis of color spaces has been done by few researchers [7], using different skin classification algorithms, generally pixel based classification, Gaussian model, mixture of Gaussian model, Multilayer Perceptron, etc [9]. This concept of skin cluster and

Gaussian model are used to make a comparative study and analysis on skin detection techniques using some of the color spaces which are rarely used for skin detection. The concept of pixel-based skin classifiers is commonly used because of its simplicity to implement and evaluate the results. Ibrahim et al. [12] has used the concept of pixel-based skin segmentation algorithm that detects the face skin region in the color images.

3 The Proposed Skin Classification Algorithms

Skin detection is a process of starting at a pixel-level that involves a pre-processing step of color space transformation followed by a classification process. A color space transformation is assumed to increase separability between skin and non-skin classes, to increase the similarity among different skin tones [1]. The following three methods were built on the pixel level skin detection concept, which deals with the pixels in the color image without looking into the shapes and contours of the object [4].

3.1 Piecewise Linear Decision Boundary Approach

It is one of the simplest and easiest approaches to implement. It builds a decision rule that will discriminate the skin pixels from the non-skin pixels of human skin region in color images. It considers each pixel of the color image sequentially, individually, and independently taking one pixel at a time and labeling it as skin or non-skin pixels based on the constraints. In this approach, skin color cluster can be explicitly defined by considering the wide range of skin samples. The process of finding the threshold values are done using the trial and error method and it is a time consuming job as it should produce better results for all kinds of the color images. The explicitly defined skin cluster methods are briefly explained in the sections that follow:

3.1.1 Skin Cluster Method Using YPbPr Color Space

Pb and Pr are the two chroma components in this color space. It can be efficiently used to define explicitly segmented skin region. The threshold values determined are given below (1). A pixel is classified as skin pixel if the values of the components fall within the range:

$$(Y > 125), (Pb > 25), and (Pr > 13) \tag{1}$$

3.1.2 Skin Cluster Method Using XYEz Color Space

It is one of the rarest color space used for the skin segmentation. It can be effectively used for the skin detection by explicitly defining the thresholds that will classify skin pixels and non-skin pixels. If a pixel value falls within the thresholds, then it is classified as skin pixel, otherwise classified as non-skin pixel.

$$(R > 160), (Y > 100), and\ (Ez < 170) \tag{2}$$

3.1.3 Skin Cluster Method Using YDbDr Color Space

This color space has been widely used for the television broadcasting in France. In this color space, Db and Dr are the chrominance components used to define the skin color cluster for skin segmentation. The skin and non-skin pixels are explicitly classified by the selection of threshold values on the basis of the observation of sample images.

$$(Y > 55), (Db > 95), and\ (Dr < 180) \tag{3}$$

3.1.4 Skin Cluster Method Using YCC Color Space

This color space is used in the digital photography system. It produces a good result in the skin segmentation area by explicitly classifying the skin and non-skin pixels.

$$(Y > 55), (C > 85), and\ (C > 145) \tag{4}$$

3.2 *Gaussian Skin Color Model*

Gaussian model is one of the commonly used parametric approaches that help in the skin classification. It is based on the probability of each pixel in the color images. It could be modeled to discriminate non-skin pixels like eyebrows, eye corners, nostrils, ear, lip corners, etc by assuming human face skin gray levels as they are of darker gray levels and could be excluded by the symmetric property of Gaussian distribution [2]. In other words, it produces the continuous output which is called skin-likelihood image. This grayscale image is transformed to skin map by using threshold technique. The general model for Gaussian joint probability density function is defined as follows [8]:

$$p(c|skin) = \frac{1}{2\pi |\Sigma_s|1/2} e^{-\frac{1}{2}(c-\mu_s)^T \Sigma_s^{-1}(c-\mu_s)} \tag{5}$$

Here, c is a color vector, μ_s and Σ_s are the mean vector and covariance vector respectively.

$$\mu_s = \frac{1}{n}\Sigma_{j=1}^n c_j \tag{6}$$

$$\Sigma_s = \frac{1}{n-1}\Sigma_{j=1}^n (c_j - \mu_s)(c_j - \mu_s)^T \tag{7}$$

where, n is the total number of skin color samples c_j. The concept of Gaussian model is also used very commonly because it is based only on the two components of the color spaces to produce skin likelihood image. The proposed skin detection algorithm based on Gaussian color model is given below:

1. Read the input RGB facial image.
2. Convert from the RGB color space into the used color spaces.
3. Using Gaussian color model the skin likelihood image is obtained. The skin likelihood image is a gray image indicating the likelihood of the each image pixel being a skin pixel.
4. Skin likelihood image is transformed into the skin region using adaptive threshold. It is a binary image showing skin and non-skin regions.
5. The skin segmented binary image is subject to postprocessing using morphological operations to remove unwanted pixels present as noise.
6. The skin segmented region is obtained as an output.

3.3 Gaussian Color Model Using Combination of Two Color Spaces

The third method uses the Gaussian model as a combination of two color spaces. This approach combines the advantages that can overcome the problems created when single color space is used. The basic goal of this approach is to find the combination of two color spaces that will efficiently classify the skin and non-skin pixels to produce human skin segmented region.

In order to determine the effect on the detection of skin region this approach is used by using the concept of combination of color spaces along with the Gaussian model. The steps that are needed to implement this approach are given below:

1. Get the binary segmented regions of skin following the Gaussian model by using color spaces.
2. Apply AND operation on the two binary images obtained to yield the final skin segmented image of the sample image.
3. The skin segmented binary image is post-processed by using morphological operations to remove unwanted pixels present as noise.
4. The skin segmented region is obtained as an output.

4 Experimental Results and Discussions

The extensive set of experimental evaluation is widely performed over the three methods using rarely used color spaces to analyze the effects of color representation on skin segmentation and to compare different classification algorithms over the commonly used face databases. The results of the methods are comparatively measured and analyzed by using different rates or metrics of the skin classifiers: detection rate or correction rate, false rejection and false acceptance ratios. The detection rate is denoted by the ratio of number of correctly identified images to the total number of images used. The false rejection rate is denoted by the ratio of number of images detected skin pixels as non-skin pixels to the total number of images used. The false acceptance rate is denoted by the ratio of number of images detected non-skin pixels as skin pixels to the total number of images used.

The problems that are commonly encountered at the time of the human skin detection during experimental evaluation are: complex background, clothes and hair having colors close to the human skin color, and identification of human facial features like eyes, mouth, nostrils, different lighting conditions etc. Various color spaces are used for the identification of human skin pixels but some of the above problems can be encountered in them. These problems are totally based on the luminance component of the color spaces as it measures the intensity of color in color images. The above explained three algorithms for skin detection are analyzed and compared over the above specified problem under normal conditions. The piecewise linear decision boundary has explicitly defined the boundaries for the skin color cluster of each color space. The characteristics of each color spaces can be figured out by the following information given in the table 1.

Table 1 Metrics Obtained by using Piecewise Linear Decision Boundary Approach

Name of the color spaces	Percentage of skin detection rate	Percentage of false rejection rate	Percentage of false acceptance rate
YPbPr	96	0.02	0.07
XYEz	77	0.11	0.18
YDbDr	77	0.05	0.2
YCC	96	0.0	0.05

The problem of classifying the skin pixels as non-skin pixels is not faced by the YCC color space. It also yielded the good result in terms of detection rate of human segmented skin region using the YPbPr color space. The major problem with XYEz color spaces is detecting the non-skin pixels as skin pixels. Some of the input samples with their corresponding binary skin segmented image are shown in figure 1.

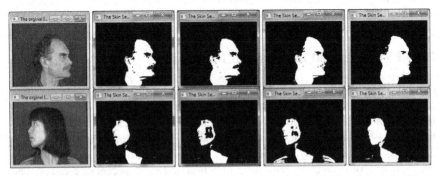

Fig. 1 Sample Images and their corresponding binary segmented images based on Piecewise Linear Decision Boundary Approach using YPbPr, XYEz, YDbDr, and YCC color spaces respectively

The second method is the Gaussian model which deals with the probability of pixels being skin color. This model yields the skin likelihood image as the intermediate product. It is a continuous gray scale image whose pixel values ranges from 0 to 1. It shows the various possible combinations of gray colors. This image clearly shows the lighter and darker region of gray levels that help in discriminating the skin and non-skin pixels. After that threshold values are used to convert it into binary segmented image. The percentage of success and failure in terms of detection of human skin region is shown in table 2. The YPbPr color space shows good result with the highest skin detection rate of 96% as compared to other ones in terms of all metrics used for skin detection.

Table 2 Metrics obtained by using Gaussian model

Name of the color spaces	Percentage of detection rate	Percentage of false rejection rate	Percentage of false acceptance rate
YPbPr	96	0.05	0.09
XYEz	82	0.11	0.16
YDbDr	77	0.07	0.2
YCC	79	0.11	0.16

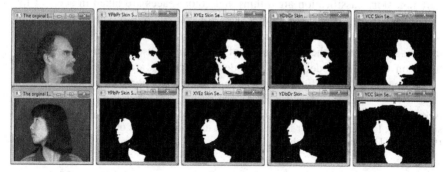

Fig. 2 Sample Images and corresponding binary segmented images based on Gaussian model using YPbPr, XYEz, YDbDr, and YCC color spaces respectively

Table 3 Metrics obtained using Gaussian model over combinations of two color spaces

Name of the color spaces	Percentage of detection rate	Percentage of false rejection	Percentage of false acceptance
YPbPr-XYEz	82	0.16	0.16
YPbPr-YDbDr	77	0.07	0.2
YPbPr-YCC	96	0.07	0.02
XYEz-YDbDr	75	0.18	0.22
XYEz-YCC	96	0.09	0.11
YDbDr-YCC	96	0.09	0.11

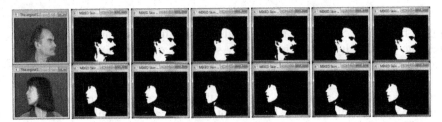

Fig. 3 Sample Images and the corresponding binary segmented images according to the order of the table using combination of two color spaces

The detection of skin region is done using single color space. In the last method, the combinations of two color spaces are taken into account. This approach helps to find out different combinations of color spaces that will overcome or override the advantages and disadvantages of the color spaces involved. All the possible combinations of two color spaces are experimented in this paper. The highest skin detection rate obtained by the combinations of two color spaces is 96%.

5   Conclusion

Different skin classification algorithms and color spaces are taken in account for human skin detection. All the color spaces considered are rarely used for the skin detection techniques. Each skin classification algorithms has shown its own results in terms of metrics. If the combination of two colors is used for skin classification based on Gaussian model, better skin detection results are obtained. Thus, they can be considered in the future applications of skin detection techniques.

Acknowledgments. This work is supported and funded by the University Grant Commission (UGC), India under Major Research Project to the department of Computer Science of Pondicherry University, Pondicherry, India.

References

[1] Shin, M.C., Chang, K.I., Tsap, L.V.: Does Colorspace Transformation Make Any Difference on Skin Detection. In: Proceedings of the Sixth IEEE Workshop on Applications of Computer Vision, WACV 2002 (2002)

[2] Hsieh, C.-C., Liou, D.-H., Lai, W.-R.: Enhanced Face-Based Adaptive Skin Color Model. Journal of Applied Science and Engineering 15(2), 167–176 (2012)

[3] Boussaid, F., Chai, D., Bouzerdoum, A.: On-Chip Skin Detection for Color CMOS Imagers. In: Proceedings of the International Conference on MEMS, NANO and Smart Systems, ICMENS 2003 (2003)

[4] Yang, Y., Wang, Z., Zhang, M., Yang, Y., Beijing, N.: Skin Region Tracking using Hybrid Color Model and Gradient vector Flow. Proceedings of the IEEE (2010)

[5] Sahdra, G.S., Kailey, K.S.: Detection of Contaminants in Cotton by using YDbDr color space. Int. J. Computer Technology & Applications 3(3), 1118–1124 (2012)

[6]   Wang, C.-X., Li, Z.-Y.: Face Detectiion based on Skin Gaussian Model and KL Transform. In: Proceeding of Ninth ACIS International Conference on Software Engineering, Artificial Intelligence, Networking, and Parallel/Distributed Computing. IEEE (2008)

[7]   Zarit, B.D., Super, B.J., Quek, F.K.H.: Comparison of Five Color Models in Skin Pixel Classification

[8]   Veznevets, V., Sazonov, V., Andreeva, A.: A Survey on Pixel-Based Skin Color Detection Techniques (2002)

[9]   Phung, S.L., Bouzerdoum, A., Chai, D.: Skin Segmentation Using Color Pixel Classification: Analysis and Comparison. IEEE Transactions on Pattern Analysis and Machine Intelligence 27(1) (January 2005)

[10]   Pasteau, F., Strauss, C., Babel, M., Deforge, O., Bedat, L.: Improved Colour Decorrelation for Lossless Colour Image Compression using the LAR Codec. In: European Signal Processing Conference, EUSIPCO 2009. Royaume-Uni, Glasgow (2009)

[11]   Ravi Subban, S., Mishra, R.: Face Detection in Color Images Based on Explicitly-Defined Skin Color Model. CCIS, vol. 361. Springer (2012)

[12]   Ibrahim, N.B., Selim, M.M., Zayed, H.H.: A Dynamic Skin Detector Based on Face Skin Tone Color. In: 8th International Conference on Informatics and Systems (INFOS 2012), May 14-16 (2012)

An Improved Local Statistics Filter for Denoising of SAR Images

Vikrant Bhateja, Anubhav Tripathi, and Anurag Gupta

Abstract. Synthetic Aperture Radar (SAR) is an active remote sensing system which is utilized for producing high-resolution images. But due to backscattering of microwave signals, these images get contaminated with speckle noise. This paper proposes an improved local statistics filter for filtering the speckle noise from the SAR images. The proposed filter is a combination of mean and hybrid median filters, employing a novel 7x7 filtering template. The performance of the proposed filter is tested against the standard Hybrid Median filters for which the evaluated values show better performs in terms of PSNR (in dB) and SSI.

Keywords: Speckle, SAR image, Detail preservation, Local statistics filter, median filter.

1 Introduction

SAR images are an outcome of coherent processing of backscattered signals received from various regions and are used in various applications like crop monitoring, navigation, military target detection etc [1]. But these images are highly contaminated with speckle noise. Speckle is a multiplicative noise which appears like a granular and degrades the quality of SAR images. For the past two decades, several speckle reduction techniques have been developed for filtering speckle and retaining edge details [2, 3]. Initially proposed filters like diffusion filters [4] and adaptive filters [5] were used for suppressing speckle noise but later on due to improper performance and less capability to reduce speckle noise other techniques [6,8] came into existence. The adaptive filtering modified the images based on statistics that are extracted from the local environment of each pixel. These filters

Vikrant Bhateja · Anubhav Tripathi · Anurag Gupta
Deptt. of Electronics and Communication Engineering, Shri Ramswaroop Memorial Group of Professional Colleges, Faizabad Road, Lucknow-227105, (U.P.), India
e-mail: {bhateja.vikrant,proabhi9,anuraggpt8}@gmail.com

S.M. Thampi et al. (eds.), *Recent Advances in Intelligent Informatics*,
Advances in Intelligent Systems and Computing 235,
DOI: 10.1007/978-3-319-01778-5_3, © Springer International Publishing Switzerland 2014

were capable of filtering noise as well as in preservation of edges. Earlier proposed Lee Filter [9] used the idea of smoothing constrain based on noise variance, but it ignored speckle noise in the areas closest to edges and lines. Kuan Filter [9] proposed after Lee Filter was designed to smooth out noise while retaining edges or shape features in the image. But it was limited in terms of its complex calculations. Frost filter [9] used an adaptive filtering algorithm which computed a set of weighted values for each pixel within the filter window surrounding each pixel. Mean filtering [10] is a simple, intuitive and easy mode to implement smoothing in images, and is also used to suppress speckle noise in SAR images. Nonlinear Local-statistics filters [11] are based on spatial ordering (ranking) of the pixels contained an image area encompassed by the filter, and then replacing the value of the center pixel with the value determined by the ranking result. Lsmv and Weiner filter [11] are examples of these types of filters which use the first order statistics such as variance and mean of the neighborhood in order to remove speckle from SAR images. In continuation of these filters, the Median filter used in the SAR image restoration was disadvantageous in terms of extra computation time needed to sort the intensity value of each set and removal of fine image details such as lines [12]. Later on, the proposal of Hybrid Median Filter [13], also called as corner preserving median filter, was a modification of median filters and gives a fourfold improvement in edge shift over that of the median filter. Hence, such detail-preserving filters improve the situation dramatically but do not completely overcome the problem. In this paper, a combination of modified hybrid median filter and mean filter is proposed to filter the speckle from SAR images effectively. The performance of proposed filter is evaluated using quality evaluation parameters: Peak Signal to Noise Ratio (*PSNR* in dB) [14] and Speckle Suppression Index (*SSI*) [15]. In the following section, the proposed Local Statistics Filter for speckle filtering is presented. Section 3 discusses obtained results followed by the conclusion in section 4.

2   Proposed Local Statistics Filter

2.1   *Background*

SAR technique proves its usability under various weather conditions and is able to penetrate clouds and soil. A SAR image is a mean intensity estimate of the radar reflectivity of the region which is being imaged [1]. Speckle noise in such system is referred as the difference between the measured and the true mean value. Since, the speckle noise is a multiplicative noise, therefore it can be mathematically modeled as:

$$G(i, j) = R(i, j).S(i, j) \qquad (1)$$

where: $R(i, j)$ and $G(i, j)$ represent original and noisy SAR image pixels respectively. Logarithmic transform is applied to the noisy image which affects the speckle noise statistics and it becomes very close to white Gaussian noise. Since, filtering is effective for additive noise, therefore, natural logarithmic transform is generally applied over the speckle noise model.

2.2 Improved Local Statistics Filter

The proposed local statistics filter for filtering the speckle noise, is a modification of median filter which combines the mean filter [10] and hybrid median filter [13]. This paper proposes a novel 7x7 template (mask) as shown in fig1. The noisy image is convolved with this improved template (mask) and the end value thus, obtained is replaced with the centre pixel of the template. In this way, the resulted denoised image is denoted as I1. The noisy image is also processed with the mean filter [10] in which the image is divided into 7x7 kernels. Then, the local mean of each kernel is calculated. The central pixel of this kernel is replaced with mean value producing image I2. In continuation of this, the mean of the two denoised images (I1 and I2) is taken in order to produce the final image. This improved mask has capability of better speckle filtering while preserving the edge content in the image. The improved hybrid median filter takes the maximum value of φ pixels to calculate φmax while it takes the median value of pixels θ to compute θmed. Then, median is calculated for the values of φmax, θmed and C using eq.(2)

$$M = median(\varphi max, \theta med, C) \tag{2}$$

where the pixel values φ are chosen at positions other than the horizontal and vertical while θ is the diagonal pixel values. Here, C denotes the central pixel values.

θ		φ		φ		θ
	θ				θ	
φ		θ		θ		φ
			C			
φ		θ		θ		φ
	θ				θ	
θ		φ		φ		θ

Fig. 1 Proposed 7x7 Filtering Template (Mask)

The results of the proposed local statistics filters are shown in fig2. Hybrid median filters have preserving properties which gives at least four fold improvement in edge shift over that of the median filter. Mean and median filters are used to reduce the intensity values in the adjacent pixel and to preserve high frequency components in the image. The proposed filter, since combines both the improved hybrid median filter and mean filter [3], therefore it has features of both of these techniques. Also, The proposal of an improved 7x7 template (mask) as shown in fig1. is advantageous in terms of speckle noise suppression while preserving the edges and other fine details of the image during filtering process.

3   Results and Discussion

The chosen SAR image of 'Glaciers' is first contaminated with speckle noise of appropriate variance level. This corrupted image is then convolved with the 7x7 filtering template of Fig1. using the methodology mentioned in section 2. The results of earlier proposed median filters[12] show that they are not much effective in terms of speckle filtering. As the tabulation of parameters reveals the relative superiority of earlier proposed filters, the proposed technique with an improved 7x7 mask is proven to be far better. The proposed filter is compared with hybrid median filter and their performance is evaluated on the basis of their values for different performance evaluation parameters such as *PSNR*(in dB)[14], and *SSI*

Fig. 2 Speckled 'Glacier' image with noise level of (a) 0.01 and (b) 0.2. (c) and (d) are results obtained with hybrid median filter at 0.01 and 0.2 respectively. Consequtive results obtained after subjection to proposed local statistics filter at (e) 0.01 and (f) 0.2.

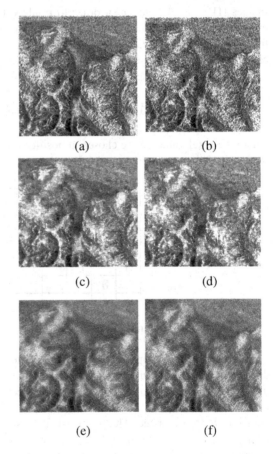

(a)   (b)

(c)   (d)

(e)   (f)

[15]. The corresponding values of these parameters at various noise variance ranging from 0.001 (low intensity noise) to 0.2 (high intensity noise) are tabulated in tables I and II. The *PSNR* (in dB)[14] values are generally supposed to be very high as it specifies peak value of signal to noise ratio in an image which are appreciably very high for the proposed method and are comparatively much improved at both the high and low levels of noise as tabulated in table I in comparison with the hybrid median filter. In SAR images, edges are to be pre-served since most of the details are present in the edges. In order to preserve the edges the vertical, horizontal, diagonal and mainly boundary pixels are considered and other pixels are ignored. Due to this reason, PSNR of hybrid Mean-Median filter is much better than many other existing filters so this may be used for reduc-ing the speckle noise in SAR images. Similarily, the speckle suppression index (*SSI*) [15] as tabulated in table II is another parameter for evaluating the performance of filtered images.

Table 1 PSNR (IN dB) for speckle filtering techniques in SAR images

Noise Level (variance)	Hybrid median filter[13]	Proposed Local Statistics filter
0.001	26.9718	28.5347
0.01	25.7614	26.3088
0.04	23.2577	25.1263
0.1	20.2689	22.7029
0.2	18.0288	21.0125

Table 2 SSI values for speckle filtering techniques in sar images

Noise Level (variance)	Hybrid median filter[13]		Proposed Statistics Local Filter	
	Noisy Image	Denoised Image	Noisy Image	Denoised Image
0.001	0.2051	0.1501	0.2051	0.1275
0.01	0.2303	0.1655	0.2303	0.1309
0.04	0.2931	0.2009	0.2931	0.2433
0.1	0.3587	0.2471	0.3587	0.2635
0.2	0.4605	0.4048	0.4605	0.3924

As the *SSI* values are the measure of average amount of speckle present in the speckled image as compared to the noisy image , lower the value of *SSI* better will be the image quality. The tabulated values of SSI in table II for the proposed method shows appreciable speckle removal results at higher level of speckle noise which is in contrast with the hybrid median filter in order to filter the speckle content from SAR images. Fig2. shows the images produced after filtering from both the hybrid median and proposed local statistics filter. The end results of filtering obtained through the proposed filter are relatively much better in terms of prevention of the image details and applicabilty in real world.

4 Conclusion

In this paper, the novel idea of suppressing the speckle content in an image with a new local statistic filter is proposed. The filtering gives more satisfactory and improved results than the earlier proposed speckle filtering techniques. The simulation results in fig2. for the proposed filter with a new 7x7 mask yield a better value of both PSNR and SSI as compared to the earlier filters. Also, this filter has a novelistic property of removing noise, alongside edge retention and details preservation. The betterment of results after filtering through this filter is appreciably up to the mark and the images obtained show their importance for various purposes of navigation and research.

References

1. Oliver, C.J.: Information from SAR Images. Journal of Applied Physics 24(5), 1493–1514 (1991)
2. Santosh, D.H.H., et al.: Efficiency Techniques for Denoising of Speckle and Highly Corrupted Impulse Noise Images. In: Proc. of the 3rd International Conference on Electronics and Computer Technology, vol. 3, pp. 253–257 (2011)
3. Lopes, A., Tauzin, R., Nezry, E.: Adaptive Speckle Filters and Scene Heterogenity. IEEE Transactions on Geoscience and Remote Sensing 28(6) (1990)
4. Perona, P., Malik, J.: Scale-Space and Edge Detection using Anisotropic Diffusion. IEEE Transactions on Pattern Analysis and Machine Intelligence 12(7), 629–639 (1990)
5. Aja-Fernandez, S., Alberola-Lopez, C.: On the Estimation of the Coefficient of Variation for Anisotropic Diffusion Speckle Filtering. IEEE Transactions on Image Processing 15(9), 2694–2701 (2006)
6. Glavin, M., Jones, E.: Echocardiographic Speckle Reduction Comparison. IEEE Transactions on Ultrasonics, Ferroelectrics and Frequency Control 58(1), 82–101 (2011)
7. Junzheng, W., Weidong, Y., Hui, B., Weiping, N.: A Despeckling Algorithm Combining Curvelet and Wavelet Transform of High Resolution SAR Images. In: International Conference on Computer Design and Application (ICCDA), vol. 1, pp. 302–305 (2010)
8. Do, M., Vetterli, M.: The contourlet transform: An efficient directional multiresolution image representation. IEEE Transactions on Image Processing 14(12), 2091–2106 (2005)
9. Weickert, J.: Coherence-enhancing diffusion filtering. International Journal of Computer Vision 31(2-3), 111–127 (1999)

10. Gonzalez, R., Woods, R.: Digital Image Processing, 3rd edn. Pearson Prentice Hall Press, New York (2009)

11. Loizou, C.P., Pattichis, C.S., Christodoulou, C.I., Istepanian, R.S., Pantziaris, M., Nicolaides, A.: Comparative Evaluation of Despeckle Filtering in Ultrasound Imaging of the Carotid Artery. IEEE Transactions on Ultrasonics, Ferroelectrics and Frequency 52(10), 1653–1669 (2005)

12. Yang, Z., Fox, M.D.: Speckle Reduction and Structure Enhancement by Multichannel Median Boosted Anisotropic Diffusion. EURASIP Journal on Applied Signal Processing 16, 2492–2502 (2004)

13. Ezhilalarasi, M., Umamaheswari, G., Vanithamani, R.: Modified Hybrid Median Filter for Effective Speckle Reduction in Ultrasound Images. In: Recent Advances Networking, Proceedings of International Conference on Networking, VLSI and Signal Processing, ICVNS 2010 (2007)

14. Wang, Z., Bovik, A.C., Sheikh, H.R., Simoncelli, E.P.: Image Quality Assessment: From Error Visibility to Structural Similarity. IEEE Transactions on Image Processing 13(4), 600–612 (2004)

15. Gupta, A., Tripathi, A., Bhateja, V.: Despeckling of SAR images via an improved anisotropic diffusion algorithm. In: Satapathy, S.C., Udgata, S.K., Biswal, B.N. (eds.) Proceedings of Int. Conf. on Front. of Intell. Comput. AISC, vol. 199, pp. 747–754. Springer, Heidelberg (2013)

Multisession Video Packet Scheduling

R. Arockia Xavier Annie, Murugesan Anitha, and P. Yogesh

Abstract. The main objective of this paper is to deliver quality video at the receiver end using the proposed scheduling schemes. The video is compressed for video streaming process, which may cause loss of packets (or) frames. Each video packet has different levels of contribution to improve the video quality at the receiver side. To overcome these problems the packet scheduling is being used. The packet scheduling is used to determine the priorities of each video packet (or) frame. The hybrid video encoder MPEG-4 is used for encoding and decoding. The importance level of video frame is based on frame types (I, P or B frames). The transmission errors are measured at finer scales for the Macro Blocks (MB) at packet-level. A simple packet scheduling by just assigning a higher and lower level priorities to the packets is tested for the video and the error scales are measured. The original input video is compared with the erroneous streamed video to determine the frame loss and packet loss. The frames which are not decoded are also considered as frame loss. With the outcome on high gain packet delivery, this scheduling scheme is better than other models.

Keywords: Video streaming, Priority Scheduling, Transmission distortion model.

1 Introduction

Video streaming is highly sensitive over mesh networks as the video packets are compressed and transmitted are erroneous with loss of packets (or) frames. A mesh network is a network that ensures that it has one of the two connection arrangements that is either, a full mesh topology or partial mesh topology.

R. Arokia Xavier Annie · Murugesan Anitha
DCSE, CEG, Anna University, Chennai – 600 025
e-mail: annie@annauniv.edu, anitha_kpm@yahoo.com

P. Yogesh
DIST, CEG, Anna University, Chennai – 600 025
e-mail: yogesh@annauniv.edu

S.M. Thampi et al. (eds.), *Recent Advances in Intelligent Informatics*,
Advances in Intelligent Systems and Computing 235,
DOI: 10.1007/978-3-319-01778-5_4, © Springer International Publishing Switzerland 2014

In a full mesh topology, each node is connected directly to each of the other nodes in the system, whereas in partial mesh topology the nodes are connected to only some of the nodes in the system. So, a mesh network utilises multisession video transmission that is more useful at the receiver side. The multisession video transmission can achieve better video quality without any delay in these network types.

In this paper, Packet Scheduling algorithm is used to schedule the packets by assigning 'higher and lower' priorities to the packets before transmission that engages in this multisession. The packets are scheduled even without setting any deadlines to packets or sometimes it may miss their deadlines which will result in loss of packets (or) frames. This erroneous video is considered as the transmission distortion model which compared with the original input video and evaluated. The evaluation is done by calculating the Peak Signal-to-Noise Ratio (PSNR) values for each frame.

Scheduling of packets involves choosing their sending order, or selecting the next packet to be sent. The basic criteria for deciding sending order is the deadline of Video Packets (VP). The sender sends the packet with Earliest Deadline First (EDF). In this case, the waiting time of packet in the receiver's buffer is mini- mized, and the minimum required buffer size at the receiver is obtained.

The remainder of this paper is structured as follows. In Section II, we summar- ize some of the important factors in the related literatures in multisession schedul- ing. Section III, we describe the system design under the action of two mesh net- work types. In Section IV we analyze the performance of the implemented scheme and analyse their results in Section V. Section VI concludes this paper.

2   Related Works

It is shown that the distortion-based utility gradients, is a simple but effective solu- tion for downlink packet scheduling in wireless video streaming applications [1]. This provides optimal solution for the case, when the video packets are indepen- dently decodable and a simple error concealment scheme is used at the decoder.

Packet scheduling algorithms as proposed in [2], [9] for video streaming over channels by applying different deadline thresholds to the video packets with dif- ferent importance. They have evaluated the performance in terms of PSNR, and observed improvements over the conventional earliest-deadline-first schemes in trace-driven simulations which are applied in our system.

X.Tong, Y.Andreopouls, and M.van der Schaar [3] has addressed the problem of robust video streaming in multi hop networks by relying on delay constrained and distortion-aware scheduling, path diversity, and retransmission of important video packets over multiple links. It is to maximize the received video quality at the destination node. And they have developed a linear model to estimate the transmission distortion of each MB. They have also observed that in wavelet based video encoders; the transmission distortion of video packets is approximate- ly an exponential function of the packet index. Their theoretical derivations demonstrate that the path diversity is not beneficial when link failures are not expected in the multi hop infrastructure.

Vander Schaar et al [4] have shown that partitioning an embedded video stream into several priority classes can improve the overall received video quality. Based on this concept he has proposed a cross-layer approach using priority queuing [5]. The essential feature behind this approach is the priority queuing [10], based on which, the most important video packet is selected and transmitted, at each intermediate node, over the most reliable link, until the transmission success or the deadline expiration.

Ehsan Maani, Peshala V. Pahalawatta, Randall Berry, Thrasyvoulos N. Pappas, and Aggelos K. Katsaggelos [6] has introduced a content-aware multi-user resource allocation and packet scheduling scheme that can be used in wireless networks where imperfect channel state information is available at the scheduler.

Z.Miao and A.Ortega [7] has proposed a new delivery method, ERDBS(Expected run-time distortion based scheduling) for the framework to solve the packet scheduling problem. The proposed algorithm is designed for the sender driven transmission system can increase the receiver quality by selecting proper packets to be transmitted at any given time during the streaming session.

A.Dua and N.Bambos [8] has proposed a modelling framework is well suited to multimedia streaming applications with soft deadline constraints, where packets which miss their deadlines are not necessarily discarded.

3   System Design Overview

In this system we provide the major network design and the packet scheduling process for the transmission.

The system design is as shown in fig 1. The input video is compressed using video codec. The compressed video contains the video packets, frames, and macro blocks.

3.1   Buffer

The intermediate and destination nodes receive the video packets from ports and store them in the buffer. Buffer size is set initially (eg: 20). When the buffer level exceeds the buffer size, the packets with the least scheduling priorities in the buffer are dropped which is considered as error. The packet scheduler decides at each time slot which packet in the buffer to transmit or drop according to the scheduling assumed.

The cause for the transmission distortion is the packet loss due to buffer overflow and delay bound violation. Once a packet is lost, it will cause distortion in the decoded video frame.

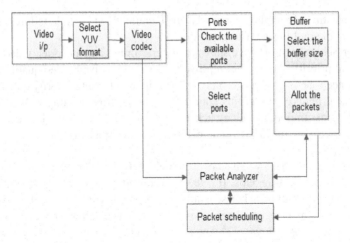

Fig. 1 Packet Priority Analysis Based Scheduling Method

3.2   Packet Scheduling Process

A frame is composed of several video packets (VP) separated by MPEG-4 codec. Considering the frame sequence {F0 F1 F2 …} to be displayed at frame rate of f frames per second. If the receiver starts to display the first frame F0 at time (t)= 0, then the n-th frame, Fn, is expected to be displayed at its deadline, i.e., at t = n/f. If a VP is not available at its expected display time at the receiver, it misses its deadline, and the receiver applies error concealment by copying corresponding macro blocks (MB) from the previous frame.

Scheduling of VPs involves determining their sending order, or selecting the next VP to be sent. One basic criterion for deciding sending order is the deadline of VPs. This means the sender sends the VP with Earliest Deadline First (EDF). In the case of EDF, the waiting time of VPs in the receiver's buffer is minimized, and as a result, minimum required buffer size at the receiver is achieved.

From the view of channel status, if the channel is in good condition without errors, then it is an advantage to use EDF criteria to send VPs in sequential order to obtain minimum average queue length in the receiver's buffer. But if the channel condition is poor with large error rates, then it is desirable to send more important VPs within GOPs first in order to achieve lower video distortion.

3.3   Packet Analyzer

The Fig-2 shows the analysis of the video packets. The erroneous video is compared with the original video in this analysis for evaluation.

3.3.1   Video Encoding

The input video is encoded using MPEG-4 codec. The video is of size 176x144. The codec basically divides this video into 400 frames. The encoded video will be stored, which is a compressed format of the video.

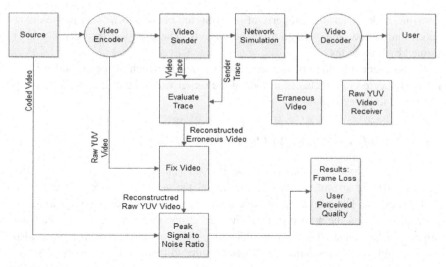

Fig. 2 Packet Analyzer

3.3.2   Video Sender

The video sender reads the compressed video file from the output of the video encoder. As shown in Fig 2, the video encoder fragments each large video into smaller segments via UDP packets over a simulated network. For each transmitted UDP packets, the framework records the timestamp, the packet-id and the packet payload-size in the sender-trace file. The video sender also generates video-trace file that contains information about every frame present in the original video file. The video-trace file and the sender-trace file will be later used for subsequent video quality evaluation.

3.3.3   System Simulation

The encoded video is simulated using NS2 simulator. Each frame will be fragmented into 1000 bytes for transmission. (Maximum packet length will be 1028 bytes, including IP header (20bytes) and UDP header (8bytes)).

3.3.4   Video Encoder

The compressed error video is decoded using MPEG-4 codec to find out the missing and the frames which are not decoded. The frames which are good and decoded correctly are also calculated.

4   Implementation

The network is designed in the Client-Server format. The Server nodes send the media streams to the client on request. The central node of the group acts as the classifier node. It has the property of ordering the packets to be sent to the client nodes and further more scheduling the packets. The group at the server side is formed by four nodes.

At the client side there is a group of four nodes, in which the central node is directly connected to the classifier node at the server side. The other three nodes are connected to the central node.

4.1   Packet Ordering and Queue Monitoring

The packets are ordered based on the flow-id (on which flow they are being transmitted), the size of the packet, the size of the queue using which they are to be transmitted. The packet-id is calculated using the flow-id in which they are present. Based on the arrival time of the packet and the id value, the packets are ordered in the queue. The total number of packets arriving in each flow is also counted and stored as separate text files in order to monitor the amount of traffic in each flow.

The queue is monitored throughout the network transmission process. Monitoring is done based on the following factors: (1) The queue which is currently transmitting the media packets, (2) the size of the queue, (3) the number of packets currently present in the queue and (4) the number of packets lost or dropped from the queue. This process of monitoring also helps in detecting the traffic and predicting the total amount of frame loss to occur.

4.2   Frame Classification

In order to classify each frame of the video file we use the trace files generated when the program starts execution. The trace files contain the following details: the frame number, the frame type, and the length of the frame. Based on the frame type the frames are classified as Type 1, Type 2, and Type 3. The "I" frames fall under the Type 1 category, "P" frames fall under Type 2, whereas "B" frames fall under Type 3. As each frame is classified the count of the frames is retrieved and stored. This procedure of classification is repeated for each and every trace file created for the video.

5   Result Analysis

ENCODING

- Total frames: 400 Frames.
- Total time taken for encoding: 55800 sec.
- Average time taken: 139.7 sec.

VIDEO SENDER

The following segment shows the packets sent and lost and the relevant frames lost. This is in consideration in the I-P-B Frames that were categorised in sent and lost section.

- Packet sent: p->nA: 549, p->nI: 173, p->nP: 109,p->nB: 266
- Packet lost: p->nA: 69, p->nI: 48, p->nP: 14,p->nB: 7
- Frame sent: f->nA: 401, f->nI: 45, f->nP: 89,f->nB: 266
- Frame lost: f->nA:43, f->nI: 23, f->nP: 13,f->nB: 266

Table 1 PSNR Calculation

Table .	PSNR VALUE
1	34.58
2	34.25
3	34.16
4	33.67
5	33.75
6	33.73
7	33.72
8	34.21
9	34.03
10	34.32
11	33.99
12	33.79
13	33.43
14	33.55
15	33.47
16	33.26
17	33.02
18	34.11
19	34.21
20	33.68
21	33.59
22	33.18
23	33.25
24	33.30
25	33.11

FRAME EVALUATION

- Good frames: 355
- Not decoded: 41
- Missing: 4

The table 3 contains the PSNR for sample of 25 frames. The system actually calculates the PSNR value for all the 400 frames. The mean and the standard deviation for those 400 frames are also calculated.

Mean of 400 frames: 24.36
Standard deviation of 400 frames: 8.92

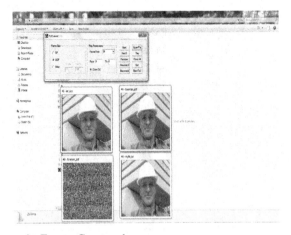

Fig. 3 Snapshot on the Frames Compared

This distortion captured is as shown in Fig. 4. The snapshot shows, the decoded video frame with normal and priority based scheduling. This graph shown in Fig 5 depicts the ratio of the number of frames to the PSNR value.

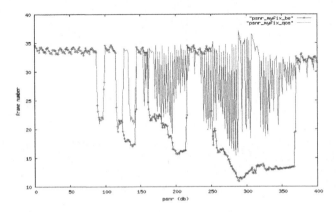

Fig. 4 Evaluated Video Frames Vs PSNR

6   Conclusion and Future Work

A fine-granularity transmission distortion model for the encoder to predict the quality degradation of the decoded videos caused by lost video packets has been developed. The frame loss and the packet loss are calculated by comparing the erroneous video and the original video. Packet loss and frame loss are being observed for all the frames (I, P and B). Packet scheduling algorithm used here, that schedules packets can be modified to use the performance evaluation of the system as a future work.

References

1. Pahalawatta, P.V., Katsggelos, A.K.: Review of content-aware resource allocation schemes for video streaming over wireless networks. Wireless Communication Mobile Computing 7(2), 131–142 (2007)
2. Kang, S., Zakhor, A.: Packet scheduling algorithm for wireless streaming. In: Proc. Packet Video (April 2002)
3. Tong, X., Andreopoulos, Y., Van der Schaar, M.: Distortion-driven video streaming over multihop wireless networks with path diversity. IEEE Transaction Mobile Computing 6(12), 1343–1356 (2007)
4. Van der Schaar, M., Anderopoulos, Y., Hu, Z.: Optimized scalable video streaming over IEEE 802.11 a/e HCCA wireless networks under delay constraints. IEEE Transaction Mobile Computing 5(6), 755–768 (2006)
5. Shiang, H., Van der Schaar, M.: Multi-user video streaming over multi-hop wireless network: A distributed, cross-layer approach based on priority queuing. IEEE J. Select. Areas. Communication 25(4), 770–786 (2007)
6. Maani, E., Pahalawatta, P.V., Berry, R., Pappas, T.N., Katsaggelos, A.K.: Resource allocation for downlink multiuser video transmission over wireless lossy networks. North western University, Electrical Engineering and Computer Science Department, 2145 Sheridan Rd, Evanston, IL 60208
7. Miao, Z., Ortega, A.: Expected run-time distortion based scheduling for delivery scalable media. In: Proc. Packet Video (April 2002)
8. Dua, A., Bambos, N.: Downlink wireless packet scheduling with deadlines. IEEE Transaction Mobile Computing 6(12), 1410–1425 (2007)
9. Goudarzinemati, A., Enokido, T., Takizawa, M.: Scheduling Algorithms for Concurrently Streaming Multimedia Objects in P2P Overlay Networks. Special issue: Frontiers in Complex, Intelligent and Software Intensive Systems. Journal of Computer Systems Science and Engineering 25(2) (March 2010)
10. Kim, H.-S., Hwang, C.-S., Lee, S.-K., Choi, S.-J., Gil, J.-M.: Priority based list scheduling for sabotage-tolerance with deadline tasks in desktop grids. Journal of Computer Systems Science and Engineering 23(2) (March 2008)

Mathematical Morphology Based Fovea Center Detection Using Retinal Fundus Images

Ganapatsingh Rajaput and Bharati Reshmi

Abstract. Exudative diabetic maculopathy is a frequent cause of visual deterioration in patients with diabetic retinopathy and represents a form of diabetic macular edema (DME), which is derived from leaking retinal vessels. The detection of the fovea center is a prerequisite for diagnosis of exudative diabetic maculopathy. In this work, a novel method for fovea center detection from color retinal fundus images is presented. With the prior knowledge of relative location of the optic disc, mathematical morphology is used to detect fovea center. The proposed method is robust to inconveniences caused by diabetic retinopathy lesions like microaneurysms, hemorrhages and exudates. Experiments were performed on local and public databases that yielded success rate of 91.38 % and 91.75 %, respectively.

Keywords: diabetic maculopathy, diabetic retinopathy, exudates, optic disk, fovea.

1 Introduction

Macula is the darkest region of the retina and macula center (fovea center) is an important anatomical landmark in automated analysis of Diabetic Maculopathy (DM) in color fundus images [1]. It is the area providing the clearest, most distinct vision. The center of the macula is called fovea, an area where all of the photoreceptors are cones; there are no rods in the fovea. The fovea is the point of sharpest, most acute vision acuity. The very center of the fovea is the foveola (Fig. 1(a)).

In physical terms, the fovea region is a circle of 0.25 mm of diameter, with its center located a number of optic disc diameters away (2 disc diameters) from the disc center, in the temporal side of the optic nerve (i.e., towards the macula center). Detection of hard exudates (bright lesions in the retinal fundus images) is not

Ganapatsingh Rajaput · Bharati Reshmi
Department of Computer Science, Gulbarga University, Gulbarga, Karnataka, India
e-mail: breshmi@yahoo.com

S.M. Thampi et al. (eds.), *Recent Advances in Intelligent Informatics*,
Advances in Intelligent Systems and Computing 235,
DOI: 10.1007/978-3-319-01778-5_5, © Springer International Publishing Switzerland 2014

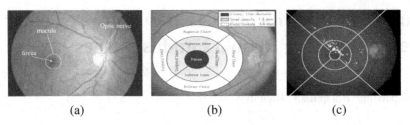

(a)   (b)   (c)

Fig. 1 (a) Retinal fundus image (b) Different sectors of retinal fundus image (c) Distribution of exudates around fovea region

sufficient to detect DM. The distribution of these exudates around the fovea also must be considered. The exudates distribution around the fovea [2] is analyzed based on the retinal sectors (Fig. 1(b)). The retinal fundus image is sub divided into 10 sectors, namely,(1) Central;   (2) Inner superior; (3) Inner temporal; (4) Inner interior; (5) Inner nasal; (6) Outer superior; (7) Outer temporal; (8) Outer inferior; (9) Outer nasal; and (10) Far temporal field, centered at the fovea center. The three fovea circles drawn are based on the optic disc diameter. The radii of the smaller, middle, and outer fovea centered circles are equal to 1/3, 1, and 2 optic disc diameters, respectively.   Exudates inside the central circle (subfield 1) and inside the inner circle (subfield 2-5) tend to affect more the patient visual acuity (clinically significant maculopathy) [3] than the exudates inside the outer circle (i.e., Non clinically significant maculopathy) (Fig. 1(c)). Analyzing the exudates distribution in this way helps the ophthalmologists to understand the severity of DM. The methods available in the literature for detection of fovea center based on optic disc location, its diameter and/or blood vessels arcade. Some of the methods highlight whether the approach is negatively affected by the diabetic lesions like exudates and hemorrhages. There are very few methods available in the literature for detecting fovea center compared to the optic disc detection. A template based method for fovea center detection is presented in [4] which uses an artificial gray-scale model of size 40x40 pixels representing a real fovea region obtained using Gaussian distribution with a fixed standard deviation. The correlation coefficients of this model and retinal image are compared with a threshold to locate fovea center. In [5] an algorithm for fovea detection based on optic disc diameter, a region of interest and adaptive threshold is described. An appearance based approach, in which the local contrast of the image is enhanced and the darkest region is selected as fovea candidate region, is presented in [6]. In [7], a method based on cost function as well as a point distribution model is proposed to detect and locate the fovea center. Hough transform based approach is presented in [8] for optic disc segmentation and for detecting fovea region and its center. In [9], first detection of the vessel pixels is performed and then macula is identified by finding the darkest cluster of pixels near the optic disc (OD). In [10], a fovea center detection method based on minimum vessel density within the search area defined from anatomical priors i.e., knowledge on the structures of the retina using MESSIDOR images is proposed. The method for locating the fovea as the centroid of the largest group of pixels with minimum intensity value is given in [11].

In this paper, we present a robust method to localize the fovea center based on optic disc center, optic disc diameter and mathematical morphology (Fig. 2). The proposed algorithm localizes the fovea center as a single pixel in the fovea candidate region (i.e., fovea centroid) that is used for classifying DM. In the first step, the optic disc center and optic disc diameter is determined. In the second step, using the location of optic disc center and optic disc diameter, the ROI for the macula region is located and then the fovea center is detected.

The rest of the paper is organized as follows. The materials used for fovea center detection are described in section 2. The proposed method for fovea detection is discussed in section 3. Section 4 describes experimental results and conclusions are given in section 5.

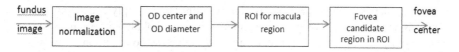

Fig. 2 Diagram for automatic detection of fovea center

2 Materials

A local database and a public database are used to test the efficacy of the proposed method. The local data set of fundus images is provided by the Department of Ophthalmology, Kasturba Medical College, and Manipal, India. The fundus images were captured from a TOPCON nonmydriatic retinal camera with model number TRC-NW200. The built-in CCD camera provides up to 3.1 megapixels of high quality imaging and the inbuilt imaging software is used to store the images in the JPEG format. The images are of 576X720 pixels in size with 8 bits per color plane. Further, the public database, namely, MESSIDOR, containing fundus color images of the posterior pole acquired by the HôpitalLariboisière Paris, the Faculté de Médecine St. Etienne and the LaTIM–CHU de Brest (France) are considered [12]. These images are of 2240X1488 pixels in size with 8 bits per color plane provided in TIFF format. Only 100 images are used for experiment purpose.

3 Proposed Method

The proposed fovea center detection method is achieved in two stages, OD detection and the fovea center detection. The following sections describe these two stages in detail.

3.1   Optic Disc Detection

The OD is a brightest component of the fundus image and is clearly visible in the red channel after inverting the image. The optic disk segmentation method is based on mathematical morphology and is explained below. In first step, the OD candidate region is localized and then the coordinates of a pixel as OD center which is located within the OD is determined. Using this information, in the second step, the circular approximation of OD boundary is localized using circular model. The operation performed in the first stage is as follows. The fundus image is normalized in order to make the image invariant with respect to background pigmentation variation between individuals. Color normalization does not aim to find the true object color, but to transform the color so as to be invariant with respect to the changes in the illumination without losing the ability to differentiate between the objects of interest. The method used for color normalization is a histogram specification [13] which transforms the red, green, and blue histograms to match the shapes of three specific histograms, rather than simply equalizing them. This has the advantage of producing more realistic looking images than those generated by equalization, and it does not exaggerate the contribution of the blue channel. In the proposed method, the reference histograms were taken from an arbitrary normal image (the reference image was chosen in agreement with the expert ophthalmologist) with good contrast and coloration. The result after color normalization is shown in the Fig.3(c). Next, the red channel is extracted from the color normalized image (Fig. 3(d)). The OD region is fragmented into multiple sub regions by blood vessels. The blood vessels enter the OD from different directions with a general tendency to concentrate around the nasal side of the OD region. Therefore, blood vessels are removed from the image by applying closing operation using octagon shape structuring element of size 15 (Fig.3(e)) to create fairly constant region. This operation contributes more towards better segmentation of OD in case of DR images. It is challenging task to identify the vessels width with precision. If the structuring element is too large, it does not ensure that all trees of the vessels are removed. A very large structuring element may deform the OD boundary. The resulting image is inverted (Fig. 3(f)). The regions with minimal intensities are identified using extended minima transform [14] (the regional minima of h minima transform) and are given in Fig. 3(e). This transformation is a thresholding technique that brings most of the valleys to zero. The h-minima transform suppresses all the minima in the intensity image whose depth is less than or equal to a predefined threshold. The output image is a binary image with the white pixels representing the regional minima in the original image. Regional minima are connected pixels with the same intensity value, whose external boundary pixels all have a higher value. The extended minima transform on the f image with threshold value h is given by the following expression

$$E = EM (f, h) \tag{1}$$

where E is the output image.

During this process, the selection of threshold is very important where the higher value of h will lower the number of regions and a lower value of h will raise the number of regions. The h value (threshold height) is selected empirically (i.e. h= 5). The result is shown in Fig 3(g). Morphological opening is applied using disk shaped structuring element to eliminate the regions that are wrongly located (Fig. 3(h)). The mean intensities of the identified regions are computed. The region with the lowest mean intensity is then selected as the optic disk region. For the selected candidate optic disk region the centroid is computed. During the second stage, for the selected OD candidate region, the equidiameter is computed using the equation given below and the circle is fitted (Fig. 3(k)).

$$\text{Equidiameter} = \sqrt{(4 \times \text{Area}/\pi)} \qquad (2)$$

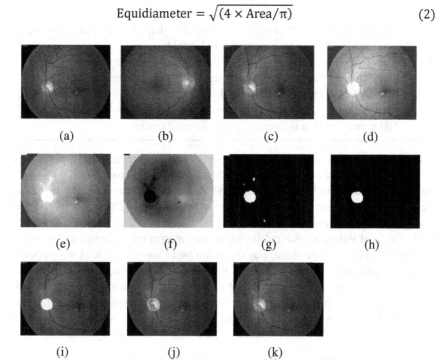

(a)   (b)   (c)   (d)

(e)   (f)   (g)   (h)

(i)   (j)   (k)

Fig. 3 (a) Original RGB retinal image.(b) Template image for color normalization. (c) Color normalized image. (d) Red channel of the image. (e) After closing operation for removal of Blood vessels. (f) Inverted image. (g) After applying Extended minima transform. (h) After applying opening operation. (i) Detected candidate OD region superimposed on the original image. (j) OD boundary and its center marked. (k) Circular approximation of the OD.

The two stages of OD detection are outlined in the algorithm given below:

Algorithm I
Input: Color Fundus Image, **Output:** Circular approximation of the OD.

Step 1: Perform color normalization on the input image.
Step 2: Retrieve the red channel of the image from the resulting image and eliminate blood vessels from the image by applying closing operation using octagon shape structuring element.
Step 3: Invert the image and identify the regions with minimal intensities in the region using Extended Minima Transform by choosing value of h empirically.
Step 4: Apply opening operation using disc shaped structuring element to eliminate regions that are wrongly located.
Step 5: Compute the mean intensities of the identified regions. The region with the lowest mean intensity is selected as the candidate optic disc region and the centroid is plotted.
Step 6: Using area of the OD, the diameter is computed and circular OD approximation is determined by fitting a circle.

The optic disc segmentation algorithm rendered an average overlapping score of 99.45% between true OD (ground truth images) and segmented OD, and success rate of 92.06% for local database. And for public database, it is resulted in an average overlapping score and success rate of 99.47% and 92%, respectively. The average difference between the diameter of the true OD and segmented OD is 5.93 pixels for local database. For public database, the average diameter difference is 24.92 pixels. For local database, our algorithm takes an average 8.21 s processing time per image and for public database; it takes 6.10 s per image. The OD detection results in normal and DR image are given in Fig. 4. The same parameter values are used for both the methods i.e., segmentation of candidate region of the OD, but in preprocessing step, for public data base blood vessels removal is not applied, but color normalization and histogram equalization are applied.

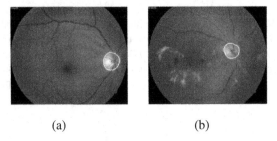

(a)   (b)

Fig. 4 Detection of OD boundary and its center (a) Normal image (b) DR image

3.2 Fovea Center Detection

After locating the optic disk center and knowing the diameter of the optic disk, the fovea center can be determined by setting an area of restriction in the vicinity of the image center, as determined by the optic disk center. The distance and position of macula with respect to the diameter of the optic disk remains relatively constant. A method for detecting ROI for the macula was referred from [15]. This approach is used to localize the ROI for macula before detecting fovea center. It is situated about 2 disk diameter temporal to the optic disk in fundus images and the mean angle between macula and the center of the optic disk against the horizon is -5.6 +or – 3.3 degrees. Since the location of macula region varies from individual to individual, a ROI for the macula is localized as shown in the Fig. 5. The width of the ROI is taken equal to 2DD as the mean angle between the fovea and the center of the optic disk to the horizontal, varying between -2.3 to -8.9.degrees.

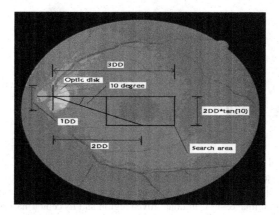

Fig. 5 Macula region search area

Macula is a darkest component of the fundus image and is clearly visible in the green channel. Histogram equalization is applied on the green channel of the image for contrast enhancement. The ROI for the macula is localized. This is the search area to detect fovea center in the macula. The regions with minimal intensities in the search region using extended minima transform (empirical threshold height, for local database, h=50 and for public database h=20 is used) are identified. Opening operation is applied using disk shaped structuring element of size 6 to eliminate the regions that are wrongly located. The mean intensities of the identified regions are computed. Then the region with the lowest mean intensity is selected as the macula region and then the centroid is computed. The approach used for fovea center detection is given in Fig. 6.

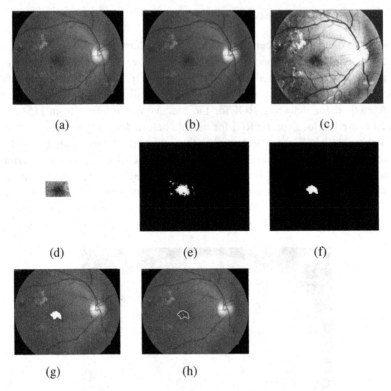

Fig. 6 (a) Original RGB image. (b) Green channel of the image. (c) Contrast enhanced using Histogram equalization. (d) Macula ROI for the candidate fovea region. (e) Extended minima transform applied. (f) After opening operation. (g) Detected fovea region superimposed on the original image. (h) Candidate fovea region and its center marked.

Thus algorithm for detection of fovea center is given below:

Algorithm II:
Input: Color fundus image,   **Output:** Fovea center pixel
Step 1: Extract the green channel of the image.
Step 2: Enhance the contrast of the image using histogram equalization.
Step 3: Localize ROI for macula region and identify the regions with minimal intensities in the ROI using extended minima transform.
Step 4: Apply morphological opening operation using disk shaped structuring element to eliminate those regions that are wrongly located.
Step 5: Compute the mean intensities of the identified regions and Select the region with the lowest mean intensity as the fovea region and compute its centro-id.

---

4   Experimental Results

Experiments were performed on the 63 images (consisting of 25 Diabetic retinopathy images and 38 normal images) of local database and on the 100 images
(consisting of 68 diabetic retinopathy images and 32 normal images) of the public
database i.e. MESSIDOR. Five images of local database and three images of
MESSIDOR database have been excluded for not presenting visually detectable
fovea region. To evaluate the algorithm performance on both available databases,
the fovea center (i.e., a single pixel coordinates) was manually delimited by an
expert ophthalmologist producing by this way a ground truth set). The Euclidean
distance between the manually marked fovea center pixel and detected fovea center pixel is used for measuring the performance of our proposed method (Fig. 7).

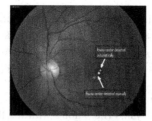

Fig. 7 The fovea center is marked manually and automatically using + and * respectively

Fig.8 shows close look up of fovea center detection. Fig. 9 gives some sample
images showing fovea region and its center marked.

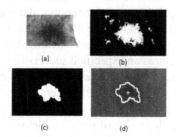

Fig. 8 (a) Macula search area. (b) After applying Extended minima transform. (c) Fovea
candidate region after opening operation applied. (d) Candidate fovea region and its center
marked.

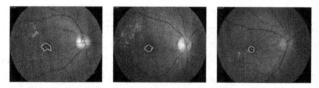

Fig. 9 Sample fovea center detection results

Fig.10 (a) and Fig. 10(b) shows that even the macula region is affected by the DR lesions; our proposed method is able to detect the fovea center successfully.

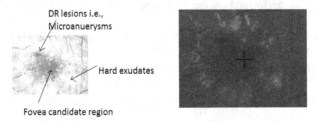

Fig. 10 (a) Macula ROI containing Diabetic retinopathy lesions like microanuerysms, hemorrhages and hard exudates. (b) Fovea center detected. (DR lesions are not affected).

In case of fundus images of the public database, one additional pre-processing step, i.e., removal of blood vessels using octagon shape structuring element with size 9, was used before applying extended minima transform. Because, it was observed that, in the images of public database, the vicinity of the thin blood vessels in the macula search area is more compared to images of local database. This contributes towards not affecting the detection of candidate fovea region by small diabetic lesions like microaneurysms (i.e., along with the thin blood vessels these lesions are also removed). The experimental results using local database and public database yielded success rate of 91.38 % and 91.75 %, respectively. Euclidean distance between manually detected fovea center and automatically detected fovea center (i.e., average error) for local and public database is 7.49 pixels and 26.38 pixels respectively. We assume tolerance rate for Euclidean error for local database (image size: 576X720 pixels) and for public database (image size: 2240X1488 pixels) 30 pixels and 60 pixels respectively. From the available literature, it is observed that most of the authors have identified only the candidate region of the fovea in the macula area and specified success rate of their proposed algorithm. Only few have used Euclidean error as performance measure for fovea detection. The algorithm results reported by [10, 11] on MESSIDOR images are used for comparison study. The algorithm proposed by [10] on MESSIDOR images performs well on images with no risk of macular edema (i.e., 80%) but less accurate on images with risk of macular edema (i.e., 59%) and it requires more processing time, because it uses 3 different stages (OD detection and vessel segmentation, search region detection and fovea center detection) to locate fovea center. Our algorithm for fovea center detection performs well on DR images (including images with risk of macular edema) and requires less processing time compared to the algorithm given in [10]. The algorithm for fovea detection which is based on ROI and pixel search in the macula region, provided in [11] is simple compared to the method in [10]. The results in [11] are directly referred from [10] and are reported the highest accuracy in MESSIDOR images (i.e., images with no risk of macular edema: 96.55%, images with risk of macular edema: 91%) compared to the results in [10]. In [10, 11], authors have not specified accuracy of their algorithms in terms of Euclidean distance error and they have given only the success rate of their algorithm.

It was observed that, our proposed OD detection algorithm did not locate OD in certain images and hence the fovea center was not correctly localized even though it was visible in the image. The failure of detecting the fovea center is attributed to uneven illuminations near the OD border that biased the OD candidates and that the myelinated retinal fibre layer (i.e., retinal lesion, which is appearing as a shiny white area of variable extent following the line of the retinal nerve fibre layer and often continuous with the OD margins and has an indistinct edge) caused by abnormal myelination of the nerve fibre of the retina. To detect fovea center, for local database, our algorithm takes an average 8.43 s per image (including OD detection and fovea center detection) and for public database, it takes 21.15 s per image (including OD detection and fovea center detection). The prototype for the experiment was implemented using MATLAB R2009a and was performed using a Laptop-based system with Intel(R) Core(TM)2 Quad CPU, clock of 2.40GHz, and4GB of RAM memory.

5 Conclusion

A novel method for detection of fovea center in the fundus eye image is presented in this paper. The proposed method is based on information of optic disc center, its diameter and mathematical morphology. The experiment results with two databases (i.e., local and MESSIDOR public database) indicate that the proposed method achieved success rate of 91.38 % for Local database and 91.75% for public database. The algorithm is robust to local disturbances introduced by diabetic retinopathy lesions like microanuerysms, hemorrhages and exudates in the color fundus images. The performance accuracy using Euclidean distance to the manually marked fovea center delimited by experts (i.e., average Euclidean distance error) is 7.49 pixels for local database and 26.38 pixels for public database indicating efficacy of the proposed method.

Acknowledgements. We acknowledge the Department of Ophthalmology, Kasturba Medical College, Manipal, Karnataka, India and MESSIDOR project for the use of their retinal images. The authors would like to thank Dr. Shivakumar. Hiremath, Department of Ophthalmology, Kumareshwara Medical College and Research Center, Bagalkot, Karnataka, India for his valuable suggestions. This work is supported by UGC, New Delhi under Major Research Project grant in Science and Technology (F. No. 40-257/2011 (SR) dated 29.06.2011).

References

1. Rema, M., Pradeepa, R.: Diabetic Retinopathy: An Indian Perspective. Journal of Medical Research 125, 297–370 (2007)
2. Daniel, W., et al.: Grading the severity of diabetic macular edema cases based on color eye images. In: Emre, M. (ed.) Color Medical Image Analysis. Lecture Notes in Computer Vision and Biomechanics, vol. 9, pp. 109–128. Springer (2013)

3. Nayak, J., Bhat, P.S., Acharya, U.R.: Automatic identification of diabetic maculopathy stages using fundus images. Journal of Medical Engineering and Technology 33(2), 119–129 (2009)
4. Sinthanayothin, C., Boyce, J.F., Cook, H.L., Williamson, T.H.: Automated localization of the optic disc, fovea and retinal blood vessels from digital color fundus images. British Journal of Ophthalmology 83, 902–910 (1999)
5. Narasimha, I.H., Can, A., Roysam, B., Stewart, C.V., Tanenbaum, H.L., Majerovics, A., Singh, H.: Robust detection and classification of longitudinal changes in color retinal fundus images for monitoring diabetic retinopathy. IEEE Transactions on Biomedical Engineering 53(6), 1084–1098 (2006)
6. Singh, J., Joshi, G.D., Sivaswamy, J.: Appearance based object detection in color retinal images. In: Proceedings of the IEEE International Conference on Image Processing, pp. 1432–1435 (2008)
7. Niemeijer, M., Abramoff, M.D., Ginneken, B.V.: Segmentation of the optic disc, macula and vascular arch in fundus photographs. IEEE Transactions on Medical Imaging 26, 116–127 (2007)
8. Sekar, S., Al- Nuaimy, W., Nandi, A.: Automatic localization of optic disc and fovea in retinal fundus images. In: Proceedings of 16th European Signal processing Conference, Lausanne, Switzerland (2008)
9. Sagar, A.V., Balasubramaniam, S., Chandrasekaran, V.: Automatic detection of anatomical structures in digital fundus retinal images. In: Proceedings of IAPR Conference on Machine Vision and Applications, Tokyo, Japan, pp. 483–486 (2007)
10. Chin, K.S., Trucco, E., Tan, L., Wilson, P.J.: Automatic fovea location in retinal images using anatomical priors and vessel density. Pattern Recognition Letters 34, 1152–1158 (2013)
11. Liang, Z., Wong, D.W.K., Liau, J.: Towards automatic detection of age related macular degeneration in retinal fundus images. In: 32nd Annual International Conference of the IEEE EMBS Buenos Aires, Argentina (2010)
12. MESSIDOR: Digital Retinal Images, http://messidor.crihan
13. Gonzalez, R.C., Woods, R.E.: Digital Image Processing, 3rd edn. Prentice Hall (2007)
14. Soille, P.: Morphological Image Analysis: Principles and Applications. Springer, New York (2003)
15. Siddalingaswamy, P.C., Gopalakrishna, P.K.: Automatic grading of diabetic maculopathy severity levels. In: Proceedings of IEEE International conference on Systems in Medicine and Biology, pp. 331–334 (2010)

Speaker Recognition Using MFCC and Hybrid Model of VQ and GMM

Dhruv Desai and Maulin Joshi

Abstract. Speaker recognition is widely used for automatic authentication of speaker's identity based on human biological features. Speaker recognition extracts, characterizes and recognizes the information about speaker identity. For feature extraction and speaker modeling many algorithms are being used. In this paper, we have proposed speaker recognition system based on hybrid approach using Mel Frequency Cepstrum Coefficient (MFCC) as feature extraction and combination of vector quantization (VQ) and Gaussian Mixture Modeling (GMM) for speaker modeling. Our approach is able to recognize speaker for both text dependent and text independent speech and uses relative index as confidence measures in case of contradiction in recognition process by GMM and VQ. Simulation results highlight the efficacy of proposed method compared to earlier work.

Keywords: Feature Extraction, Feature Matching, Mel Frequency Cepstral Coefficient (MFCC), Gaussian mixture modeling.

1 Introduction

Speaker recognition automatically identifies the speaker on the basis of individual information incorporated in speech waves and divided into two classifications namely a) speaker identification and b) speaker verification. Speaker identification system once properly trained can recognize authenticated speaker. Speaker

Dhruv Desai
Department of Electronics enginnering,
Sarvajanik College of Engineering and Technology, Surat, 395001, India
e-mail: dhruvdesai_09@rediffmail.com

Maulin Joshi
Department of Electronics & Communication,
Sarvajanik College of Engineering and Technology, Surat, 395001, India
e-mail: maulin.joshi@scet.ac.in

S.M. Thampi et al. (eds.), *Recent Advances in Intelligent Informatics*,
Advances in Intelligent Systems and Computing 235,
DOI: 10.1007/978-3-319-01778-5_6, © Springer International Publishing Switzerland 2014

identification can further be divided into text- dependent and text independent methods. In text dependent method, speaker utters key words or sentences having the same text for both training and testing trials. In case of text independent method, identification does not rely on a specific text being spoken. Speaker verification is the process of rejecting or accepting the identity claim of a speaker. In many applications, voice is used as the key to confirm the identity of a speaker. Speaker recognition is widely applicable to verify user's identity and control access to services such as banking by telephone, database access services, voice dialing telephone shopping, information services, security control for secret information areas.

In spite of impressive advances in the field of speaker recognition in recent years, it is still the area of an active research because of uncertainties involved due to unknown environments in real world scenarios. These uncertainties are due to following reasons like mimicking voice, background noise, recording, stress condition of individuals.

Among many approaches proposed to solve the above mentioned challenges for feature extraction, MFCC is found more accurate [2-9] compared to LPC which has been discussed in. [1] Effects of number of filters and types of windows are discussed in [1][3]. Feature modeling is discussed with GMM in [4-5] while; the same is emphasized by vector quantization in [3-4]. However, disadvantage of VQ lies with ignorance of the possibility of a specific training vector that may also belong to another cluster. [8] This disadvantage is overcome by using GMM method. This approach provides good identification rate [6-9] due to the fact that individual Gaussian classes are interpreted to represents set of acoustic classes. These acoustic classes represent vocal tract information. Gaussian mixture density provides smooth approximation to distribution of feature vectors in multi-dimensional feature space but the disadvantage of this approach is that it requires sufficient data to model the speaker well also outliers are there in this modeling also.

In this paper, we propose model that combines the two modeling methods i.e. Vector quantization (VQ) and Gaussian mixture modeling (GMM) along with MFCC as a feature extraction technique. In our approach, overall decision is based on agreement or disagreement by individual models. In case of agreement, speaker identification is simple but in case of difference of opinion, we have used confidence ratio- ratio of best score to the second best score - as a secondary measure that shows the confidence of the given model for particular recognition task. This overcomes the disadvantage of individual VQ and GMM methods. In the proposed system, parameters for feature extraction like filter type and size, number of MFC coefficients and for modeling like number of Gaussians and codebook size are fine tuned by performing experimentation for different users, different conditions and correcting errors. Proposed system is capable for recognition in case of both text dependent and text independent speech.

The rest of paper is organized as follows. In Section 2, discuses proposed algorithm including details of feature extraction and speaker modeling techniques. In section 3, highlights comparison of different techniques of speaker recognition. Finally, we conclude in Section 4.

2   Proposed Algorithm

2.1   Basic Model for Speaker Recognition

Speaker recognition systems involve two phases namely, training and testing in similar ways to other pattern recognition systems Training is the process of familiarizing the system with the voice characteristics of the speakers registering, testing is the actual recognition task. Speaker recognition system once been trained with number of speakers should be able to identify speaker. The basic block diagram of such system is shown in fig.1.

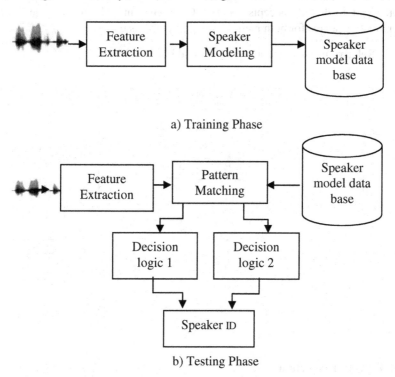

a) Training Phase

b) Testing Phase

Fig. 1 Schematic diagram of the speaker identification system

During training phase, signal processing part converts the sampled speech signal into set of feature vectors. These feature vectors characterize the properties of speech unique for different speakers. This part is common in training and testing phases. In speaker modeling, feature vectors are modeled by speaker modeling. Different speaker models are stored in speaker database during the training phase. During the testing phase, after feature extraction, pattern matching block compares the extracted feature with the data base of known speakers. Finally, decision about the identity of the speaker is given by decision logic block.

2.2   Proposed Algorithm

Gaussian mixture density provides smooth approximation to distribution of feature vectors in multi-dimensional feature space. Vector quantization allows the modeling of probability density functions by the distribution of prototype vectors and also suitable for lossy data compression. As described earlier, in the proposed work as shown in fig.2 combination of both are used to provide more security and robustness to the system.

Once extracted features are modeled using both VQ and GMM blocks, they are compared with database of known speaker to identify the user. If both models agree, the system directly accepts speaker. Otherwise, in order to take the final decision relative scores for both methods are computed as under:

Relative index = ((best score – second best sore)/best score)*100

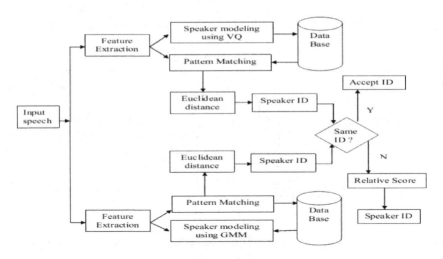

Fig. 2 Proposed Hybrid method using GMM and VQ

2.2.1   Feature Extraction

To recognize the speaker, extraction of the features from speaker's speech is required. Features can be extracted either directly from the time domain signal or from a transformation domain depending upon the choice of the signal analysis approach. Some of the techniques successfully been used for audio classification include Mel-frequency cepstral coefficients (MFCC), Linear predictive coding (LPC) etc. We have selected MFCC as recently, the majority of systems [2-6] have converged to the use of a cepstral vector derived from a filter bank designed according to some model of the auditory system.

Mel Frequency Cepstrum Coefficient (MFCC)

A block diagram of an MFCC processor is given in Fig.3. The speech input is recorded at sampling rate f_s Hz. Sampling frequency is chosen to minimize the effects of aliasing in the analog-to-digital conversion process. The process of segmenting the speech samples into a small frame with the length within the range of 20 to 40 msec. The voice signal is divided into frames of N samples. Adjacent frames are being separated by M. Hamming window is used in feature extraction processing chain and integrates all the closest frequency lines. The Hamming window is given as:

$$h(n) = 0.54 - 0.46 \cos (2\pi n/N\text{-}1) \qquad (1)$$

Where, N = number of samples in each frame. Frequency wrapping has been used in order to maintain the range or scale. In the case of speaker recognition, looking at complexities associated with human speech range; we use mel scale as it gives linear frequency spacing below 1KHz and a logarithmic spacing above 1 KHz. We have used following formula to compute the mels for a given frequency f in Hz [3].

$$Mel (f) = 2595*\log_{10} (1+f/700) \qquad (2)$$

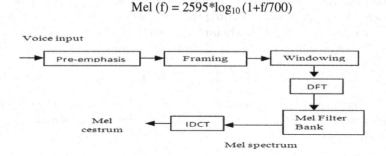

Fig. 3 Block diagram of the MFCC processor [3]

Filter bank is used to simulating the subjective Spectrum, one filter for each desired mel frequency component. The triangular filter bank has a triangular band pass frequency response, and the spacing as well as the bandwidth is determined by a constant mel- frequency interval [1] [3].Once log mel spectrum has been computed, it has to be converted back to time domain by using Inverse Discrete Cosine Transform (IDCT). The result is called the mel frequency cepstrum coefficients (MFCCs). Using the same procedure, a set of mel-frequency cepstrum coefficients are computed for each speech frame of about 20-40 ms with overlapping manner. These sets of coefficients are called acoustic vectors. These acoustic vectors are different for different speakers and can be used to represent the voice characteristic of the speaker.

2.2.2   Speaker Modeling

Speaker modeling algorithms have been used for speaker recognition by compress feature vectors but retain most prominent characteristics. Generally, speaker modeling techniques used in speaker recognition include, Dynamic Time Warping (DTW) [6,7], Vector Quantization (VQ) [3], Gaussian mixture modeling (GMM) [4,5] etc.

Vector Quantization (VQ)
Vector Quantization (VQ) is a pattern classification technique applied to speech data to form a representative set of features. It works by dividing a feature vector (large set of points) into groups having approximately the same number of points closest to them. Each group is represented by its point which is known as centroid point. Codebooks contain the numerical representation of features that are speaker specific. The speaker specific codebook is generated in the training phase by clustering the feature vectors of each speaker. We have used standard LBG algorithm for clustering a set of L training vectors into a set of M codebook vectors.

Gaussian mixture modeling (GMM)
The Gaussian mixture model (GMM) is a density estimator and is one of the most commonly used types of classifier[6],[7]. In this method, the distribution of the feature vector x is modeled clearly using a mixture of M Gaussians. A gaussian mixture density is a weighted sum of M component densities, and is given by

$$P\ (X|\lambda) = \sum_{i=1}^{M} P_i B_i\ B_i(X) \tag{3}$$

Where X is N- dimensional random vector, $B_i(X)$, i=1... M are the component densities and $P_{i,}$ i=1,...M, are the mixture weights. Each component density is a D-variate Gaussian function of the form:

$$B_i B_i(X) = \frac{1}{(2\pi)^{\frac{D}{2}}|\Sigma_i|^{1/2}} \exp\{-\frac{1}{2}(\bar{X} - \bar{\mu}_i)'\Sigma_i^{-1}(\bar{X} - \bar{\mu}_i)\}\ ) \} \tag{4}$$

With mean vector $\bar{\mu}_i$ and covariance matrix Σ_i. The mixture weight satisfy the constraint that $\sum_{i=1}^{M} P_i = 1$. The complete Gaussian mixture density is parameterized by the mean vectors, covariance matrices and mixture weights from all component densities. These parameters are collectively represented by the notation

$$\lambda = \{P_i, \bar{\mu}_i, \Sigma_i \Sigma_i\} \qquad\qquad i = 1,....,M. \tag{5}$$

For Speaker identification, each speaker is represented by a GMM and is referred to by his/her model λ. GMM uses the expectation –maximization (EM) algorithm

On each EM iteration the following re estimation formulas are used which guarantee a monotonic increase in the model's likelihood value:

Mixture Weights

$$\bar{P}_i = \frac{1}{T} \sum_{t=1}^{T} p(i|\bar{x_t}, \lambda) \tag{6}$$

Weighted Means

$$\bar{\mu}_i = \frac{\sum_{t=1}^{T} p(i|\bar{x_t}, \lambda) \bar{x_t}}{\sum_{t=1}^{T} p(i|\bar{x_t}, \lambda)} \tag{7}$$

Variances

$$\bar{\sigma i}^2 = \frac{\sum_{t=1}^{T} p(i|\bar{x_t}, \lambda) \bar{x_t}^2}{\sum_{t=1}^{T} p(i|\bar{x_t}, \lambda)} - \bar{\mu}_i^2 \tag{8}$$

Assuming we have S speakers in a closed-set which is different from each other. For a given speech feature vector $\{X_t\}$, t = 1,2,...,T. The purpose of speaker recognition is to find the speaker k in the closed-set k ∈ {1,2,...,S}, whose corresponding model λ_k will obtain the largest posterior probability $P(\lambda_k | X)$.

3   Simulation Results

3.1   Text Dependent Speaker Identification

As described earlier, during training phase MFCC is used to generate feature vectors and vector quantization and Gaussian mixture modeling are separately used for speaker modeling and all data are stored in data base. During testing phase, speaker identification is done using pattern matching block in which Euclidean distances are calculated. Decisions by individual GMM and VQ are compared and if both agree, the system directly accepts the speaker. If both models differ in identification, relative score are computed for both method and based on this score system takes decision. In our final modeling, identification rate affecting parameters like codebook size (number of Gaussians), number of MFC coefficient and number of Filters in filter bank are fine tuned after performing several experiments. Table 1 shows number of Gaussians/no of centroid Vs identification rate for noisy environment with 30 speakers for both training and testing with 12 cepstral coefficient and 40 filters. Table 2 illustrates effects of number of filters on identification rate for noisy environment with 30 speakers for both training and testing with 12 cepstral coefficient and 16 Gaussians. Table 3 demonstrates effects of number of MFC coefficient on identification rate for noisy environment with 30 speakers for both training and testing with 40 filters.

Table 1 Number of Gaussians/no of centroid Vs. Identification Rate using different approaches

No of Gaussians/no of centroid	Identification Rate (%)		
	VQ [4]	GMM [5]	Proposed GMM+VQ
2	70	70	70
4	76.66	83.33	80
8	83.33	86.67	93.33
16	90	93.33	93.33
32	86.67	83.33	96.67

Table 2 Number of filters Vs. Identification Rate using different approaches

No of filters	Identification Rate (%)		
	VQ [4]	GMM [5]	Proposed GMM+VQ
12	70.66	83.33	83.33
16	83.33	86.67	86.67
20	83.33	86.67	93.33
40	90	93.33	96.67

3.2 Text Independent Speaker Identification

During training phase, MFCC is used to generate feature vectors and Gaussian mixture modeling is used in speaker modeling block. During the training phase, input speech for around 8 sec with the sampling rate of 8000 Hz is used. During the testing phase, input speech for around 4 sec with the sampling rate of 8000 Hz is used. Being text independent system, each speech samples are taken twice ones for training and other for testing but the content of speech is different for all the speakers for both during training and testing. Training database is created for 20 different speakers. Table 1 shows number of Gaussians Vs identification rate.

Table 3 Number of MFC coefficient Vs. Identification Rate using different approaches

No of MFCC	Identification Rate (%)		
	VQ [3]	GMM [6]	Proposed GMM+VQ
1	26.66	26.66	26.66
2	50	50	50
4	50	66.67	76.67
8	76.67	76.67	76.67
12	90	93.33	96.67

Table 4 No of Gaussians Vs. Identification Rate

No Of Gaussian	Identification Rate (%)
2	55
4	65
8	70
16	85
32	85

Table 5 Performance evaluation of speaker identification using proposed method Vs. GMM and VQ

Test speaker	Method	Identified Speaker and Score		Second best speaker and Score		Relative index
s1	**GMM**	**s5**	**7.72**	**s1**	**8.1**	**0.049**
	VQ	**s1**	**7.51**	**s5**	**8.06**	**0.073**
s2	GMM	s2	5.91	s6	6.97	0.179
	VQ	s2	6.14	s7	7.48	0.218
s3	GMM	s3	5.11	s5	7.45	0.457
	VQ	s3	5.82	s7	7.25	0.245
s4	GMM	s4	5.11	s5	8.55	0.673
	VQ	s4	5.03	s5	7.8	0.550
s5	GMM	s5	7.87	s3	9.14	0.161
	VQ	s3	8.3	s6	8.62	0.038
s6	GMM	s6	6.75	s10	7.7	0.140
	VQ	s6	6.94	s2	7.42	0.069
s7	**GMM**	**s7**	**5.86**	**s15**	**8.08**	**0.378**
	VQ	**S6**	**6.42**	**s6**	**6.98**	**0.087**
s8	GMM	s8	6.62	s10	10.5	0.586
	VQ	s8	7.09	s6	8.75	0.234
s9	GMM	s9	5.49	s25	17.06	2.107
	VQ	s9	7.56	s26	12.92	0.708
s10	GMM	s10	5.18	s6	8.39	0.619
	VQ	s10	6.24	s12	7.12	0.141

Table.5 shows performance evaluation of speaker identification using proposed method Vs. GMM and VQ. Identification success rate is an obvious choice as evaluation parameter. From the test case of s1 and s7 it is clear that GMM and VQ methods are outliers, respectively. However, with our combined hybrid approach in order to take the final decision (when two methods differ in decision) relative scores for both methods are computed. For s1 relative index of VQ is higher than GMM while for s7 relative index of GMM is higher. In effect, a system takes decisions of VQ for s1 and GMM for s7 and hence false identification is avoided. The effectiveness of proposed system is observed while performing different experimentations.

4   Conclusions

In this paper, speaker recognition using MFCC and hybrid modeling methods of VQ and GMM is proposed for the text dependent and text independent speech. The techniques used in this paper were able to identify particular user based on individual information stored in database during training. From the performance evaluation it is clear that for speaker identification, combination (GMM+VQ) gives better identification rate than individual models. It is proven that in addition to careful selection of various performance parameters like number of Gaussians/code book size, number of MFC coefficient and numbers of filters consideration of relative scoring as a evaluation measures has strengthen the system capacity for increased success rate.

References

1. Hai, J., Joo, E.M.: Improved linear predictive coding method for speech recognition. In: Information, Communications and Signal Processing and Fourth Pacific Rim Conference on Multimedia. Proceedings of the Joint Conference of the Fourth International Conference, vol. 3, pp. 1614–1618 (2003)
2. Muda, L., Begam, M., Elamvazuthi, I.: Voice Recognition Algorithms using Mel Frequency Cepstral Coefficient (MFCC) and Dynamic Time Warping (DTW) Techniques. Journal of Computing 2(3), 138–141 (2010)
3. Hasan, R., Jamil, M., Rahman, G.R.S.: Speaker Identification Using Mel Frequency Cepstral Coefficients. In: 3rd International Conference on Electrical & Computer Engineering ICECE, pp. 565–568 (2004)
4. Tiwari, V.: MFCC and its applications in speaker recognition. International Journal on Emerging Technologies 1, 19–22 (2010)
5. Shende, A., Mishra, S., Kumar, S.: Comparison of Different Parameters Used In GMM Based Automatic Speaker Recognition. International Journal of Soft Computing and Engineering (IJSCE) 1(3), 14–18 (2011) ISSN: 2231-2307
6. Reynolds, D.A., Rose, R.C.: Robust Text-Independent Speaker Identification using Gaussian Mixture Speaker Models. IEEE Transactions on Speech and Audio Processing 3, 72–83 (1995)

7. Bagul, S.G., Shastri, R.K.: Text Independent Speaker Recognition System using GMM. International Journal of Scientific and Research Publications 2(10), 1–5 (2012)
8. Jayana, H.S., Mahadeva Prasana, S. R.: Analysis, Feature Extraction, Modeling and Testing Techniques for Speaker Recognition. International Journal of Institution of Electronics and Telecommunication Engineers (IETE) 26(3), 181–190 (2009)
9. Kumar, P., Jakhanwal, N., Chandra, M.: Text Dependent Speaker Identification in Noisy Environment. In: International Conference on Device and Communication (ICDeCom), pp. 1–4 (2011)

Image Restoration Based on Scene Adaptive Patch In-painting for Tampered Natural Scenes

Ravi Subban, Subramanyam Muthukumar, and P. Pasupathi

Abstract. Many Researchers proposed algorithms which restored damaged images. These methods cause textures broken while inpainting texture image with complex structure. Most of the existing inpainting techniques require knowing beforehand where those damaged pixels are, either given as a priori or detected by some preprocessing. However, in certain applications, such information is neither available nor can be reliably pre-detected, like noise from archived photographs. This paper propose a patch based adaptive inpainting model to solve these types of problems, i.e., a model of simultaneously identifying and recovering damaged pixels of the given image. The proposed inpainting method is applied to various challenging image restoration tasks, including recovering images that are blurry and damaged by scratches. The experimental result shows that it is effective in inpainting complex texture images.

Keywords: Image Inpainting, Image Decomposition, Restoration, Texture Segmentation, Texture Synthesis, Boundary Restoration, Image Reconstruction, Occlusion Removal.

1   Introduction

Reconstruction of missing or scratched portion of images is an ancient practice used widely in artwork restoration. It consists of filling in the missing areas,

Ravi Subban
Dept. of Computer Science, Pondicherry University, Pondicherry, India
e-mail: sravicite@gmail.com

Subramanyam Muthukumar
Dept. of Computer Sci. and Engg., National Institute of Technology, Puducherry, India
e-mail: sm.cite.msu@gmail.com

P. Pasupathi
Centre for Information Tech. and Engg., M.S. University, Tirunelveli, India
e-mail: pp.cite.msu@gmail.com

S.M. Thampi et al. (eds.), *Recent Advances in Intelligent Informatics*,   65
Advances in Intelligent Systems and Computing 235,
DOI: 10.1007/978-3-319-01778-5_7, © Springer International Publishing Switzerland 2014

retouching, modifying the damaged ones in such a manner imperceptible to an observer who is not familiar with the original images. This activity is known as inpainting or retouching [19]. The goal of inpainting algorithms varies, depending on the application and idea chosen for making the inpaint parts such as restoration of damaged old photographs, error recovery of images and videos, computer assisted multimedia editing, transmission loss and replacing large regions in an image or video for privacy protection [17].

The natural images are unruffled of structures and textures in which the structures comprise the primitive sketches of an image such as edge and corners. The textures are image regions with homogeneous features statistics. Pure textures synthesis technique cannot handle the missing region with composite textures and structures [18]. The input may be static or dynamic images. Initially, the structure and texture values of the given images are calculated and the target area is removed. Edges of the object to be removed are obtained through the edge detection and the similarities of the patch with its neighboring patches in the source regions are computed [20] [22]. Then, the patch priority is obtained by multiplying the transformed structure sparsity term with patch confidence term. The patch with highest priority is selected for further in-painting. The above process is repeated until the missing region is completely filled by the known values of the neighboring patches. The sequence of display of the entire individual inpainted image frame is referred to as the video inpainting [21].

2   Related Work

The main aim of the image inpainting is to modify the damaged regions of an image or video, so that the in-painted region is undetectable to a neutral observer. The in-painting strateregy is split into three categories viz., structure, texture and patch based in-painting algorithms. Mumford and Shah (1989) proposed inpainting model which takes care of the edges explicitly on the functional to be minimized [1]. J.S.D. Bonet proposed the Multi resolution sampling procedure for analysis and synthesis of texture images [2]. Masnou and Morel (1998) proposed the word disocclusion rather than inpainting [3]. Efros and Leung (1999) use the same texture synthesis techniques [4]. This algorithm based on Markov Random Field and texture is synthesized in a pixel by pixel way, by picking existing pixels with similar neighborhoods in a randomized fashion. This algorithm performs very well, it is very slow since the filling-in is being done pixel by pixel. Bertalmio et.al (2000) proposed a method that propagates the isophotes arrive at the boundary of the region, smoothly inside the region while preserving the arrival angle. In the same context of mimicking natural processes [5], Bertalmio et.al (2001) suggested another similar model, where the evolution of the isophotes is based on the Navier Stokes equations that govern the evolution of fluid dynamics [6]. Chan and Shen (2001) derived an in-painting model by considering the image as an element of the space of Bounded Variation (BV) images, endowed with the

Total Variation (TV) norm [7]. The solution to the in-painting problem comes from the minimization of an appropriate functional. Meyer (2002) used the technique of inpainting using image denoising [8]. Esedoglu and Shen (2002) use extension for curvature with Euler's elastic [9]. Texture synthesis used to complete regions where the texture is stationary or structured areas. Reconstructing methods can be used to fill in large scale missing regions by interpolation. Drori et.al (2003) proposed an inpainting which is suitable for relatively small, smooth and non-textured regions. This approach focuses on image based completion without the knowledge of the underlying scene [10].

Criminisi et.al (2004) presented an algorithm specifically meant for texture in-painting. This algorithm uses the texture synthesis technique. The only difference is that the pixels that are placed along the edges of the image are filled in with high priority [11]. Aujol et.al (2005) discriminates between texture and noise [12]. Brennan (2007) implemented using some experimental model for simultaneous structure and texture image in-painting [13]. Zhang Hong-bin, Wang Jia-wen (2007) proposed an approach for image inpainting by integrating both structure and texture features [14]. Zhongyu Xu et al (2008) presented a novel algorithm for structure inpainting. This algorithm revealed the concepts in manual inpainting and obtained good inpainting results, especially for images with high contrast edges [15]. Exemplar based problem of image inpainting method was proposed by Muthukumar et al [16]. The keen interest in maintaining the structures, in and out of the mask region since the structures makes an impact in every plane of the indoor environment. There are quite a good number of research studies in the area of structure propagation inpainting. Jiying Wu et.al (2010) proposed the exemplar-based in-painting method that needs no user intervention to construct missing edge information [18]. S. Ravi et.al analyzed various techniques such as Texture Synthesis, Semi-automatic and Fast Digital In-painting, PDE based inpainting, Exemplar and Search based inpainting and Hybrid inpainting [17,23].

2.1   Adaptive Patch Based Methodology

In-painting techniques can be used for modifying/repairing the lost or deteriorated image parts in an undetectable manner [17]. Here, inpainting technique employed for making up for the occluded regions with computer generated texture patches. Initially target region is identified like exemplar based in-painting [16], identifying the boundary for the textured regions. It is observed that the results were to be improved with the structure propagation, since the exemplar based techniques does not handle the structures globally. The patch selection process in [19] increases the overall computation time but the limitations are the structures. This turns out to be a non-trivial task. In this paper, a new novel Scene adaptive in-painting technique is proposed that combines the merits of these methods and overcome their limitations. This method produces better results as compared to some of the techniques proposed by some other researchers in inpainting.

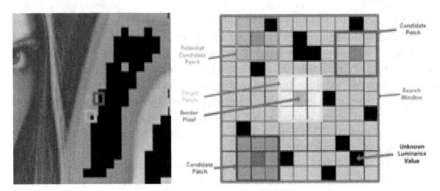

Fig. 1 a) Area Selection b) Methodology for Scene Adaptive Patch based

The proposed algorithm for Scene adaptive in-painting technique is given as follows and its corresponding flowchart is shown in fig 2.

The image restoration based on scene adaptive patch in-painting for tampered natural scenes algorithm:

Input: Image f(x,y) with target Region Φ.

Repeat until all the pixels in Ω are filled.

1. Calculate patch priority for each pixel p on the boundary $\delta\Omega$, and select the patch with the highest priority as the target.
2. Find N-1 candidate patches almost similar to the target patch, and create the incomplete data matrix according to steps 3 through 6.
3. Recuperate the incomplete data matrix using scene adaptive patch based in-painting.
4. Copy the pixel values in the target region of selected patch from the improved data matrix.
5. Update the target region Ω and boundary $\delta\Omega$
6. Finally produce the in-painted image f '(x,y).

The patch priority is calculated by the multiplicative of the data term and the confidence term. The patch Ψp with the highest patch priority on the fill-front is selected. Then, in-painting is carried out based on the patch propagation by inwardly propagating the image patches from the base region to the target region, patch by patch. The selected patch Ψp is in-painted by the corresponding pixels in the sparse linear combination of exemplars to infer the patch in a frame work of sparse representation. The fill front $\delta\Omega$ and the missing region Ω are updated iteratively. For each newly-apparent pixel on the fill-front, its patch similarities are computed with the neighboring patches and its patch priority is fixed.

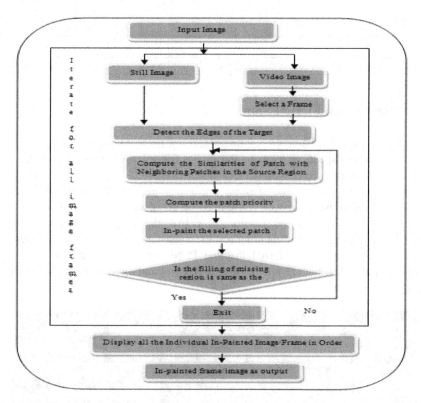

Fig. 2 The Flowchart for the Proposed Algorithm

Table 1 Restoration Efficiency In Psnr (Db) & Bitrate Savings (Bpp)

Test Cases	Structural	Textural	Exemplar	Proposed	Results
	26.56 (0.1)	28.35 (0.648)	32.43 (1.156)	34.16 (0.919)	
	30.43 (0.85)	30.55 (0.989)	31.12 (1.112)	33.13 (1.008)	
	34.66 (0.487)	35.26 (0.313)	36.72 (0.913)	35.98 (0.512)	
	35.95 (0.69)	36.56 (0.736)	36.97 (1.217)	37.15 (1.765)	
	30.28 (0.547)	32.64 (0.833)	36.65 (1.058)	36.9 (0.909)	
	36.59 (0.52)	36.67 (0.73)	36.89 (0.895)	37.08 (1.408)	

Table 2 Comparison of Relative Computation Time (Sec)

Method	Structural	Textural	Exemplar	Proposed
Time	0.793	0.845	1.012	1

3   Results and Discussion

Novel adaptive patch based in-painting algorithm for restoration of natural scene
focus on the occlusion removal in the image frames with appropriate patch to
recover back from the patch similarities the structure with larger structure sparsity
is given higher priority. The sparsest linear combination of candidate patches
under the local consistency was synthesized by the display of the sequence of the
image frames. Experimental results show that the proposed adaptive patch
algorithm produces sharp in-painting results, consistent with surrounding textures.
From the analysis, it is inferred that the PSNR taken for proposed image in
painting approach is better than the other image in-painting methods and also time
computation is normal.

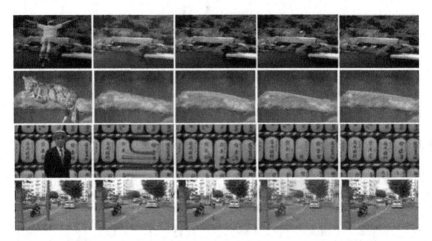

Fig. 3 a) Test Image b) Structural c) Textural d) Exemplar e) Proposed Method

4   Conclusion

This paper presents a novel algorithm for filling the image for the removed large
objects from digital images / frames. In the resultant image, the selected area has
been replaced by a visually plausible a background that mimics the appearance of
the source region. This approach employs an adaptive patch-based technique
modulated by a unified scheme for determining the fill order of the target region.
Pixels maintain a better confidence value, which influence their fill priority. The
technique is capable of propagating both linear structure and two dimensional
texture target regions with a single and simple algorithm.

The experimental results show that a simple selection of the fill order is necessary and sufficient to handle target regions. This proposed method performs better than the other techniques designed for the restoration of scratches and in the instances in which larger objects are removed. It dramatically outperforms earlier work in terms of both perceptual quality and computational efficiency. Moreover, robustness towards changes in shape and topology of the target region has been demonstrated, together with other advantageous properties such as: (i) preservation of edge sharpness, (ii) no dependency on image segmentation and (iii) balanced region filling to avoid over-shooting artifacts. Adaptive scene based patch filling method helps to achieve: (i) visually more pleasing (ii) accuracy in the synthesis of texture (less garbage growing) and (iii) accurate propagation of linear structures. Currently, the investigation of the possibility of constructing ground-truth data and designing evaluation tests that would fine-tune the performance of the proposed algorithm is taken as the future work. In addition, the investigating of the extensions to the current algorithm to handle accurate propagation of curved structures in still photographs and removing objects from video is also taken as the future enhancement to this currently proposed method.

References

[1]   Mumford, Shah: Optimal approximations by piecewise smooth functions and associated Variational problems. Comm. Pure Appl. Math. 42(5), 577–685 (1989)

[2]   Bonet, J.S.D.: Multi resolution sampling procedure for analysis and synthesis of texture images. In: Computer Graphics. Annual conference Series, vol. 31, pp. 361–368 (1997)

[3]   Masnou, Morel: Level lines based disocclusion. In: Proc. IEEE-ICIP, pp. 259–263 (1998)

[4]   Efors, A.A., Leung, T.K.: Texture synthesis by non-parametric sampling. In: ICCV (2), pp. 1033–1038 (1999)

[5]   Bertalmio, et al.: Image in-painting. In: Siggraph, Computer Graphics Proceedings, pp. 417–424. ACM Press/ACM SIGGRAPH (2000)

[6]   Bertalmio, M.: Processing of flat and non-flat image information on arbitrary manifolds using partial Differential Equations. Computer Eng. Program (2001)

[7]   Chan, T.F., Shen, J.: Non Texture in-painting by curvature-driven diffusions (CDD). Journal of Vis. Comm. Image Rep. 4(12), 436–449 (2001)

[8]   Meyer: Oscillating Patterns in Image Processing and Nonlinear Evolution Equations. University Lecture Series, vol. 22. AMS (2002)

[9]   Esedoglu, S., Shen, J.: Digital inpainting based on Mumford shaheuler image model. Eur. J. Appl. Math. 13, 353–370 (2002)

[10] Drori, et al.: Fragment Based Image Completion. In: Proceedings of ACM SIGGRAPH (2003)

[11] Criminisi, A., Perez, P., Toyama, K.: Region filling and object removal by exemplar based image in-painting. IEEE Trans. on Image Processing 13, 1200–1212 (2004)

[12] Aujol, et al.: Image decomposition into a bounded variation component and an oscillating component. J. MIV 22, 71–88 (2005)

[13]   Brennan: Simultaneous structure and texture image inpainting. Department of Computer Engineering, University of California at Santa Cruz, EE264 (2007)

[14]   Zhang, H.-B., Wang, J.-W.: Image In Painting by Integrating Structure and Texture Features. Journal of Beijing University of Technology 33(8), 864–869 (2007)

[15]   Xu, Z., et al.: Image Inpainting Algorithm Based on Partial Differential Equation. In: International Colloquium on CCCM (2008)

[16]   Muthukumar, S., et al.: Analysis of Image Inpainting Techniques with Exemplar, Poisson, Successive Elimination and 8 Pixel Neighborhood Methods. International Journal of Computer Applications 9(11), 15–18 (2010)

[17]   Faizal, M., Fauzi, A., Lewis, P.H.: A multi-scale approach to texture-based image retrieval. Pattern Analysis and Applications 11, 141–157 (2007)

[18]   Xu, Z., Sun, J.: Image in-painting by patch propagation Using Patch Sparsity. IEEE Transactions on Image Processing 19(5) (2010)

[19]   Li, S., Zhao, M.: Image inpainting with salient structure completion and Texture propagation. Pattern Recognition, 0167-8655 (2011)

[20]   Du, X., et al.: Image segmentation and inpainting using hierarchical level set and texture mapping (2011)

[21]   Zhong, Z., Wang: Image inpainting-based edge enhancement using the eikonal equation (2011) 978-1-4577-0539-7/11 IEEE

[22]   Vidhya, B., Valarmathy, S.: Novel Video In-painting Using Patch Sparsity. In: IEEE – International Conference on Recent Trends in Information Technology, ICRTIT, 978-1-4577-0590- 8/11, IEEE, AnnaUniversity, Chennai (2011)

[23]   Ravi, S., et al.: Image Inpainting Techniques – A Survey And Analysis. In: International Conference on IIT, 978-1- 4673-6203-0/13© IEEE (2013)

An Impact of Complex Hybrid Color Space in Image Segmentation

K. Mahantesh, V.N. Manjunath Aradhya, and S.K. Niranjan

Abstract. Image segmentation is a crucial stage in image processing and pattern recognition. In this paper, color uniformity is considered as a significant criterion for partioning the image into considerable multiple disjoint regions and the distribution of the pixel intensities are investigated in different color spaces. A study of single component and hybrid color components is performed. As a result, it is noticed that different color spaces can be created and the performance of an image segmentation procedure is known to be very much dependent on the choice of the color space. In this study, a novel complex hybrid color space HCbCr is derived from the basic primary color spaces and then transformed it into LUV color space. Further, an unsupervised k-means clustering has been applied which significantly describes the relationship between the color space and the impact on color image segmentation. We experiment our proposed color space image segmentation model with the standard human segmented images of Berkeley dataset, results proved to be very promising compared to conventional and existing color space models.

Keywords: Color space models, k-means, Hybrid color space, Image segmentation.

1 Introduction

Image segmentation is a prerequisite stage in image and video processing and in the field of computer vision applications. In this stage an image is segmented into different regions corresponding to different real world objects. In content analysis and image understanding, segmentation stage is considered as one of the critical

K. Mahantesh
Department of ECE, Sri Jagadguru Balagangadhara Institute of Technology, Bangalore, India
e-mail: mahantesh.sjbit@gmail.com

V.N. Manjunath Aradhya · S.K. Niranjan
Department of MCA, Sri Jayachamarajendra College of Engineering, Mysore, India
e-mail: {aradhya.mysore,sriniranjan}@gmail.com

S.M. Thampi et al. (eds.), *Recent Advances in Intelligent Informatics*,
Advances in Intelligent Systems and Computing 235,
DOI: 10.1007/978-3-319-01778-5_8, © Springer International Publishing Switzerland 2014

step. Many image segmentation algorithms have been proposed over last decade, can be found in literature [1] & [2]. Felzenszwalb and Huttenlocher [3] describe an efficient graph theoretic algorithm for image segmentation by partitioning image into pair of regions such that the variation across neighboring regions should be larger than the variation within each individual region. Shi and Malik [4] proposed a general image segmentation approach based on normalized cut by solving an eigen system of equations. Mean shift and normalized cut can be applied to segment the color images [5]. The discontinuity property of the images can be maintained by the mean shift algorithm to form the segmented regions. Finally, the normalized cut has been used on the segmented regions to reduce the complexity of the process. Wang and Siskind [6] developed a cost function, named as ratio cut, based on the graph reduction method to segment color images efficiently. The pixels belonging to the same object have gray levels within a specific range defined by two or several thresholds.

Chang and Wang [7] used a lowpass/highpass filter repeatedly for adjusting the peaks or valleys to a desired number of classes. The valleys in the filtered histogram are then considered as threshold values. A Genetic Algorithm (GA) was combined with a wavelet transform [8]. The wavelet transform is used for reducing the length of the histogram while the genetic algorithm allows finding the threshold number and the optimal threshold values. A multilevel threshold method using a multiphase level set technique is used for determining the number & the values of the thresholds for segmentation [9]. Pixel-level color feature extracted considering local human visual sensitivity in HSV color space and then image pixel's texture features, Maximum local energy, Maximum gradient, and Maximum second moment matrix, are represented via Gabor filter. Then, LS-SVM model is used for classification [10]. Fuzzy and physics-based methods were investigated in [11] for usage of different color space representation and color models in image segmentation. Comaniciu and Meer [12], describe a segmentation method based on the mean-shift algorithm which will converge from an initial location in feature space to a region of locally maximal density and regions smaller than a user defined threshold are eliminated. An implementation of FCM is also analyzed [13] to classify the image into different clusters.

Several approaches combine the analysis of the spatial interaction between pixels and of the color distribution in order to improve the segmentation results. The main highlight focuses on the study of color space models and its complex hybrid approach with the application of k-means algorithm in an effective segmentation process. The remaining paper is structured as: section 2 describes the proposed method, section 3 describes experiment results, finally conclusions are drawn at the end.

2   Proposed Method

This section describes the study of color spaces such as single component analysis and hybrid component analysis of different color spaces. And also describes an application of k-means clustering and derivation of complex hybrid color space for image segmentation.

2.1 Color Spaces

Color space is a geometrical representation defined by means of three components in a space, numerical values of which define a specific color. Earlier color of human perception was considered to be color spaces, later according to trichromatic theory, three primary color known as Red (R), Green (G) & Blue (B) are necessary and sufficient to match any color by mixture in primary spaces. Given an input image and it is transformed into different color spaces like YCbCr [19], YUV, Y'PbPr, Y'CbCr, Y'DbDr, Y'UV, Y'IQ [20], HSI [17], XYZ, CIE XYZ, CIE L*a*b*, L*u*v*, and L*C*H* [18]. All these color space adopt nonlinear transformation in order to use euclidian distance to compare color in different spaces. After the transformation process, we considered only single plane (eg. Y component from YCbCr color space) and later two significant planes from different color spaces (eg. HCbCr) were taken to generate hybrid color space and were analyzed using k-means [25]. Here in this study, we choose K=3 due to the property of discriminating the foreground pixels from background very effectively (please refer Fig. 6.).

2.2 Complex Hybrid Color Space Model

Instead of searching the best classical color space for segmentation, we propose an original approach 'pixel classification' in order to improve the results of image segmentation. In this we introduce a complex hybrid color space model by choosing color components belonging to classical color spaces.The proposed model is derived by considering prominent components from the early described different color space models. In this method, chrominance and human color vision perception components are extracted, which is crucial and found to be very effective for

Table 1 Proposed Algorithm

Algorithm: Proposed complex hybrid color space model
Input: RGB color image   **Output: Segmented image**   **Method**   Step 1: Transform RGB image into YCbCr and HSI Color spaces.   Step 2: Consider CbCr of YCbCr & H of HSI.   Step 3: Compute high dimensional hybrid HCbCr three 2-D matrices.   Step 4: Transform hybrid HCbCr to LUV color space.   Step 5: Apply k-means (k=3) for U & V components.   Step 6: End.   **Method ends**

image segmentation. Algorithm of the derived complex hybrid color space model is as shown in Table 1.

The derived color spaces have neither psycho-visual nor physical color significance, named hybrid color spaces in the proposed work. In this, pixels are discriminated between the pixel classes in the hybrid color space for better classification. The resultant space is called the complex hybrid color space, built by means of a sequential supervised feature selection scheme.

3 Experimental Results and Performance Analysis

The proposed method is implemented in MATLAB 7.0 on an intel core2duo 2.20 GHz with 3GB RAM. For the purpose of experiment we considered a standard image segmentation datasets such as Berkely BSDS300 dataset, Caltech-101 and Corel datasets. Berkeley segmentation Dataset [14], the BSDS300 consists of 200 training and 100 test images, each with multiple ground-truth segmentations. Caltech 101 dataset contains 9,197 images comprising 101 different object categories, includeing a background category, collected via Google image search by Fei-Fei et. al. [15]. Corel dataset consists of 10,800 images from the Corel Photo Gallery [16], classified into 80 concept groups, e.g., autumn, aviation, bonsai, castle, cloud, dog, elephant, iceberg, primates, ship, stalactite, steam-engine, tiger, train, and waterfall etc.

3.1 Segmentation in different color spaces

In this experiment, we considered an image of butterfly from Berkeley dataset comprising complex background and analyzed few comprehensive observations based on the single component analysis as shown in Fig. 1. In this green leaves and white petal flowers are considered as a background along with butterfly resulted in some images. The white colored flower petals are partially visible with prominent green colored leaf and vice versa. It is observed that, background is eliminated completely but shape of an object is not continuous & well defined and appears to be partially blended with the background.

Fig. 2. shows segmentation in hybrid color space in which two or more than two components from different color spaces are selected to overcome the problems of single component analysis. Here the green colored background is completely removed retaining white colored flower petals scarcely visible as spots. But, in HCbCr color space, we obtained the object's structure and its prominent features along with well defined color information.

Complex hybrid color space is derived based on the above obtained set of results. In this, first we considered CbCr components of YCbCr color space, H component of HSI color space to yield an HCbCr as three 2-dimensional matrices. Then HCbCr is transformed to LUV color space to obtain a complex hybrid color space model. Further k-means clustering is applied to segment the object efficiently by setting k=3. The resultant segmented image is shown in Fig. 3.

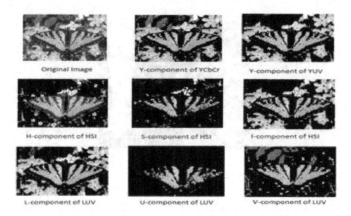

Fig. 1 Original image and resulted segmented images (considered single component in respective color spaces)

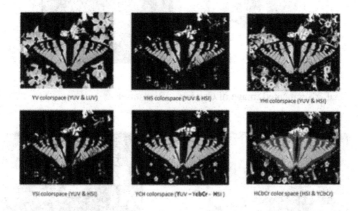

Fig. 2 Results of Hybrid color model (two and more than two components of color spaces)

The image segmentation results of caltech-101 and corel datasets are shown in Fig. 4. and Fig. 5. along with its input images respectively

3.2 Performance Analysis

To evaluate the performance of the proposed system, we compared the results of the present system with the well known existing methods such as Edison's mean shift [12] and fuzzy c-means [13] using popular standard segmentation evaluation metrics, and are explained in detail with this session.

A multi-stage algorithm canny edge detection operator is applied on to the resultant segmented images of the proposed system to evaluate the performance with

Fig. 3 Segmentation result of complex hybrid color space

Fig. 4 Original & segmented image of Caltech - 101 dataset

Fig. 5 Original & segmented image of Corel dataset

human segmented edges of Berkeley dataset. The obtained edge detected image is shown in Fig. 7. aong with its proposed segmented image.

The proposed system's edge detected efficiency is compared with the edges obtained for Edison's mean shift & fuzzy c-means methods and can be seen in Fig. 8.

In this, the proposed segmented edges are compared with the human segmented edges of Berkeley dataset to obtain the performance measures such as correlation [21], Hausdorff distance [22], Jaccard & Dice coefficients [23], root mean square error (RMSE) [24] and are computed using equations 1, 2, 4, 5 & 6 respectively.

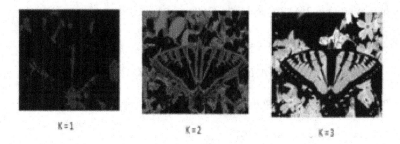

K=1   K=2   K=3

Fig. 6 Segmented results performing k-means for K=1, 2 & 3 respectively

Fig. 7 Proposed segmented and edge deteceted image

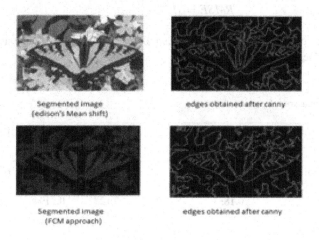

Segmented image
(edison's Mean shift)   edges obtained after canny

Segmented image
(FCM approach)   edges obtained after canny

Fig. 8 Mean shift and Fuzzy c-Means results

Correlation is given by:

$$r = \frac{\sum_m \sum_n (A_{mn} - \bar{A})(B_{mn} - \bar{B})}{\sqrt{\sum_m \sum_n (A_{mn} - \bar{A})^2 \sum_m \sum_n (B_{mn} - \bar{B})^2}} \qquad (1)$$

where $\bar{A} = mean2(A)$ & $\bar{B} = mean2(B)$.

Hausdorff distance can be calculated using:

$$H(A,B) = max(h(A,B), h(B,A)) \tag{2}$$

where $h(A,B) = max_{a \in A} min_{b \in B} \|a - b\|$

Jaccard co-efficient is given by:

$$J(A,B) = \frac{|A \cap B|}{|A \cup B|} \tag{3}$$

$$J(A,B) = 1 - J(A,B) = \frac{|A \cup B| - |A \cap B|}{|A \cup B|} \tag{4}$$

Dice co-efficient can be written as:

$$Dice(A,B) = \frac{2\|AB\|}{|A| + |B|} \tag{5}$$

And finally root mean square error can be computed using:

$$RMSE = \sqrt{\frac{\Sigma_{i=1}^{h}(A - B)^2}{n}} \tag{6}$$

A & B in above equations are matrices indicating segmented image and ground truth/human segmented image respectively.

Table 2 Comparison of proposed method

Method	Correlation	Hausdorff distance	Jaccard	Dice	RMSE
Edison's mean shift [12]	0.1652	8.2462	0.1150	0.2063	0.2808
Fuzzy c-means [13]	0.1635	8.2462	0.1128	0.2028	0.2738
Proposed	**0.1851**	**8.0623**	**0.1232**	**0.2194**	**0.2599**

With reference to Table 2, analysis on Berkeley dataset of 110 images is listed as follows:

- Correlation refers to the statistical dependency between two images lies in the range [0, 1] and higher value of 0.1851 of proposed method shows better segmentation result than 0.1652 & 0.1635 of methods [12] & [13].
- Hausdorff distance of value 8.0623 measures the extent to which each point of proposed segmented image lies minimally near some point of ground-truth segmentations.

- Jaccard and dice should be in the range of [0, 1], shows 0.1232 greater than the related values revealed extremely close results compared to other two existing methods.
- Finally, RMSE takes segmented and ground truth images as input and produces real valued output in the range [0, 1], the obtained value 0.2599 signifies lesser error with improved result of 0.2808 & 0.2738 of Edison's mean shift and fuzzy c-means respectively.

4 Conclusion and Future Work

In the present work, we have conducted image analysis describing the relationship between the color spaces and its impact on color image segmentation considering different color space models and derived an efficient complex hybrid color space model resulting in the best discrimination by means of a sequential procedure in a supervised context. Experimental results based on the standard segmentation metrics shows that the image is efficiently segmented into sub regions with clearly defined edges based on human color visual perception and retaining low frequency components which efficiently describes the shape, color and texture of the object. The proposed system is experimented on Berkeley, Caltech-101 and corel datasets and is evaluated on Berkeley dataset. In future, we are planning to use combination of four components to extract discriminating color texture features for better segmentation of multiple objects in images with more complex background.

References

1. Gonzalez, R.C., Woods, R.E.: Digital Image Processing. Prentice Hall (2002)
2. Cheng, H.D., Jiang, X.H., Sun, Y., Wang, J.: Color image segmentation: advances and prospects. Pattern Recognition 34(12), 2259–2281 (2001)
3. Felzenszwalb, P., Huttenlocher, D.: Image segmentation using local variation. In: Proceedings of IEEE Conference on Computer Vision and Pattern Recognition, pp. 98–104 (1998)
4. Shi, J., Malik, J.: Normalized cuts and image segmentation. IEEE Transaction Pattern Analysis Machine Intelligence 22(8), 888–905 (2000)
5. Wenbing, T., Hai, J., Yimin, Z.: Color image segmentation based on mean shift and normalized cuts. IEEE Transactions on Systems, Man, and Cybernetics-Part B: Cybernetics 37(5), 1382–1389 (2007)
6. Wang, S., Siskind, J.M.: Image segmentation with ratio cut. IEEE Transactions on Pattern Analysis and Machine Intelligence 25(6), 675–690 (2003)
7. Chang, C.C., Wang, L.L.: A fast multilevel thresholding method based on lowpass and highpass filtering. Pattern Recognition Letters 18, 1469–1478 (1977)
8. Hammouche, K., Diaf, M., Siarry, P.: A multilevel automatic thresholding method based on a genetic algorithm for a fast image segmentation. Computer Vision and Image Understanding 109, 163–175 (2008)
9. Dirami, A., Hammouche, K., Diaf, M., Siarry, P.: Fast multilevel thresholding for image segmentation through a multiphase level set method. Signal Processing 93, 139–153 (2013)

10. Yang, H.-Y., Wang, X.-Y., Wanga, Q.-Y., Zhang, X.-J.: LS-SVM based image segmentation using color and texture information. J. Vis. Commun. Image R. 23, 1095–1112 (2012)
11. Cheng, H.D., Jiang, X.H., Sun, Y., Wang, J.: Color image segmentation: advances and prospects. Pattern Recognition 34(12), 2259–2281 (2001)
12. Comaniciu, D., Meer, P.: Mean shift: A robust approach toward feature space analysis. IEEE Transactions on Pattern Analysis and Machine Intelligence 24(5), 603–619 (2002)
13. Wang, X.-Y., Zhang, X.-J., Yang, H.-Y., Bu, J.: A pixel-based color image segmentation using support vector machine and fuzzy C-means. Neural Networks 33, 148–159 (2012)
14. Martin, D., Fowlkes, C.: The Berkeley segmentation database and benchmark. Computer Science Department, Berkeley University (2001), http://www.eecs.berkeley.edu/Research/Projects/CS/vision/bsds/
15. Fei-Fei, L., Fergus, R., Perona, P.: Learning generative visual models from few training examples: an incremental bayesian approach tested on 101 object categories. In: CVPR Workshop on Generative-Model Based Vision (2004)
16. Rui, Y., Huang, T.S., Ortega, M., Mehrotra, S.: Relevance feedback: a power tool for interactive content-based image retrieval. IEEE Transactions on Circuits and Systems for Video Technology 8(5), 644–655 (1998)
17. Liu, G.H., Li, Z.Y., Zhang, L., Xu, Y.: Image retrieval based on micro-structure descriptor. Pattern Recognition 44(9), 2123–2133 (2011)
18. Burger, W., Burge, M.J.: Principles of Digital image processing: Core Algorithms. Springer (2009)
19. Shih, F.Y., Cheng, S.: Automatic seeded region growing for color image segmentation. Image Vision Comput. 23(10), 877–886 (2005)
20. ITU-R BT.601-7, Studio encoding parameters of digital television for standard 4:3 and wide screen 16:9 aspect ratios. Tech. rep., International Telecommunication Union (2007)
21. Szekely, G.J., Rizzo, M.L., Bakirov, N.K.: Measuring and testing independence by correlation of distances. Annals of Statistics 35(6), 2769–2794 (2007)
22. Tyrrell Rockafellar, R., Wets, R.J.-B.: Variational Analysis, p. 117. Springer (2005) ISBN 3-540-62772-3, ISBN 978-3-540-62772-2
23. Jackson, A.D., Somers, M.K., Harvey, H.H.: Similarity coefficients: measures for co-occurrence and association or simply measures of occurrence? Am. Natur. 133(3), 436–453 (1989)
24. Parmar, K., Kher, R.: A Comparative Analysis of Multimodality Medical Image Fusion Methods. In: 2012 Sixth Asia IEEE, Modelling Symposium (AMS), May 29-31, pp. 93–97 (2012)
25. MacQueen, J.B.: Some Methods for classification and Analysis of Multivariate Observations. In: Proceedings of 5th Berkeley Symposium on Mathematical Statistics and Probability, pp. 281–297. University of California Press (1967)

Natural Color Image Enhancement Based on Modified Multiscale Retinex Algorithm and Performance Evaluation Using Wavelet Energy

M.C. Hanumantharaju, M. Ravishankar, and D.R. Rameshbabu

Abstract. This paper presents a new color image enhancement technique based on modified modified MultiScale Retinex (MSR) algorithm and visual quality of the enhanced images are evaluated using a new metric, namely, Wavelet Energy (WE). The color image enhancement is achieved by downsampling the value component of HSV color space converted image into three scales (normal, medium and fine) following the contrast stretching operation. These downsampled value components are enhanced using the MSR algorithm. The value component is reconstructed by averaging each pixels of the lower scale image with that of the upper scale image subsequent to upsampling the lower scale image. This process replaces dark pixel by the average pixels of both the lower scale and upper scale, while retaining the bright pixels. The quality of the reconstructed images in the proposed method is found to be good and far better then the other researchers method. The performance of the proposed scheme is evaluated using new wavelet domain based assessment criterion, referred as WE. This scheme computes the energy of both original and enhanced image in wavelet domain. The number of edge details as well as WE is less in a poor quality image compared with naturally enhanced image. Experimental results presented confirms that the proposed wavelet energy based color image quality assessment technique efficiently characterizes both the local and global details of enhanced image.

Keywords: Color Image Enhancement, Sampling, Multiscale Retinex, Image Quality Assessment, HSV.

M.C. Hanumantharaju · M. Ravishankar
Department of ISE, Dayananda Sagar College of Engineering,
Shavige Malleshwara Hills, Bangalore
e-mail: {mchanumantharaju,ravishankarmcn}@gmail.com

D.R. Rameshbabu
Department of CSE, Dayananda Sagar College of Engineering,
Shavige Malleshwara Hills, Bangalore
e-mail: bobrammysore@gmail.com

S.M. Thampi et al. (eds.), *Recent Advances in Intelligent Informatics,*   83
Advances in Intelligent Systems and Computing 235,
DOI: 10.1007/978-3-319-01778-5_9, © Springer International Publishing Switzerland 2014

1 Introduction

Image processing is a 2D-signal processing which improves the characteristics, properties and parameters of an input image in order to produce a true output picture more suitable than the input image. The important image processing operation includes enhancement, reconstruction and compression. Among these operations, image enhancement is the key step which modifies the attributes of an image to make it more appropriate for display, analysis and further processing in an image processing system. The realm of image enhancement wraps up restoration, reconstruction, filtering, segmentation, compression and transmission. Image enhancement algorithms are mainly used to pick up some important features in an image. For instance, image sharpening is done in order to bring out the details such as car license plate number, edge or line enhancement to reconstruct the objects in an aerial image and highlighting the region of interest in medical images for pathology detection of various lesions etc. These enhancement operations needs highly efficient, integrated algorithm with less number of parameters to specify. The applications such as high definition telivision, video conferencing, remote sensing etc., handles huge volumes of image data owing to increased complexity in processing. Development of image enhancement algorithm for these applications are imperative. It is of paramount importance to design an image enhancement algorithm suitable for the applications handling large image data and offer better enhancement.

Image enhancement techniques include a filtering operation for reducing the noise present in images, contrast stretching to stretch the range of intensity values, Histogram Equalization (HE) operation to increase the contrast of an image by increasing the dynamic range of intensity values etc., The goal of the image enhancement is to extract the true image of the recorded scene. The discrepancies present in the recorded pictures described earlier are overcome in the present work by using efficient image enhancement techniques. The image enhancement algorithms mainly used in spatial domain are HE [1], Adaptive HE [2], intensity transformations [3], homomorphic filtering [4] and MSR with Color Restoration (MSRCR) [5]. Rahman et al. [6] proposed state-of-the-art image enhancement techniques for most commonly used image enhancement techniques and validated other enhancement schemes with the MSRCR.

This paper is organized as follows: Section 2 gives a brief review of existing work. Section 3 describes the proposed modified multiscale retinex algorithm. Section 4 provides experimental results and discussions. Finally conclusion arrived at is presented in Section 5.

2 Existing Work

Faming et al. [7] proposed a new pixel based variational model for remote sensing multisource image fusion using gradient features. Although multisource image fusion method adapted in this work offers integration of multiple sources, visual inspection of the reconstructed image reveals distortion in the spectral information while merging the multispectral data. Chan et al. [8] proposed fast MSR algorithm using dominant Single Scale Retinex (SSR) in weight selection. This scheme describes an approach to reduce the computational complexity in the conventional MSR algorithm. Authors claim that the quality reconstructed pictures are similar to that of conventional MSR method. Although the algorithm offers fast enhancement, subjective evaluation of the obtained results indicate the presence of blocking artifacts, owing to poor visual quality.

Qingyuan et al. [9] proposed an improved MSR algorithm for medical image enhancement. In this scheme, Y-component of medical image is separated into edge and non-edge area subsequent to RGB to YIQ color space conversion. The MSR technique has been used for the non-edge area in order to accomplish the medical image enhancement. However, this method provides satisfactory results for immunohistochemistry images but this approach may not provide inevitable results for other medical images. Real time modified retinex image enhancement algorithm with hardware implementation has been proposed by Hiroshi et al. [10], in order to reduce the halo artifacts. This method adaptively adjust the parameter of the cost function owing to reduced halo artifacts with improved contrast. As is seen from the experimental results presented, reconstructed images using this method are generally not satisfactory. In addition, authors have not provided the information about quality assessment of the reconstructed images.

The drawbacks of image enhancement methods specified earlier are overcome in the proposed method. The input image is first downsampled into three versions namely, normal, medium and fine scale. This downsampled images are contrast stretched to increase the picture element range and enhanced by the popular MSR algorithm. Subsequently, lower scale is upsampled to the size of next upper scale version and then combined with the next upper scale image. While combining these images, if the upsampled image has a zero pixel, then the upper scale pixel is retained otherwise, the pixel average is computed. The proposed method removes the black spots present in the Chao et al. [11] technique in an efficient way. The design developed here is much faster compared to other MSR methods since the image is downsampled into three versions. The proposed work is validated with various images of different environmental conditions.

3 Proposed Modified Multiscale Retinex Algorithm

This section presents the proposed modified Multiscale Retinex based color image enhancement. The input image of resolution 256×256 pixels read from RGB color space is converted into Hue-Saturation-Value (HSV) color space since HSV space separates color from intensity. The value channel of HSV is scaled into three versions namely, medium scale (64×64 pixels), fine scale (128×128 pixels) and normal scale (256×256 pixels) in order to speed up the MSR enhancement process. The hue and saturation are preserved to avoid distortion. Each of these scaled image versions may have a random pixel range. Therefore, contrast stretching operation is accomplished for each of the scaled versions of the value channel in order to translate the pixels in the display range of 0 to 255. The summary of the proposed algorithm is outlined as follows:

1. Read the poor quality color image which needs image enhancement.
2. Convert the image in RGB color space to HSV space to separate color from intensity.
3. Scale the value Component of HSV into three versions, namely, medium, fine, and normal.
4. Apply contrast stretching operation on each of the scaled versions to translate pixels into the display range of 0 to 255.
5. Apply MSR based color image enhancement of Ref. [12]
6. Reconstructed the value component is obtained by upsampling and combining the fine, medium and normal scale. The upsampled fine scale is combined with the medium scale by eliminating dark pixel. Similarly, the upsampled medium scale is combined with the normal scale by eliminating the dark pixel.
7. The reconstructed value component is combined with the preserved hue and saturation component.
8. Convert the image from HSV to RGB color space.
9. Display the enhanced Image.

3.1 Multi-scale Retinex Algorithm

Numerous MSR based image enhancement algorithms have been reported by many researchers. However, the most popular one is the Jobson et al. [5] MSR image enhancement algorithm since this scheme offers better image enhancement compared to other methods. The new version of the MSR algorithm developed by the Hanumantharaju et al. [12] reduces the halo artifacts of Jobson method. Therefore, in this work MSR technique proposed by Hanumantharaju et al. has been adapted in order achieve natural color image enhancement. The core part of the MSR algorithm is the design of 2D Gaussian surround function. The Gaussian surround functions are scaled in accordance with the size of the scaled value component. The size of Gaussian function employed is 64×64 for the medium version, 128×128 for the fine version and 256×256 for the normal version of the value channel, respectively. The general expression for the Gaussian surround function is given by Eqn. (1)

$$G_n(x,y) = K_n \times e^{-\frac{x^2+y^2}{2\sigma^2}} \qquad (1)$$

and K_n is given by the Eqn. (2)

$$K_n = \frac{1}{\sum_{i=1}^{M} \sum_{j=1}^{N} e^{-\frac{x^2+y^2}{2\sigma^2}}} \qquad (2)$$

where x and y signify the spatial coordinates, $M \times N$ represents the image size, n is preferred as 1, 2 and 3 since the three Gaussian scales are used for each downsampled versions of the image.

Next, in order to accomplish color image enhancement, the SSR algorithm follows the MSR technique. The SSR for the value channel is given by Eqn. (3)

$$R_{SSRi}(x,y) = \log_2\left[V_i(x,y)\right] - \log_2\left[G_n(x,y) \otimes V_i(x,y)\right] \tag{3}$$

where $R_{SSRi}(x,y)$ shows SSR output, $V_i(x,y)$ represents value channel of HSV, $G_n(x,y)$ indicated Gaussian Surround function, $\otimes$ denotes convolution operation.

The MSR operation on a 2-D image is carried out by using Eqn. (4)

$$R_{MSRi}(x,y) = \sum_{n=1}^{N} W_n \times R_{SSRni}(x,y) \tag{4}$$

where $R_{MSRi}(x,y)$ shows MSR output, W_n is a weighting factor which is assumed as $\frac{1}{3}$ and N indicates number of scales.

The color image enhancement is achieved by applying the MSR algorithm for each downsampled versions subsequent to SSR operation. The new value channel is reconstructed from the individual enhanced images by combining medium, fine and normal versions of the image in an efficient way. The MSR enhanced image of medium version with resolution of 64×64 pixels is upsampled by two in order to match with the resolution of 128×128 pixels of the fine version. However, the upsampling and reconstruction operations adapted by Chao et al. [11] technique introduces zeros between alternative pixels. Although an image enhanced by this scheme is satisfactory, it has actually resulted in appearance of dots in the enhanced image and thus affects overall image quality. The present work overcomes this difficulty in a proficient way. The new fine scale version of the image is obtained as follows. The pixel of the medium scale version is retained for the zeros encountered in the upsampled medium version of the image. If there are no zeros in the upsampled medium version image than the pixel average is computed between upsampled medium version and fine version. This is illustrated by the detailed flow chart presented in Fig. 1. Finally, the composite enhanced image is reconstructed by combining new value channel with that of hue and saturation channels and converting back into RGB color space.

3.2   *Image Quality Assessment Using Wavelet Energy*

The wavelet domain is a powerful and efficient technique for analyzing, decomposing, denoising, and compressing signals. In particular, the Discrete Wavelet Transform (DWT) breaks a signal into several time-frequency components that enables the extraction of features desirable for signal identification and recognition. The DWT and wavelet theory have been developing rapidly over the past few years. In the paper, DWT and its energy computation is exploited for visual quality assessment of an enhanced color image. The extraction of Detailed Wavelet Energy

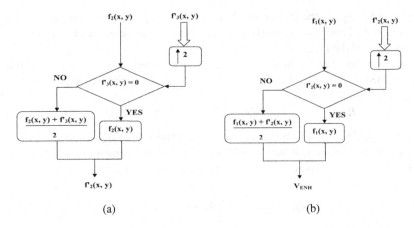

Fig. 1 Detailed Flow Sequence of the Proposed Method : (a) Flow Chart for Obtaining New Fine Scale (Resolution of 128×128 pixels) of Enhanced image (b) Flow Chart for Obtaining New Normal Scale (Resolution of 256×256 pixels) of Enhanced image

(DWE) coefficient from an image provides information about image details and the extraction of Approximate Wavelet Energy (AWE) coefficients offer the global contrast information of an image.

3.3 Wavelet Energy

A Continuous Wavelet Transform (CWT) maps a given function in the time domain into two dimensional function of s and t. The parameter 's' represents the scale and corresponds to frequency in Fourier transform and 't' indicates the translation of the wavelet function. The CWT is defined by Eqn. (5)

$$CWT(x,y) = \frac{1}{\sqrt{s}} \int S(T)\varphi\left(\frac{T-t}{s}\right) dt \qquad (5)$$

where S(T) is the signal and $\varphi(T)$ is the basic wavelet and $\varphi\left(\frac{T-t}{s}\right)\frac{1}{\sqrt{s}}$ is the wavelet basis function. The DWT for a signal is given by Eqn. (6)

$$DWT(m,n) = \frac{1}{2^m} \sum_{i=1}^{N} S(I,i)\phi\left[2^{-m}(i-n)\right] \qquad (6)$$

Wavelet energy is a method for finding wavelet energy for 1-D wavelet decomposition. The WE provides percentage of energy corresponding to the approximation and the vector containing the percentage of energy corresponding details. The WE is computed as follows

$$WE = \frac{1}{2^{-m/2}} \sum_{i=1}^{N} S(I,i)\phi \left[2^{-m}(i-n)\right] \tag{7}$$

The WE is a Full Reference (FR) image quality assessment algorithm uses subband characteristics in wavelet domain. The existing metrics are analyzed and limitations are investigated. Image quality evaluation using WE computation uses a linear combination of high frequency coefficients after a Daubechies wavelet transform. The probability density function of the enhanced image has relatively higher energy in wavelet domain compared to other transforms. Therefore, the wavelet energy metric is an effective and efficient metric to evaluate the quality of the enhanced image. The approximate wavelet energy coefficients provide the information on the global image enhancement and detailed WE coefficients provide statistics on the image details.

4 Experimental Results and Comparative Study

The algorithm presented in the earlier section was coded using Matlab Version 8.0. The experiment was conducted by considering poor quality images of different environmental conditions downloaded from various databases. The first column of Fig. 2 presents the original image. The same images enhanced using NASA's MSR scheme

Fig. 2 First Column: Original Image of Resolution 256 × 256 pixels, **Second Column:** Image Enhanced using Multiscale Retinex with Color Restoration (MSRCR) of Ref. [5], **Third Column:** Improved MSRCR of Ref. [13], **Fourth Column:** Proposed Method

Fig. 3 First Column: Original Image of Resolution 256×256 pixels, **Second Column:** Image Enhanced using Multiscale Retinex with Color Restoration (MSRCR) of Ref. [5], **Third Column:** Improved MSRCR of Ref. [13], **Fourth Column:** Proposed Method

of Ref. [5] is shown in second column of Fig. 2. The improved MSR scheme of Ref. [13] is presented in third column of Fig. 2. Finally, reconstructed images using proposed modified MSR based scheme is shown in last column of Fig. 2.

In order to show the efficiency of the proposed method in more detail, the algorithm is tested with other test images. The first column of Fig. 3 shows the original image. The second column of Fig. 3 shows the same image enhanced using MSRCR of Ref. jobson. The third column presents the image enhanced using improved MSRCR of Ref. shen. The last column of Fig. 3 shows the image enhanced using proposed modified MSR method.

The image enhancement achieved using the proposed modified MSR algorithm is validated using the WE metric described earlier. Although, numerous metrics for color image quality assessment were reported by various researchers, measurement do not correlate well with image perception. Therefore, WE based IQA has been exploited in this work in order to verify the performance of the image enhancement algorithm. Table 1 shows the image quality assessment based on AWE and DWE. It may noted that the Average WE calculated for enhanced image of the proposed method is in close proximity to that of the original image. However, the image enhanced using MSRCR of Ref. [5] and Improved MSRCR of Ref. [13] are large compared to proposed method. Similarly, it may be analyzed from the DWE presented that the proposed modified MSR algorithm has DWE close to that of original image. However, DWE calculated for other enhancement methods are smaller than the original image. This shows that the proposed modified MSR based image enhancement method offers a natural enhancement which may be perceived by the Human Visual System (HVS).

Table 1 Approximate (A) and Detailed (D) WE Metric for Color Image Quality Assessment, **Note:** The highlighted values in the table shows better enhancement method based on the proposed wavelet energy metric. In the evaluation, first preference is given for details followed by overall enhancement.

	Org. Image		MSRCR		Improved MSRCR		Proposed Method	
Test Images	AWE	DWE	AWE	DWE	AWE	DWE	AWE	DWE
Swan	99.57	0.423	99.84	0.154	99.76	0.2375	99.62	0.374
Port	99.60	0.394	99.87	0.126	99.83	0.163	99.68	0.317
Girl	99.75	0.244	99.91	0.087	99.87	0.127	99.82	0.175

5 Conclusion

A new color enhancement approach based on modified multiscale retinex algorithm as well as the wavelet energy metric for image quality assessment has been proposed. HSV color space is exploited since this domain separates color from intensity. The value component is downsampled into three versions, namely, normal, medium and fine. The contrast stretching operation is performed on the each downsampled value components. Subsequently, MSR algorithm used for value component enhancement. As is evident from the experimental results presented that the proposed modified MSR based color image enhancement offers natural quality images. It is found that the enhanced images are more vivid, brilliant, and correlates to the HVS. Research is in progress to implement the proposed algorithm on reconfigurable hardware device such as Field Programmable Gate Arrays (FPGAs).

References

1. Cheng, H.D., Shi, X.J.: A Simple and Effective Histogram Equalization Approach to Image Enhancement. Digital Signal Processing 14(2), 158–170 (2004)
2. Zhu, Y., Huang, C.: An Adaptive Histogram Equalization Algorithm on the Image Gray Level Mapping. Physics Procedia 25, 601–608 (2012)
3. Lee, E., Kim, S., Kang, W., Seo, D., Paik, J.: Contrast Enhancement Using Dominant Brightness Level Analysis and Adaptive Intensity Transformation for Remote Sensing Images, pp. 62–66 (2013)
4. Jie, X., LiNa, H., GuoHua, G., MingQuan, Z.: Based on hsv space real color image enhanced by multiscale homomorphic filters in two channels. In: Proceedings of WRI Global Congress on Intelligent Systems, vol. 3, pp. 160–165 (2009)

5. Jobson, D.J., Rahman, Z.U., Woodell, G.A.: A Multiscale Retinex for Bridging the Gap Between Color Images and the Human Observation of Scenes. IEEE Transactions on Image Processing 6(7), 965–976 (1997)
6. Rahman, Z.U., Woodell, G.A., Jobson, D.J.: A Comparison of the Multiscale Retinex with other Image Enhancement Techniques. In: Proceedings of IS and T Annual Conference, pp. 426–431. Citeseer (1997)
7. Fang, F., Li, F., Zhang, G., Shen, C.: A variational method for multisource remote-sensing image fusion. International Journal of Remote Sensing 34(7), 2470–2486 (2013)
8. Jang, C.Y., Lim, J.H., Kim, Y.H.: A Fast Multi-scale Retinex Algorithm using Dominant SSR in Weights Selection. In: Proceedings of International SoC Design Conference (ISOCC), pp. 37–40 (2012)
9. Meng, Q., Bian, D., Guo, M., Lu, F., Liu, D.: Improved Multiscale Retinex Algorithm for Medical Image Enhancement. In: Proceedings of Information Engineering and Applications, vol. 154, pp. 930–937 (2012)
10. Tsutsui, H., Yoshikawa, S., Okuhata, H., Onoye, T.: Halo Artifacts Reduction Method for Variational based Real-time Retinex Image Enhancement. In: Proceedings of Asia-Pacific Signal & Information Processing Association Annual Summit and Conference (APSIPA ASC), pp. 1–6 (2012)
11. An, C., Yu, M.: Fast Color Image Enhancement based on Fuzzy Multiple-Scale Retinex. In: 6th International Forum on Strategic Technology (IFOST 2011), vol. 2, pp. 1065–1069 (2011)
12. Hanumantharaju, M.C., Ravishankar, M., Rameshbabu, D.R., Ramachandran, S.: Color Image Enhancement using Multiscale Retinex with Modified Color Restoration Technique. In: Second IEEE International Conference on Emerging Applications of Information Technology (EAIT 2011), pp. 93–97 (2011)
13. Shen, C.T., Hwang, W.L.: Color Image Enhancement using Retinex with Robust Envelope. In: Proceedings of 16th IEEE International Conference on Image Processing (ICIP 2009), pp. 3141–3144 (2009)

Multiple Moving Object Recognitions in Video Based on Log Gabor-PCA Approach

M.T. Gopalakrishna, M. Ravishankar, and D.R. Rameshbabu

Abstract. Object recognition in the video sequence or images is one of the sub-field of computer vision. Moving object recognition from a video sequence is an appealing topic with applications in various areas such as airport safety, intrusion surveillance, video monitoring, intelligent highway, etc. Moving object recognition is the most challenging task in intelligent video surveillance system. In this regard, many techniques have been proposed based on different methods. Despite of its importance, moving object recognition in complex environments is still far from being completely solved for low resolution videos, foggy videos, and also dim video sequences. All in all, these make it necessary to develop exceedingly robust techniques. This paper introduces multiple moving object recognition in the video sequence based on LoG Gabor-PCA approach and Angle based distance Similarity measures techniques used to recognize the object as a human, vehicle etc. Number of experiments are conducted for indoor and outdoor video sequences of standard datasets and also our own collection of video sequences comprising of partial night vision video sequences. Experimental results show that our proposed approach achieves an excellent recognition rate. Results obtained are satisfactory and competent.

Keywords: Moving object recognition, LoG Gabor-PCA, Intelligent Video Surveillance.

M.T. Gopalakrishna · M. Ravishankar
Department of ISE, Dayananda Sagar College of Engineering,
Shavige Malleshwara Hills, Bangalore
e-mail: {gopalmtm,ravishankarmcn}@gmail.com

D.R. Rameshbabu
Department of CSE, Dayananda Sagar College of Engineering,
Shavige Malleshwara Hills, Bangalore
e-mail: bobrammysore@gmail.com

S.M. Thampi et al. (eds.), *Recent Advances in Intelligent Informatics*,
Advances in Intelligent Systems and Computing 235,
DOI: 10.1007/978-3-319-01778-5_10, © Springer International Publishing Switzerland 2014

1 Introductions

Moving object recognition from a video sequence is a fascinating topic with applications in various areas such as airport safety, intrusion surveillance, video monitoring, intelligent highway, etc [1]. Moving object recognition in the video sequence has been considered as one of the most fascinating and challenging areas in computer vision and pattern recognition, in recent years. In last one decade, researchers have proposed a variety of approaches for moving object detection and classification, most of them are based on motion and shape features. For example, an end-to-end method for extracting moving targets from a real-time video stream has been presented by [2]. This method is applicable to human and vehicle classification with shapes that are remarkably different. Petrovic et al [3] extract gradient features from reference patches in images of car fronts, and recognition is performed in two stages. Gradient-based feature vectors are used to produce a ranked list of possible classes of the candidate. A novel match refinement algorithm is used to refine the obtained result. There are many other moving objects classification methods based on multi-feature fusion [4, 5, 6, 7, 8]. Image features are arguably the most fundamental task in moving object recognition. Generally, there are two categories of feature representation: appearance feature based and geometric feature based. Appearance features have been demonstrated to be better than geometric features, for the superior insensitivity to noises, especially illumination changes, foggy weathers etc. Gabor wavelets are reasonable models of visual processing in primary visual cortex and are one of the most successful approaches to describe local appearance of the human face, exhibiting powerful characteristics of spatial locality, scale and orientation selectivity [9]. However, they fail to provide excellent simultaneous localization of the spatial and frequency information due to the constraints of the narrow spectral bandwidth, which is crucial to the analysis of highly complex scene. Jamie Cook et al. [10] proposed a system for 3D Face Recognition using Log-Gabor Filter to obtain a simultaneous response that is Gaussian when viewed on a logarithmic frequency scale instead of a linear one. As an alternative to traditional common approaches, D. Field [11] proposed a Log Gabor Filter to perform DC compensation and to overcome the bandwidth limitations of traditional Gabor Filter banks. However, the dimensionality of the resulting data is exceptionally high. For this reason, a computationally effective approach is needed. One common choice would be Principle Component Analysis (PCA). As per the review reports, it is clear that research on moving object recognitions in complex background, low contrast, foggy videos and cluttered are still a challenging problem. Motivated by the above facts, in the present work, an idea of LoG-Gabor-PCA approach is explored to obtain desirable high pass characteristics as well to capture more information in high frequency areas for efficient moving objects recognitions in the video sequences.

2 Proposed Method

The proposed system comprises of three main steps. The primary step is moving object detection by using Tensor Locality Preserving Projections (Ten-LoPP). The second step is moving object tracking based on the Centroid and Area of detected object. Finally, moving object recognition is performed by using Log-Gabor-PCA approach. The entire processing of the proposed system is illustrated in Fig. 1.

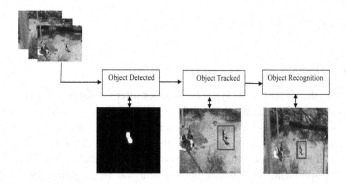

Fig. 1 Block Diagram of the Proposed System

2.1 Moving Object Detection and Tracking

In the proposed method of moving object detection and tracking, the method proposed by us in system [12] is considered. Here, moving object detection is performed by using Ten-LoPP and moving object tracking is done by considering the Centroid and Area of detected object. Results of Detection and Tracking is shown in Fig. 2(a), & 2(b). In this work concentration is mainly given towards recognitions of multiple moving object in video sequences and hence recognitions steps described in details.

2.2 Moving Object Recognition

This section explains the process of moving object recognition in video sequences. The detected moving object features are extracted by applying LoG-Gabor filter. The feature dimension reduction is performed using PCA and are stored in the library.

(a) Successfull detection of moving   (b) Moving Object Tracking
object

Fig. 2 Results of Detection and Tracking

2.2.1   LoG-Gabor

In this section, the reason of applying Log-Gabor filter for the moving object recognition is illustrated. Excellent spatial and frequency information is provided by the Gabor filters for the object localization in scenes, and the description is given about the characteristics of scale, orientation and spatial locality selectivity [11]. The main drawback of Gabor filter is that it doesn't provide the excellent simultaneous localization of the spatial and frequency information and its limitation being that the maximum bandwidth captured by a Gabor filter cannot exceed approximately to one octave. This drawback can be overcome by using Log-Gabor function proposed by field [11]. It is possible to vary the bandwidth form one to three octaves using Log-Gabor filters. Log-Gabor filters have features like null DC component, which reinforces the contrast ridges and edges of images and can be constructed with an arbitrary bandwidth which can be optimized to produce filter with minimal spatial extent. In linear frequency scale, co-ordinates by the transfer function H (f, θ) in the Log-Gabor filters defined in the frequency domain using polar can be represented in a polar form as

$$H(f,\theta) = H_f X H_\theta = exp\left\{\frac{1}{2}\frac{\left(ln\frac{f}{f_0}\right)^2}{\left(ln\frac{\sigma_f}{f}\right)^2}\right\} exp\left\{-\frac{1}{2}\frac{(\theta-\theta_0)^2}{\sigma_\theta{}^2}\right\} \qquad (1)$$

The radial component H_f controlling the bandwidth and the angular component H_θ controlling the spatial orientation that the filter responds [13]. The resultant image of applied LoG Gabor filter is shown in Fig. 3(b). The obtained LoG Gabor filter features whose dimensionality is exceptionally high. For this reason, one of the computationally effective subspace approach is needed. Which is Principle Component Analysis (PCA) have been used which reduce the dimensions space significantly.

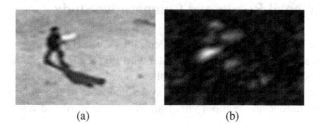

(a) (b)

Fig. 3 Results of Gabor filtering process:(a) original image.(b) Log Gabor filtered Image

2.2.2 Combination of LoG Gabor-PCA

The accuracy is significantly high for the object recognition from Log-Gabor filter based PCA for the standard and our own collected Datasets. This is capable of recognizing objects in video with small far captured objects, diffuse glow effect, dark objects, dark backgrounds etc. The construction of Log-Gabor filters are done with arbitrary bandwidth and the optimization can be done to produce filters with minimal spatial extent. There are two prominent features: First, Log-Gabor functions always have no DC component, and second, the transfer function of the Log-Gabor function has an extended tail at high frequency end [11].

Feature vectors of moving objects are extracted for a given set of training images through Log-Gabor approach. For finding the Principle Components and to reduce the dimensional space of the image to store in the library the reduced image data is further processed by PCA. Feature vectors are extracted through appropriate Log-Gabor approach for testing purpose of moving object image sequence. Further reduction of the dimensional space of the image is done and features are extracted using PCA, and the better classification of the moving object is done by measuring the angle based distance of mean values of training image sequences in each class and the testing image sequences. The recognition results obtained are shown in Fig. 4.

(a) (b)

Fig. 4 Results of Moving Objects Recognition in test video sequences

3   Experimental Results and Comparative Study

The proposed system is implemented in Pentium IV 1 GHz processor with MAT-LAB 10. The system is experimented on standard PETS, OTCBVS, and Videoweb Activities datasets and also on our own collected video sequences comprising various environmental scenes. The proposed system capable of recognizing the moving objects of indoor and outdoor environments of standard and also of our own

Table 1 Percentage of Recognition Accuracy

Input Sequences	Correctly Recognized	Incorrectly nized	Recog-	Recognition racy	Accu-
Arial Fig 3	25	03		88	
OTCBVS Fig 7(a)	43	07		86	
Fig 8(a)	23	02		92	
Fig 8(b)	24	01		96	

Fig. 5 Results of Moving Object Recognition of standard Video Sequences using proposed LoG-Gabor-PCA approach

collected video sequences efficiently by using LoG-Gabor-PCA Approach. The Fig. 5, 6 & 7 shows successive moving object recognitions of a scene of standard dataset and our own datasets respectively. The LoG-Gabor-PCA approach is successfully experimented on standard and our own collected datasets. Recognition accuracy is also tabulated for different standard and our own collected datasets which is shown in Table 1. From the table, it is clear that the percentage of recognition accuracy 90.5%.

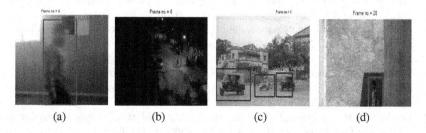

(a)   (b)   (c)   (d)

Fig. 6 Results of Moving Object Recognition of our own video sequences using proposed LoG-Gabor-PCA approach

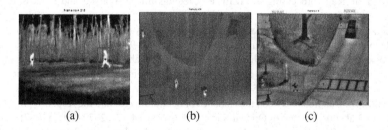

(a)   (b)   (c)

Fig. 7 Results of Moving Object Recognition of standard OTCBVS datasets (Infrared Images) Video Sequences using proposed LoG- Gabor-PCA approach.

4   Conclusion

The Proposed Method is successfully implemented and experimented on standard and our own collected datasets of different environments. A LoG-Gabor-PCA based moving object recognition scheme has been developed to overcome the drawback of the original PCA and Gabor Filter Approaches. Applying LoG-Gabor-PCA approach on detected moving objects is a significant advantage in achieving dimensionality reduction and insensitive texture feature extraction. The LoG-Gabor-PCA approach gives better recognition accuracy in recognizing moving objects of video sequences. The performance of the moving object detection and recognition can be improved in specific circumstances such as occlusion between objects and will be considered in our future work.

References

1. Hsu, Wallace: An industrial network flow information integration model for supply chain management and intelligent transportation. Enterprise Information Systems 1(3), 327–351 (2007)
2. Lipton, A.J., Fujiyoshi, H., Patil, R.S.: Moving target classification and tracking from real-time video. In: Proceedings of the IEEE Applications of Computer Vision, WACV 1998, pp. 8–14 (1998)
3. Petrovic, V.S., Cootes, T.F.: Vehicle type recognition with match refinement. In: International Conference on Pattern Recognition, vol. 3(8), pp. 95–98 (2004)
4. Lin, Y., Bhanu, B.: Evolutionary feature synthesis for object recognition. IEEE Transactions on Systems, Man, and Cybernetics, Part C: Applications and Reviews 35(2), 156–171 (2005)
5. Sullivan, G.D., Baker, K.D., Worrall, A.D., Attwood, C.I., Remagnino, P.M.: Model-based vehicle detection and classification using orthographic approximations. Image and Vision Computing 15(8), 649–654 (1997)
6. Bergboer, N.H., Postma, E.O., van den Herik, H.J.: Context-based object detection in still images. Image and Vision Computing 24(9), 987–1000 (2006)
7. Takano, S., Minamoto, T., Niijima, K.: Moving object recognition using wavelets and learning of eigenspaces, p.151 (1998)
8. Zanin, M., Messelodi, S., Modena, C.M.: An efficient vehicle queue detection system based on image processing. In: IEEE Proceedings Image Analysis and Processing, pp. 232–237 (2003)
9. Liu, C., Wechsler, H.: Independent Component Analysis of Gabor Features for Face Recognition. IEEE Trans. Neural Networks 14(4), 919–928 (2003)
10. Cook, J., Chandran, V., Sridharan, S., Fookes, C.: Gabor Filter Bank Representation for 3D Face Recognition. In: Proc. IEEE Digital Imaging Computing: Techniques and Applications, pp. 16–23 (2005)
11. Fields, D.: Relations between the statistics of natural images and the response properties of cortical cells. Journal of Optical Society of America 4(12), 2379–2394 (1987)
12. krishna, M.T.G., Ravishankar, M., Rameshbabu, D.R.: Ten-loPP: Tensor locality preserving projections approach for moving object detection and tracking. In: Meesad, P., Unger, H., Boonkrong, S. (eds.) IC2IT2013. AISC, vol. 209, pp. 291–300. Springer, Heidelberg (2013)
13. Lajevardi, S.M., Hussain, Z.M.: Facial Expression Recognition Using Log-Gabor Filters and Local Binary Pattern Operators. In: Proceedings of the International Conference on Communication, Computer and Power, pp. 349–353 (2009)

Off-Line Signature Verification Based on Principal Component Analysis and Multi-Layer Perceptrons

B.H. Shekar and R.K. Bharathi

Abstract. The off-line signature verification approach based on Principal Component Analysis (PCA) and Muti-Layer Perceptrons (MLP) is proposed in this work. The proposed approach involves three major phases. In the first phase, the signature image is subjected to preprocessing, such as binarization, noise elimination, skew correction followed by normalising the image by a series of thinning and dilating with a structuring element of size $3X3$. The principal component analysis is employed on the preprocessed signature samples and the features extracted are termed as eigen-sign feature vectors, in second phase. The multilayer perceptrons, a neural network based approach is trained with eigen-sign feature vector and subsequently used for verification of signature samples. Extensive experimentation has been conducted on the publicly available signature datasets namely, CEDAR, GPDS-100 and MUKOS, a regional language dataset. The state-of-art off-line signature verification methods are considered for comparative study and objective analysis through experimental results is provided to justify the accuracy of the proposed approach.

Keywords: Principal Component Analysis (PCA), eigen-sign, Multi-layer perceptrons (MLP), Classification, Off-line signature verification.

1 Introduction

Handwritten signature is one of the oldest accepted mode of authenticating a person in many of the business transactions. In spite of the existence of many behavioural

B.H. Shekar
Department of Computer Science, Mangalore University, Mangalore
e-mail: bhshekar@gmail.com

R.K. Bharathi
Department of Master of Computer Applications, S.J. College of Engineering,
Mysore, Karnataka, India
e-mail: rkbharathi@hotmail.com

S.M. Thampi et al. (eds.), *Recent Advances in Intelligent Informatics*,   101
Advances in Intelligent Systems and Computing 235,
DOI: 10.1007/978-3-319-01778-5_11, © Springer International Publishing Switzerland 2014

biometrics, handwritten signatures plays a vital role in society, because of its advantages: it is one of the natural established mode, non-presence of signer for authentication, minimal disruption of the signer, invasive measurement of the signature characteristics. Generally, the signature is also the word with which a person identifies himself, and as such will have a greater personal significance than any other word he/she writes. Based on the ownership, the signature is classified as genuine and forge. Genuine signature is the representative of a person through a few strokes which is stable and unique for a long period of time. The genuine signature, above all others, is the design of the stokes/word which is written automatically and without concious thought about the mechanics of its reproduction for any number of times.

One of the major threat in signature verification for authentication is the forged signatures. A forged signature (forgery) is the imitation of the genuine signature to the level of acceptance without the knowledge of the genuine signer. Based on the knowledge the forger has about the signature and the signer, forgery can be broadly classified into three types, such as: *skilled, random and simple forgery*. In simple forgery, the forger knows the name of the signer but not the genuine signature pattern and hence produces his/her own pattern of stokes. Random forgeries occur when the forger neither knows the name of the signer nor the signature pattern, where as the skilled forger will have the access to the genuine signature sample pattern and also the name of the signer, hence resulting as the major threat for verification and authentication of a person through signature. Apart from the forge threat, there are many other instances, such as the intra class deviation of the signature sample, i.e variation of the signature by the genuine signer due to age, illness, orientation of the document used to sign, pen width, deteriorated signatures, illegible signatures and so on which needs greater attention in signature verification.

There are basically two acquisition modes of the signatures, namely On-line and Off-line signature. On-line signatures are acquired at the instance of its registration beholding the dynamic details viz: velocity, acceleration, duration, pen lifts, direction of pen movement, pressure applied, where as off-line signatures are the static image of the registered signature. Off-line signature lacks the above mentioned dynamic details, still possess global and local features viz: signature image area, height, width, zonal information, important points such as end points, cross points, cusps, loops, and so on.

Although the human ability to recognise patterns is generally far superior to that of computer, there are many real-life activities where machine recognition is more efficient and convenient. Efficient selection of feature and classifier plays important role in off-line signature verification. Randhawa et al. [10] exploited the use of Hu's moment in extracting the feature and support vector machine (SVM) as classifier. Kisku et al. [6] concentrates on the fusion of classifiers, where Gaussian empirical rule, Euclidean and Mahalanobis distance based classifiers are fused using SVM. Malik et al. tried to fuse both local and global features and examined the performance on ICDAR2009 dataset [8]. Solar et al. [11] concentrated on local interest points and descriptors for off-line signature verification. Kumar et al. [7] presents a novel set of features based on surroundedness property of a signature image to provide a measure of texture through the correlation among signature pixels. To examine the efficacy they used two popular classifiers, SVM and MLP on CEDAR and GPDS dataset. Very few work are done on non-English signature verification. Although the general model underlying the signature generation process is invariant in terms of cultural habits and language differences among signers, the enormous diversity in the signatures of people from different countries has suggested the development of specifically designed solutions [3]. For instance, Pal et al. [9] have concentrated on signature in Chinese, Japanese and Arabic. Shekar and Bharathi [12] , [13] have attempted on signatures in Kannada, a regional language of India.

Selection of dominant and important features for the representation of the sample is crucial in any of the pattern recognition approach. The idea behind most dimension reduction methods is to transform the original set of variables in such a way that only a few of the new transformed variables incorporates most of the information contained in the original ones. The most popular technique is the principal component analysis (PCA). Principal component analysis, projects images into a subspace such that the first orthogonal dimension of the subspace captures the greatest amount of variance among the images and the last dimension of the subspace captures the least amount of variance among the images. Ismail et al. [4] presents a novel off-line signature identification based on chain codes and Fourier descriptors. In their work, for the compact representation of feature vector, PCA is employed and classifier is designed based on the feed forward artificial neural network. Shekar and Bharathi [12] concentrated on reducing the dimension of feature vectors, preserving the discriminative features obtained through principal component analysis on shape based signature and extended approach is given based on Kernel-PCA [13].

The major contribution of this paper is to pre-process the signature image suitable for further processing and extraction of dominant/prominent features by applying the concept of principal component analysis. The classification / recognition of genuine sample from forge is achieved through a back propagation neural network approach, multi-layer perceptrons (MLP). The paper is organised as follows: Section 2 demonstrates the process of pre-processing and feature extraction through PCA. Section 3 discusses the experimental set-up and result analysis with a comparative study on the state-of-art methods followed by conclusion and future work in section 4.

2   Proposed Approach

The proposed PCA-MLP approach for off-line signature verification is of three phases: pre-processing, feature extraction through PCA and verification by MLP. In pre-processing, the signature image is binarized, and subjected to morphological filter operation to eliminate the intruded noise, and finally the thickness of the input image is normalised once the skew is corrected by rotating the image parallel to x-axis. The pre-processed sample signature image is fed into PCA for compact representation and hence the dominant feature representative vector is extracted. These dominant features are thus trained by MLP and tested to classify the genuine from the forge, and the performance accuracy is evaluated.

2.1 Pre-processing

In this phase, the given signature image is binarized using Otsu's binarization method. The noise intruded due to binarization is eliminated using morphological filter operations. We have employed the steps discussed in [5] to rotate back the signature image parallel to X-axis of the Cartesian co-ordinate to correct the skew. We later normalise the thickness of the strokes in the skew corrected signature image by a series of thinning and dilating the sample image. The structuring element of size $3X3$ is considered for normalising the sample thickness. The results obtained in different pre-processing steps are shown on a sample image from GPDS-100 corpus in figure 1.

Fig. 1 Results of preprocessing on an signature sample image from GPDS-100 dataset (a) Input signature image (b) Binarized (c) Noised removed image (d) Rotated by an angle of 29.101 degree and thinned image (e) Normalised image

2.2 Feature Extraction

The preprocessed signature image is subjected to principal component analysis (PCA) for the purpose of feature extraction. Each of the preprocessed signature image results in a matrix with the pixel intensity values either 0 or 1. Now, each signature image, being represented in a matrix form, is transformed into a vector, resulting in a feature vector (but with high dimension). Obviously, the length of this preprocessed image's feature vector is very high and hence we employ PCA for the compact representation. All the signature samples considered during training forms a training matrix from which we obtain the feature vectors. The prominent eigen vectors are considered to project the training data that results in *eigen-valued features* knowledge base.

More formally:

- Let Img is a preprocessed signature image of size $R \times C$.
- Let each training image matrix is transformed into a vector form of size $[(R \times C) \times 1]$.
- Let S_i $(i = 1....N)$ be the number of such training signatures (Total no. of training samples be N).

- Let *Avg* be the average signature image of size $R \times C$, defined as:

$$Avg = \sum_{i=1}^{N} S_i, \tag{1}$$

  and is represented in a vector form.
- Let T be a training matrix obtained by concatenating all the preprocessed signature image vectors, resulting in a matrix of size $[(R \times C) \times N]$ i.e.,

$$T = [Img_1, Img_2, ..., Img_N], \tag{2}$$

  where each Img_i is the column vector representation of training sample.
- The training matrix T is mean centred by subtracting average image *Avg* from each of the training sample image, i.e.,

$$MC = [Img_1 - Avg, Img_2 - Avg, ..., Img_N - Avg]. \tag{3}$$

- The eigenvectors, e_i and the corresponding eigenvalues λ_i of MC are determined by solving the well known eigen structure decomposition problem:

$$\lambda_i.e_i = MC.e_i. \tag{4}$$

However, the L eigenvectors, corresponding to the L largest eigen values, are good enough to capture the dominant characteristics of any signature. Thus, all the signatures considered for training are projected onto these L eigenvectors that results in ***eigen-sign feature vector*** knowledge base. This knowledge base consists of all signer's genuine signature feature set as well as their skilled forgery and this compact representation of knowledge base substantiate the original feature vector for further processing.

3 Experimental Results and Comparative Analysis

This section presents the analysis of the experimental results conducted on the well known publicly available datasets. The proposed approach is experimented on standard off-line English signature datasets namely: CEDAR and GPDS-100 (A sub-corpus of GPDS-300). In addition, we have also extended the experiments on our regional language off-line signature corpus called MUKOS (Mangalore University Kannada Off-line Signature). Each dataset has varying number of signers, genuine and forge samples. All experiments are developed using MATLAB tool and tested on Pentium(R) dual core CPU with 3GB RAM on windows-7. The knowledge base created is the accumulation of both genuine sample features and skilled forge sample features of the respective datasets. The Multi-layer perceptrons are trained with the combination of genuine and forge samples of the datasets. For each dataset, the signature samples are divided into two groups: training sample set and testing sample set with varying number of samples. As per the test configurations in state-of-art works on off-line signature verification,We have carried out two sets of experiments. In Set-1, first ten genuine and first ten skilled forgeries are considered as training samples and tested against the remaining samples of the respective datasets, where as in Set-2, we have taken first 15 samples of genuine and first 15 samples of skilled forgery for training the MLP and tested with remaining samples. As we have used principal component analysis for compact representation of the feature vector, we have carried out the experimentation with varying dimensional feature vectors for projection purpose, say, 20,30,40 and 50 on all datasets. Practically the skilled forgery will not be at hand and cannot train the system, still, its the great challenge to segregate the genuine from skilled forge sample. More over, negative sample is needed to train the MLP (a bi-classifier), and hence, here we have considered the skilled forge samples as the negative samples.

Selection of optimal parameters for MLP is always a challenging task as it is solely application (feature value) based and can be decided only after trial experimentations. Here we have employed 2 hidden layers, with Sum of Squared Error (SSE) as an error performance function with a goal of 0.001 and a maximum of 3000 epochs. The training is carried out by creating feed-forward back propagation network and later a simulator is used to test the data along with the network created while training. In order to estimate the accuracy of the proposed approach, we have considered two well-known evaluation metrics: FAR (False Acceptance Rate) and FRR (False Recognition Rate). The FAR is defined as the percentile ratio of the total number of accepted forgeries to the total number of tested forgeries. The FRR is defined as the percentile ratio of total numbers of genuine rejected versus total number of genuine tested.

3.1 Experimental Result Analysis on CEDAR Dataset

The Centre of Excellence for Document Analysis and Recognition (CEDAR), at SUNY Buffalo, has built the off-line signature dataset with 55 signers. From each signer, 24 genuine and 24 forge samples were collected resulting a total of 2640 signature samples. The CEDAR signature dataset is available on $http : //www.cedar.buffalo.edu/NIJ/publications.html/$.

We started experimenting with set-1 configuration, where we trained the MLP with first 10 genuine and first 10 skilled forge sample features and tested with the remaining 14 genuine and 14 skilled forge samples. Set-2 test configuration had first 15 genuine and first 15 skilled forge sample features for training the MLP and tested against the remaining 9 genuine and 9 forge samples. The metrics FAR and FRR obtained for above experiments on varying dimensional feature vectors are given in Table 3.

From the literature we observed that, Kalera et al. [5], Chen and Shrihari [1] and Kumar et al. [7] have experimented on CEDAR dataset and hence a comparative analysis is given in Table 1.

Table 1 Experimental Results obtained for CEDAR Dataset - A comparison

Proposed by	Feature type	Classifier	Accuracy	FAR	FRR
Kalera et al. [5]	Word Shape	PDF	78.50	19.50	22.45
Chen and Shrihari [1]	Zernike moments	DTW	83.60	16.30	16.60
Kumar et al. [7]	Signature morphology	SVM	88.41	11.59	11.59
Proposed Approach	**PCA-MLP**	**MLP**	**96.88**	**3.22**	**3.06**

3.2 Experimental Result Analysis on GPDS-100 Dataset

Digital Signal Processing Group (GPDS) of the Universidad de Las Palmas de Gran Canaria, has comeout with a good scale dataset called GPDS-300 corpus. GPDS-300 is a dataset of 300 signers signature samples with 24 genuine and 30 forge of each, summing to a total of 16200 samples. For our experimentation, a subset of 100 signers, starting from the first signer to 100th signer is extracted from the corpus and named GPDS-100 with 5400 signature samples. GPDS-300 corpus is available on $http: //www.gpds.ulpgs.cs/download/index.htm/$.

Here we have conducted experimentation with set-1 test configuration where we considered first 10 genuine and first 10 skilled forgery sample features to train the MLP and tested with remaining 14 genuine and 20 skilled forge. Extending the experimentation, set-2, considering first 15 genuine and 15 skilled forge sample features to train the MLP and tested with remaining 9 genuine and 15 skilled forge sample features of all 100 signers in the corpus. The experiment was conducted with varying dimensional eigen feature vectors. Thus, the results in terms of FAR and FRR on GPDS-100 dataset on varying dimensional feature vectors are tabulated in Table 3. A comparative analysis with the state-of-art work is given in Table 2.

Table 2 Experimental result obtained for GPDS-300/100 dataset: A comparative analysis

Model Proposed	Feature type	Classifier type	Accuracy	FAR	FRR
Ferrar et al. [2]	Geometric features	SVM	86.65	13.12	15.41
Vargas et al. [14]	GLCM + LBP	SVM	87.28	6.17	22.49
Solar et al. [11]	Local interest points	Bayseian	84.70	14.20	16.40
Proposed Approach	**PCA-MLP**	**MLP**	**95.67**	**4.44**	**5.33**

3.3 Experimental Result Analysis on MUKOS Dataset

MUKOS [Mangalore University Kannada Off-line Signature] is a corpus with signatures in Kannada, a regional language in south India . We have collected 30 genuine signatures and 15 skilled forgeries from 30 signers resulting in 1350 signatures [12].

The experimentation is conducted considering set-1 test configuration with first 10 genuine and first 10 skilled forgery samples by training the MLP and further classified with the same set of feature vectors. The remaining 15 genuine and 5 skilled forgery samples of each signer is considered for testing. Similar experimentation is carried out with set-2, where we have considered first 15 genuine and 15 skilled forge samples to yield the feature vectors of varying dimensions. Here the remaining 15 genuine and all 15 skilled forge samples of all signer's are considered for testing. The classification accuracy due to set-1 and set-2 experimental configurations are tabulated in Table 3 along with varying dimensional feature vectors. A comparative analysis on the MUKOS dataset with our earlier work is tabulated in Table 4.

Table 3 Experimental Results obtained for CEDAR, GPDS-100 and MUKOS Dataset

Dataset	CEDAR				GPDS-100				MUKOS			
Experimental set-up	SET-1		SET-2		SET-1		SET-2		SET-1		SET-2	
Dimension / Accuracy	FAR	FRR	FAR	FRR	FAR	FRR	FAR	FRR	FAR	FRR	FAR	FRR
20	11.07	8.21	12.62	7.22	11.07	8.21	12.62	7.22	5.00	5.60	5.44	3.20
30	7.93	7.46	3.22	3.06	10.10	8.92	9.22	8.64	6.32	5.60	3.62	2.40
40	8.63	6.94	3.94	4.02	9.78	9.28	8.44	6.11	5.60	6.10	4.74	3.62
50	10.32	7.41	4.62	3.92	6.43	7.56	4.44	5.33	6.46	6.24	5.40	3.96

Table 4 Experimental Results for MUKOS dataset- A comparative analysis

Proposed by	Feature Type	Classifier	Accuracy	FAR	FRR
Shekar and Bharathi [12]	Shape based eigen signature	Euclidean Distance	93.00	11.07	6.40
Proposed Approach	**PCA-MLP**	**MLP**	**96.38**	**3.62**	**2.40**

4 Conclusions and Future Works

In this paper we explored the application of PCA on off-line signatures to extract the dominant features followed by verification using MLP. Here the input image is well pre-processed and PCA is employed to obtain the eigen feature vectors representing the sample image which are used for training and testing using MLP. Extensive experimentation is conducted on well known publicly available signature dataset :CEDAR and GPDS-100 (a sub-corpus of GPDS-300) and a regional language signature dataset called MUKOS. In order to highlight the superiority of the proposed approach, a comparative analysis is provided with the state-of-the-art off-line signature methods. It is found that the proposed approach is simple to implement, computationally efficient and accurate in terms of verification.

References

1. Chen, S., Srihari, S.: Use of exterior contours and shape features in off-line signature verification. In: ICDAR, pp. 1280–1284 (2005)
2. Ferrer, M., Alonso, J., Travieso, C.: Offline geometric parameters for automatic signature verification using fixed-point arithmetic. IEEE Transactions on Pattern Analysis and Machine Intelligence 27(6), 993–997 (2005)
3. Impedovo, D., Pirlo, G., Plamondon, R.: Handwritten signature verification: New advancements and open issues. In: International Conference on Frontiers in Handwriting Recognition, pp. 367–372. IEEE (2012)
4. Ismail, I.A., Ramadan, M.A., El-Danaf, T.S., Samak, A.H.: An efficient off line signature identification method based on fourier descriptor and chain codes. IJCSNS 10(5), 29 (2010)
5. Kalera, M.K., Srihari, S., Xu, A.: Off-line signature verification and identification using distance statistics. International Journal of Pattern Recognition and Artificial Intelligence 18, 228–232 (2004)
6. Kisku, D.R., Gupta, P., Sing, J.K.: Fusion of multiple matchers using svm for offline signature identification. abs/1002.0416 (2010)
7. Kumar, R., Kundu, L., Chanda, B., Sharma, J.D.: A writer-independent off-line signature verification system based on signature morphology. In: First International Conference on Intelligent Interactive Technologies and Multimedia, pp. 261–265. ACM, New York (2010)
8. Malik, M.I., Liwicki, M., Dengel, A.: Evaluation of local and global features for off-line signature verification. In: Proc. Int. Workshop on Automated Forensic Handwriting Analysis, Beijing, China, pp. 26–30 (2011)
9. Pal, S., Blumenstein, M., Pal, U.: Recognition Unit. Non-english and non-latin signature verification systems: a survey. In: Proc. of First International Workshop on Automated Forensic Handwriting Analysis, pp. 1–5 (2011)
10. Randhawa, M.K., Sharma, A.K., Sharma, R.K.: Off-line signature verification based on hu's moment invatiants and zone features using support vector machine. International Journal of Latest Trens in Engineering and Technology 1, 16–23 (2012)
11. Ruiz-del-Solar, J., Devia, C., Loncomilla, P., Concha, F.: Offline signature verification using local interest points and descriptors. In: Ruiz-Shulcloper, J., Kropatsch, W.G. (eds.) CIARP 2008. LNCS, vol. 5197, pp. 22–29. Springer, Heidelberg (2008)
12. Shekar, B.H., Bharathi, R.K.: Eigen-signature: A robust and an efficient offline signature verification algorithm. In: International Conference on Recent Trends in Information Technology, pp. 134–138 (June 2011)
13. Shekar, B.H., Bharathi, R.K., Sharmilakumari, M.: Kernel eigen-signature: An offline signature verification technique based on kernel principal component analysis. In: Emerging Applications of Computer Vision, EACV 2011 Bilateral Russian-Indian Scientific Workshop, pp. 37–44 (November 2011)
14. Vargas, J.F., Ferrer, M.A., Travieso, C.M., Alonso, J.B.: Off-line signature verification based on grey level information using texture features. Pattern Recognition 44(2), 375–385 (2011)

Gradual Transition Detection Based on Fuzzy Logic Using Visual Attention Model

Amudha Joseph and P. Naresh Kumar

Abstract. Shot boundary detection (SBD) is the process of automatically detecting the boundaries between shots in video. It is a problem which has attracted much attention since video became available in digital form as it is an essential pre-processing step to almost all video analysis, indexing, summarization, search, and other content based operations. The existing SBD algorithms are sensitive to video object motion and there are no reliable solutions to detect gradual transitions (GT). GT is difficult to detect because of the following reasons. First, GT include various special editing effects, including dissolve, wipe, Fade Out/In. Each effect results in a distinct temporal pattern over the continuity signal curve. Secondly, GT exhibit varying temporal duration and also the temporal patterns of GT are similar to those caused by object/camera movement, since both of them are essentially processes of gradual visual content variation. The proposed approach uses Fuzzy rule based system to detect the Gradual Transitions based on the features derived from visual attention model which detects the gradual transition better than the existing approaches.

1   Introduction

Digital videos are the major type of data in the new generation databases. With increase in amount of videos online, huge amount of digital videos are available for previewing. This huge amount of digital data poses new challenges of storage and access. The key step to manage the data easily is to segment the videos into shots. This makes the video data more manageable by imposing a hierarchy. This segmentation process is generally referred as shot boundary detection (SBD). A shot is generally defined as a series of interrelated pictures taken by a camera continuously in space and time.

Amudha Joseph · P. Naresh Kumar
Amrita Vishwa Vidyapeetham, Amrita School of Engg., Bangalore-35
e-mail: j_amudha@blr.amrita.edu, nareshchowdary463@gmail.com

S.M. Thampi et al. (eds.), *Recent Advances in Intelligent Informatics*,   111
Advances in Intelligent Systems and Computing 235,
DOI: 10.1007/978-3-319-01778-5_12, © Springer International Publishing Switzerland 2014

The transition from one shot to next shot is broadly categorized as Abrupt Transition (AT) and Gradual Transition (GT). Abrupt Transition, also known as Cut, denotes instantaneous transition from one shot to another. On the other hand, a Gradual Transition is obtained by incorporating digital effects usually through editing. The Abrupt Transition however has been tackled easily in most of the shot boundary detection, but the detection of GT remains a difficult problem. The detection of GT is more difficult than that of CUT in the perspective of the temporal and spatial interrelation of the two adjacent shots. Here, from a different point of view, we summarize three reasons why it is difficult. First, GT include various special editing effects, including dissolve, wipe, Fade Out/In, etc. Each effect results in a distinct temporal pattern over the continuity signal curve. Second, GT exhibit varying temporal duration, probably from three to dozens of frames. Finally, the temporal patterns of GT are similar to those caused by object/camera movement, since both of them are essentially processes of gradual visual content variation.

In this paper gradual transition detection is mainly focused and visual attention model is used for feature extraction and fuzzy logic is used for classification of shots. The major advantage of using the Visual attention model is using the saliency map which captures the discriminate information in each frame. The features obtained in visual attention model can be reused in the higher level abstractions like key frames extraction [1], annotation [2] and for object recognition [3].

The rest of the paper is organized as follows. A brief survey of previous work is explained in section 2. Section 3 explains the feature extraction for the proposed system; the shot boundary detection classification is explained detail in section 4. Experimental results are presented in section 5 and conclusion is given in section 6.

2   Related Works

As mentioned, the detection of Gradual Transition is one of the major challenges to the proposed formal framework. So far, no techniques of Gradual Transition detection have been able to achieve the result comparable to that of abrupt transition detection. Some of the existing methods are designed to detect one specific editing effect, such as Fade-in and Fade-out, wipe and dissolve, while others are developed to detect several types of editing effects simultaneously. Here, we present a brief overview of the existing methods.

2.1   Features Used in SBD

Almost all shot boundary detection algorithms reduce the large dimensionality of the video domain by extracting a small number of features from each video frame. These are extracted either from the whole frame or from a subset of it. The pixel-based [4] method is the simplest method of constructing the mapping, which maps each image to itself. Obviously, this is the most sensitive method, since it has captured all details of the frame. Color histograms [5] methods are more invariant to local or small global movements than pixel-based methods. However, it is not expressive enough to distinguish the shots within the same scene. An obvious choice of feature is edge information in a ROI [6]. Edges can be used as it is, be combined to form objects or used to extract ROI statistics. They are invariant to illumination changes and most motion but their main disadvantage is computational cost, noise sensitivity and high dimensionality. Statistical methods expand on the idea of pixel differences by dividing the images into regions and comparing statistical measures of the pixels in those regions [7]. These methods are reasonably tolerant of noise, but are slow due the complexity of the statistical formulas. It also generates many false positives. Visual Attention Model is based on the extended frequency-tuned saliency model [1]. Saliency maps are produced from the color, luminance and orientation features of the image, Major advantage of the algorithm is that it is robust to dissolving digital video effects used during shot transition. There are many more algorithms [8] which use multiple features like edges, motion vectors, histogram, and statistical measures etc., and use them in combination for better results [9].

2.2   Spatial Feature Domain

The size of the region from which individual features are extracted plays a great role in the performance of shot boundary detection. A small region tends to reduce detection invariance with respect to motion, while a large region tends to miss transitions between similar shots.

Single frame: Some algorithms use a single frame per feature. This feature can be luminance [4], edge strength [6] or other. However, such an approach results in a very large feature vector and is very sensitive to motion.

Rectangular block: Another method is to segment each frame into equal-sized blocks, and extract a set of features [10]. This approach is invariant to small camera and object motion. By computing block motion it is possible to enhance motion invariance, or to use the motion vector itself as a feature.

Arbitrarily shaped region: Feature extraction can also be applied to arbitrarily shaped and sized regions [11]. This exploits the most homogeneous regions, enabling better detection of discontinuities. Object-based feature extraction is also included in this category. The main disadvantage is high computational complexity and instability due to the complexity of the algorithms involved.

2.3   Temporal Domain of Continuity Metric

Another important aspect of shot boundary detection algorithms is the temporal window of frames which is used to perform shot boundary detection. These can be one of the following:

Two frames: The simplest way to detect discontinuity is to look for a high value of the discontinuity metric between two successive frames [5, 8]. However, such an approach fails when there is significant variation in activity among different parts of the video, or when certain shots contain events that cause short-lived discontinuities (e.g. photographic flashes).

N-frame window: The most common technique for alleviating the above problems.It is based on detecting the discontinuity by using the features of all frames within a temporal window [4, 12]. This is either by computing a dynamic threshold against which a frame-by-frame discontinuity metric is compared or by computing the discontinuity metric directly on the window.

Entire video: The most accurate method is to take the characteristics of the whole video into consideration when detecting a shot boundary, as in [14]. Again, the problem is that the video can have great variability within and between shots.

2.4   Shot Boundary Detection Method

*Rule Based Classifier:*In Rule-based classifier [4, 5, 6], the classification function is usually defined as

$$w_t = \begin{cases} 1, & \text{if } s_t > T \\ 0, & \text{otherwise} \end{cases} \tag{1}$$

Where T is predefined threshold, as mentioned in the above eq. (1), s_t hold the computed (absolute difference of successive frames) value from the frames. If it exceeds the threshold 'T', theclassifier gives an output '1' which indicates that there is a 'Shot' detected between the I_t and I_{t+1} frames, else there is no transition between the pair of frames. In the early work, heuristically chosen global thresholds were used. It is difficult to select a threshold 'T' appropriate for various genres of videos. To address this drawback, various local adaptive thresholds were proposed [4]

Statistical Machine Learning: There has been some recent efforts treating SBD as a pattern recognition problem and turning to the tools of machine learning. Various discriminative approaches, including K-means [15], KNN[14], and support vector machines (SVMs) [16], have been employed to perform SBD. With the statistical machine learning methods, the parameters of the models are chosen via cross validation processes and the shapes of classification hyper plane are constructed automatically during the training procedure. Compared with the thresholding schemes, the machine learning methods make decisions via the recognition of the shot transition patterns instead of the evaluation of the amplitude of content variations. It is expected that full use of contextual information can be made by machine learning methods.

Extensive research has been done in the area of Shot Boundary Detection and each of the above techniques has its own pros and cons. However, the performance of the system depends on the major factors like how much complexity is involved in the model. The major difficulty in detecting gradual transition lies due to occurrence of gradual transition over a set of varying number of frames. So, we present a unique model for detecting the Gradual transition using features extracted from Visual attention model, which is able to identify all the transitions frames. Moreover experimental results show that the proposed method works well in detecting the Gradual transitions measured by both recall and precision.

3   Feature Extraction

In the proposed method, Visual attention model is used for feature extraction. Visual attention is an important biological mechanism which can rapidly help human to capture the interested region within eye view and filter out the minor part of image. By means of visual attention, checking for every detail in image is unnecessary due to the property of selective processing.

3.1   Visual Attention Model

A Visual attention system [13] approach is a bottom-up approach where two different features are computed like color and intensity. For each feature, the saliencies are computed on different scales and for different feature types e.g. red, green, blue, yellow and intensity. Thereafter, the maps are fused step-by-step, thereby strengthening important aspects and ignoring others. For each feature, wefirst compute an image pyramid from which we compute scale maps. These maps are fused into feature maps representing different feature types and these again are combined to a single saliency map. Fig. 1 shows the Visual attention model.

Fig. 1 Flow of visual attention Module

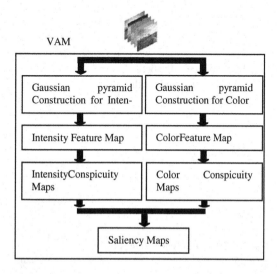

3.2   Calculation of Covariance and Cumulative Distribution Function

After receiving all the saliency maps from the Video file, calculate the covariance for the successive frames of the saliency map as shown in the eq. (2).The covariance value between i^{th} and $(i+1)^{th}$ frames from the saliency map are considered as feature for the Gradual Transition detection. Fig. 2 shows the values obtained for the space video. In the proposed method fuzzy system is used to find Gradual Transitions, the inputs to the fuzzy system is a covariance values but the covariance value range should be from '0' to '1', so to convert the covariance values to a range from 0 to 1 we used Normal Cumulative distribution function (N-CDF) as shown in the eq. (3) and the difference between successive values of N-CDF are calculated and shown in the Fig. 3 for the space video.

$$COV(X,Y) = \frac{1}{n-1} \sum_{i=1}^{n} (X_i - E(X)) (Y_i - E(Y)) \tag{2}$$

where

$E(X)$ = Expected Value of X
$X = i^{th}$ frame of Saliency Map
$Y = (i+1)^{th}$ frame of Saliency Map

$$F(x, \mu, \sigma^2) = \frac{1}{\sqrt{2\pi}} \int_{-\infty}^{\frac{x-\mu}{\sigma}} e^{\frac{-t^2}{2}} dt \tag{3}$$

where

x = Covariance values
μ = Mean of the Covariance values
σ^2 = Variance of the Covariance values

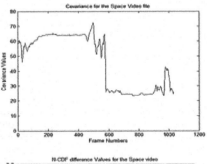

Fig. 2 Covariance value between every two frames computed for the 1200 frames of a space video

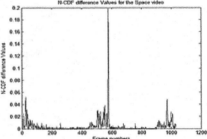

Fig. 3 N-CDF difference Values for the Space Video

The Space video has one gradual transition between the frame numbers 574-581.

4   Shot Boundary Detection

Here the proposed system is using Fuzzy logic to find the shot boundaries in a video.The framework is characterized by appropriate formulation of relevant fuzzy variables, fuzzy rules and inference methods. We know that an abrupt transition occurs due to a sudden and large change among neighbouring frames. On the other hand the gradual transition can occur due to small as well as large change occurring over long time which makes Fuzzy Logic system a better choice in place of thresholding methods.

4.1   Fuzzifier

To use fuzzy logic and fuzzy system for problem solving, the problem must be represented in fuzzy terms. The purpose here is to represent input and output values as linguistic variable. A linguistic variable is a variable which takes fuzzy value and has a linguistic meaning.

The difference in N-CDF for consecutive frames –variables- is fuzzified by using triangular membership functions for each component so that it can be labelled as "negligible", "small", "medium" and "large". We classified the class boundaries for these membership functions as shown in Table 1and the Fuzzy membership functions for the inputs are shown in Fig. 4. Theoutput is categorized as "No transition", "Abrupt transition" and "Gradual transition" using triangular membership functions as shown in Fig. 5 and these components are linked in a Mamdani-style fuzzy inference system.

Table 1 Class boundaries for fuzzy Memberships functions

Sl.No	Class	Class Boundaries
1	Negligible	0.0-0.15
2	Small	0.04-0.3
3	Medium	0.2-0.6
4	Large	0.5-1.0

4.2   Inference Engine

The Inference engine uses Fuzzy rules to convert fuzzy inputs to fuzzy outputs. Twenty-five fuzzy rules are derived for detecting the abrupt transition and the gradual transition as shown in the Table 2. In this the first four rules are for abrupt transition and remaining 21 rules are for gradual transition. Here $F_{(i-1)}$, $F_{(i)}$ and $F_{(i+1)}$ are considered as previous, current and next frame to frame difference of N-CDF values as inputs to the fuzzy system. A set of 3 consecutive frames are classified by the Fuzzy system as Abrupt transition, No transition or gradual transition and the middle frame is labelled as the transition frame.

Fig. 4 Fuzzy membership functions
for the input variables

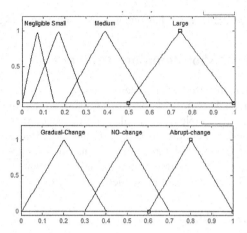

Fig. 5 Fuzzy membership functions
for the output variables

4.3 Defuzzifier

Here the fuzzy rules are evaluated and their outputs are combined by MAX aggregation operator. After that the resulting fuzzy set is defuzzified to produce the crisp decision value, which classifies the frames as Abrupt Transition, No Transition and Gradual Transition. We pass all the frames in a video file to the fuzzy logic and record the frame numbers where the transition is taking place. After recording all the transition frames number, ignore the gradual transition frames which have less than 4 consecutive frames because we know that gradual transition occurs in general more than 3 frames.

The proposed system is implemented and executed on Windows 7 operating system with Intel core2 Duo processor using MATLAB 7.10.0 (R2010a). The dataset used in the experiment has more than 60,000 frames collected from various video files present in TRECVID 2001 test database as well as some other videos. The dataset contains both types of transitions: Abrupt (Cut) transition and Gradual transition which includesfade-in, fade-out, wipes and dissolve. The entire dataset consists of 8 videos with 358 shots.

Performance of the system is measured following the TRECVID evaluation protocols in terms of Recall, Precision and F-measure respectively defined and shown in the equations (4)-(6)

$$\text{Recall}(R) = \frac{\text{Correct}}{\text{Correct} + \text{Miss}} \times 100 \tag{4}$$

$$\text{Precision}(P) = \frac{\text{Correct}}{\text{Correct} + \text{False}} \times 100 \tag{5}$$

$$\text{F} - \text{Measure}(F) = \frac{2 \times \text{Precision} \times \text{Recall}}{\text{Precision} + \text{Recall}} \times 100 \tag{6}$$

Table 2 Fuzzy Rules

Sl.No	Rules for detecting the Transitions
1	If ($F_{(i-1)}$ is Negligible) and ($F_{(i)}$ is large) and ($F_{(i+1)}$ is Negligible) then (output is AT)
2	If ($F_{(i-1)}$ is Negligible) and ($F_{(i)}$ is large) and ($F_{(i+1)}$ is Small) then (output is AT)
3	If($F_{(i-1)}$ is Negligible)and ($F_{(i)}$ is Medium)and($F_{(i+1)}$ is Negligible) then (output is AT)
4	If ($F_{(i-1)}$ is Negligible) and ($F_{(i)}$ is Medium) and ($F_{(i+1)}$ is Small) then (output is AT)
5	If ($F_{(i-1)}$ is Negligible) and ($F_{(i)}$ is Small) and ($F_{(i+1)}$ is Medium) then (output is GT)
6	If ($F_{(i-1)}$ is Negligible) and ($F_{(i)}$ is Small) and ($F_{(i+1)}$ is Small) then (output is GT)
7	If ($F_{(i-1)}$ is Negligible) and ($F_{(i)}$ is Small) and ($F_{(i+1)}$ is large) then (output is GT)
8	If ($F_{(i-1)}$ is Small) and ($F_{(i)}$ is Small) and ($F_{(i+1)}$ is Small) then (output is GT)
9	If ($F_{(i-1)}$ is Small) and ($F_{(i)}$ is Small) and ($F_{(i+1)}$ is Medium) then (output is GT)
10	If ($F_{(i-1)}$ is Small) and ($F_{(i)}$ is Medium) and ($F_{(i+1)}$ is Medium) then (output is GT)
11	If ($F_{(i-1)}$ is Small) and ($F_{(i)}$ is Small) and ($F_{(i+1)}$ is Negligible) then (output is GT)
12	If ($F_{(i-1)}$ is Small) and ($F_{(i)}$ is Medium) and ($F_{(i+1)}$ is Small) then (output is GT)
13	If ($F_{(i-1)}$ is Small) and ($F_{(i)}$ is large) and ($F_{(i+1)}$ is Small) then (output is GT)
14	If ($F_{(i-1)}$ is Medium) and ($F_{(i)}$ is Medium) and ($F_{(i+1)}$ is Medium) then (output is GT)
15	If ($F_{(i-1)}$ is Medium) and ($F_{(i)}$ is Medium) and ($F_{(i+1)}$ is Small) then (output is GT)
16	If ($F_{(i-1)}$ is Medium) and ($F_{(i)}$ is Small) and ($F_{(i+1)}$ is Negligible) then (output is GT)
17	If ($F_{(i-1)}$ is Medium) and ($F_{(i)}$ is Small) and ($F_{(i+1)}$ is Small) then (output is GT)
18	If ($F_{(i-1)}$ is Medium) and ($F_{(i)}$ is Medium) and ($F_{(i+1)}$ is Large) then (output is GT)
19	If ($F_{(i-1)}$ is Medium) and ($F_{(i)}$ is Large) and ($F_{(i+1)}$ is Medium) then (output is GT)
20	If ($F_{(i-1)}$ is Medium) and ($F_{(i)}$ is Large) and ($F_{(i+1)}$ is Small) then (output is GT)
21	If ($F_{(i-1)}$ is Medium) and ($F_{(i)}$ is Large) and ($F_{(i+1)}$ is Large) then (output is GT)
22	If ($F_{(i-1)}$ is Large) and ($F_{(i)}$ is Large) and ($F_{(i+1)}$ is Medium) then (output is GT)
23	If ($F_{(i-1)}$ is Large) and ($F_{(i)}$ is Medium) and ($F_{(i+1)}$ is Medium) then (output is GT)
24	If ($F_{(i-1)}$ is Large) and ($F_{(i)}$ is small) and ($F_{(i+1)}$ is Medium) then (output is GT)
25	If ($F_{(i-1)}$ is Large) and ($F_{(i)}$ is Medium) and ($F_{(i+1)}$ is Small) then (output is GT)

Table 3 Performance Values for video files

Sl.No	File Name	No of Shots	R	P	F
1	Space	1	1.00	1.00	1.00
2	Ksc_launch	3	1.00	1.00	1.00
3	You sang to me	15	0.93	1.00	0.96
4	GLCN	139	0.89	0.95	0.92
5	Anni005	38	0.92	0.97	0.94
6	Anni006	41	0.93	0.97	0.95
7	Anni009	38	0.90	0.92	0.91
8	NAD53	83	0.95	0.89	0.92
	Total	385	0.94	0.96	0.95

Recall reflects, in a way, the rate of miss-classification and is high when miss-classification is low. Precision, on the other hand, reveals the false alarm rate, the lower the false alarm higher is the precision. F-measure is a combination of recall and precision, gives the overall performance. Table 3 lists the performance of the proposed algorithm for the test video files. It is clear from the Table 3 that proposed systems misclassification rate for the given dataset is high and false alarm rate is low.

4.4   Comparative Study of Shot Boundary Detection

The proposed system performance has been compared with Video Shot Detection using Saliency Measure for a set of video and compared with other three approaches for the TRECVID 2001 dataset.In case of Video Shot Detection using Saliency Measure [13] which uses Visual Attention model for extracting the features and then finding Saliency Measure for detecting Shot boundary detection by thresholding, the major disadvantage is it fails to identify all the frames involved in the occurrence of gradualtransition. The proposed system performance measures are better as shown in the Table 4.Secondly, the proposed system iscompared with three other techniques [8,17,18] for the TRECVID 2001 dataset of videos. First Technique is Cooper's system [18] which used YUV color histogram features to detect shot boundary detection using KNN Classifier. Second approach is Yoo's System [17] which used variance distribution of edge information in the frame sequences. The lowest variance in the local frame sequence is chosen as a gradual detection. In the third technique Frame Transition Method[8], which used both global and local features for boundary detection, scatter matrix of edge strength and motion matrix. Finally the frames are classified using multilayer perceptron network. The performance measure for these three technique with the proposed system is shown in the below Table 5.

Table 4 Performance measure Comparison between Proposed system & Saliency Measure [13]

Video File	Proposed Method			Saliency Measure[13]		
	P	R	F	P	R	F
Ksc-launch	1.00	1.00	1.00	1.00	0.80	0.89
Space	1.00	1.00	1.00	1.00	1.00	1.00
You sang to me	0.93	1.00	0.96	0.60	0.90	0.72
GLCN	0.89	0.95	0.92	0.61	0.83	0.70
Total	0.96	0.99	0.97	0.80	0.88	0.83

The proposed method is better than Cooper's system in all the performance measures and when recall is compared the proposed method is very close to Yoo's system and Frame transition model. But in the proposed method, the false detection of shots is less than compared to Yoo's method and almost equal to Frame transition method and the overall performance of the system is equal to Yoo's method and close to Frame transition method.

However Yoo's and Frame transitions methods are slower than proposed method because of the complexity in their systems. In Yoo's method, dividing each frame into 9 blocks and comparing these blocks with the next frames and in the Frame transition model, using total of 183 features-Local and Global- for the neural network with two level of classification for detecting the transition frames, increases the complexity of the model.

Table 5 Performance measure Comparison between Proposed system & other methods [8, 17, 18]

Video Files	Cooper's system[18]			Yoo's system[17]			Frame Transition system[8]			Proposed method		
	R	P	F	R	P	F	R	P	F	R	P	F
Anni005	0.64	0.72	0.68	0.89	0.87	0.88	0.90	0.97	0.93	0.92	0.97	0.94
Anni006	0.69	0.79	0.75	0.88	0.84	0.86	0.93	0.95	0.94	0.93	0.97	0.95
Anni009	0.81	0.83	0.82	0.94	0.86	0.90	0.92	0.94	0.93	0.90	0.92	0.91
NAD53	0.80	0.75	0.77	0.94	0.79	0.86	0.93	0.86	0.89	0.95	0.89	0.92
Total	0.74	0.77	0.76	0.91	0.84	0.88	0.92	0.93	0.92	0.92	0.94	0.93

5   Conclusion

We proposed a model of shot boundary detections which represents various types of shot boundaries. Visual attention model is used to capture the discriminate information in the frames. In the proposed system the difference between successive frames of the covariance value is used as a feature for shot detection.For classification we have employed a fuzzy system which classifies the frames into one of the three categories: no transition, gradual transition and abrupt transition. Thus, the proposed system is not dependent on the thresholds or sliding windowsize. Finally, a simple but effective shot boundary detection system is proposed to reduce the false transition and misclassification error. The method is found to be efficient and works better for a variety of benchmark videos compared to the existing approaches.

References

[1] Mendi, E., Bayrak, C.: Shot Boundary Detection and key Frame Extraction using Salient Region Detection and Structural similarity. In: ACMSE 2010, Oxford, MS,USA, April 15-17 (2010)

[2] Amudha, J., Soman, K.P., Vasanth, K.: Video Annotation. Using Saliency. In: International Conference on Image Processing Computer Vision and Pattern Recognition, vol. 1, pp. 191–195 (2008)

[3] Amudha, J., Soman, K.P., Kiran, Y.: Feature Selection in Top Down Visual Attention Model with WEKA. International Journal of Computer Application, Foundation of Computer Sciences 24(4), 38–43 (2011)

[4] Campisi, P., Neri, A., Sorgi, L.: Automatic dissolve and fade detection for video sequences. In: Proc. Int. Conf. on Digital Signal Processing (July 2002)

[5] Cernekova, Z., Kotropoulos, C., Pitas, I.: Video shot segmentation using singularvalue decomposition. In: Proc. 2003 IEEE Int. Conf. on Multimedia and Expo., Baltimore, Maryland, USA, vol. II, pp. 301–302 (July 2003)

[6]   Heng, W.J., Ngan, K.N.: An object-based shot boundary detection using edge trac-
       ing and tracking. Journal of Visual Communication and Image Representation 12(3),
       217–239 (2001)

[7]   Kasturi, R., Jain, R.: Dynamic vision. In: Kasturi, R., Jain, R. (eds.) Computer
       Vision: Principles. IEEE Computer Society Press, Washington (1991)

[8]   Mohanta, P.P., Saha, S.K., Chanda, B.: A Model-Based Shot Boundary Detection
       Technique Using Frame Transition Parameters. IEEE Transactions on Multime-
       dia 14(1), 223–233 (2012)

[9]   Qi, Y., Hauptmann, A., Liu, T.: Supervised classification for video shot segmenta-
       tion. In: Proc. 2003 IEEE Int. Conf. on Multimedia and Expo., Baltimore, Maryland,
       USA, vol. II, pp. 689–692 (July 2003)

[10]   Lelescu, D., Schonfeld, D.: Statistical sequential analysis for real-time video scene
       change detection" on compressed multimedia bitstream. IEEE Trans. on Multime-
       dia 5(1), 106–117 (2003)

[11]   Sanchez, J.M., Binefa, X., Vitria, J., Radeva, P.: Local color analysis for scene break
       detection applied to tv commercials recognition. In: Proc. Third Int. Conf. on Visual
       Information and Information Systems, Amsterdam, The Netherlands, pp. 237–244
       (June 1999)

[12]   Huang, C.-L., Liao, B.-Y.: A robust scene-change detection method for video seg-
       mentation. IEEE Trans. on Circuits and Systems for Video Technology, 1281–1288
       (December 2001)

[13]   Amudha, J., Radha, D., Naresh Kumar, P.: Video Shot Detection using Saliency
       Measure. International Journal of Computer Applications 45, 17–24 (2012)

[14]   Bescós, J., Cisneros, G., Martínez, J.M., Menendez, J.M., Cabrera, J.: A unified
       model for techniques on video shot transition detection. IEEE Trans. Multime-
       dia 7(2), 293–307 (2005)

[15]   Naphade, M.R., Mehrotra, R., Ferman, A., Warnick, J., Huang, T.S., Tekalp, A.M.:
       A high-performance shot boundary detection algorithm uses multiple cues. In: IEEE
       Inte. Conf. Image Process, pp. 884–887 (1998)

[16]   Yuan, J., Li, F., Zhang, B.: A unified shot boundary detection framework based on
       graph partition model. In: Proc. ACM Multimedia 2005, pp. 539–542 (November
       2005)

[17]   Yoo, H.W., Ryoo, H.J., Jang, D.S.: Gradual shot boundary detection using localized
       edge blocks. Multimedia Tools and Classification 28, 283–300 (2006)

[18]   Cooper, M., Liu, T., Rieffel, E.: Video segmentation via temporal pattern classifica-
       tion. IEEE Trans. on Multimedia 9(3), 610–618 (2007)

Some Constructions of $\mathscr{T}$-*Direct* Codes over $GF(2^n)$

R.S. Raja Durai and Meenakshi Devi

Abstract. The class of $\mathscr{T}$-*Direct* codes are an extension to the class of *linear codes with complementary duals*. In this paper, a construction procedure that constructs an n^2-*Direct* code from an n-*Direct* code is described. Further, the construction procedure is employed recursively to construct $n^{2^{m+1}}$-*Direct* codes for $m \geq 0$. Finally, $\mathscr{T}^2$-*Direct* codes are obtained from arbitrary $\mathscr{T}$-*Direct* codes with *constituent* codes of variable rates. The proposed construction procedure, when employed on an existing $\mathscr{T}$-*Direct* code, in fact increases the number of *constituent* codes (users), thereby supporting more users in a multi-user environment.

1 Introduction

A $\mathscr{T}$-*Direct* code $(\Gamma_1, \Gamma_2, \ldots, \Gamma_\mathscr{T})$ is constituted by a set of $\mathscr{T}$ F-ary linear codes $\Gamma_1, \Gamma_2, \ldots, \Gamma_\mathscr{T}$ with $\Gamma_i \cap \Gamma_i^\perp = \{\mathbf{0}\}$, where $\Gamma_i^\perp = \Gamma_1 \oplus \Gamma_2 \oplus \cdots \oplus \Gamma_{i-1} \oplus \Gamma_{i+1} \oplus \cdots \oplus \Gamma_\mathscr{T}$ is the dual of Γ_i with respect to the direct sum $\Lambda = \Gamma_1 \oplus \Gamma_2 \oplus \cdots \oplus \Gamma_\mathscr{T}$ for each $i = 1, 2, \ldots, \mathscr{T}$. The class of $\mathscr{T}$-*Direct* codes are in fact an extension to the class of *linear codes with complementary duals* (*LCD* codes) [2]. An interesting algebraic characterization of the class of $\mathscr{T}$-*Direct* codes is that $G_i G_i^{\mathbf{T}}$ is non-singular [5, Theorem 1], where G_i is the generator matrix of Γ_i for $i = 1, 2, \ldots, \mathscr{T}$. These class of multi-user codes can be used in a multiple access channel environment for encoding and decoding (or separating the user codewords). A coding problem for the $\mathscr{T}$-user Binary Adder Channel and $\mathscr{T}$-user F-Adder Channel is addressed via the class

R.S. Raja Durai
Department of Mathematics, Jaypee University of Information Technology Waknaghat,
Solan-173 234, Himachal Pradesh, India
e-mail: rsraja.durai@juit.ac.in

Meenakshi Devi
Department of Mathematics, Bahra University, Shimla Hills, Waknaghat,
Solan - 173 234, India
e-mail: meenakshi_juit@yahoo.co.in

S.M. Thampi et al. (eds.), *Recent Advances in Intelligent Informatics*,
Advances in Intelligent Systems and Computing 235,
DOI: 10.1007/978-3-319-01778-5_13, © Springer International Publishing Switzerland 2014

of $\mathscr{T}$-*Direct* codes in [5, 6], where the F-Adder Channel is a non-binary channel model (analogues to the so-called binary adder channel) introduced by Urbanke and Rimoldi [3, 4] to model transmission of message symbols, possibly from a finite field F, of non-binary type. Unlike the *binary adder channels* which are real-number adder channels, arithmetic operations on the channel alphabets in F-Adder Channel are performed under the operations defined in the underlying field F.

A construction method (termed as the *distance* construction) for the class of $\mathscr{T}$-*Direct* codes (defined over $GF(2^n)$) that can increase the minimum distance of the *constituent* codes is proposed in [8], where it is shown that the proposed construction method also increases the number of *constituent* codes from n to $2n-1$ [7], and further to $\frac{n(n+1)}{2}$. This paper generalizes the constructions given in [7, 8] to the constructions of n^2-*Direct* codes and $\mathscr{T}^2$-*Direct* codes. While the former class of n^2-*Direct* codes comprises of constant-rate *constituent* codes, the later class of codes are equipped with variable-rate *constituent* codes. In all the constructions, Kronecker product represented by the symbol $\otimes$ is used as a basic tool. The class of Maximum Rank Distance (MRD) codes introduced by Gabidulin [1] are used to facilitate the results obtained in this paper.

The remaining part of the paper is organized as follows. By making use of the non-commutative nature of Kronecker product, section 2 constructs the class of n^2-*Direct* codes from the class of n-*Direct* codes. Section 3 generalizes the construction procedure outlined in section 2 and obtains $n^{2^{m+1}}$-*Direct* codes from n^{2^m}-*Direct* codes for $m \geq 0$. A construction of $\mathscr{T}^2$-*Direct* codes which are equipped with variable-rate *constituent* codes is presented in section 4. Final section draws the conclusion based on the results.

Notation and abbreviation: We use the notation $(\{n_1, n_2, \ldots, n_{\mathscr{T}}\}, \{k_1, k_2, \ldots, k_{\mathscr{T}}\}, \{d_1, d_2, \ldots, d_{\mathscr{T}}\})$ to denote a $\mathscr{T}$-*Direct* code constituted by the *constituent* codes $(n_1, k_1, d_1), (n_2, k_2, d_2), \ldots, (n_{\mathscr{T}}, k_{\mathscr{T}}, d_{\mathscr{T}})$ and is abbreviated as $(\{n_i\}, \{k_i\}, \{d_i\})$. In particular, a $\mathscr{T}$-*Direct* code with $k_1 = k_2 = \cdots = k_{\mathscr{T}} = k$ (say) is denoted by $(\{n_i\}, \{k\}, \{d_i\})$ rather than $(\{n_i\}, k, \{d_i\})$ - to distinguish a $\mathscr{T}$-*Direct* code from a conventional single user code, namely (n, k, d).

2   Construction of n^2-*Direct* Codes

Consider an $(\{n\}, \{1\}, \{n\})$ n-*Direct* code $(\Gamma_1, \Gamma_2, \ldots, \Gamma_n)$ along with the generator matrices of the *constituent* codes:

$$G_1 = \left[\alpha_1^{[1]} \ \alpha_2^{[1]} \ \cdots \ \alpha_n^{[1]} \right] \tag{1}$$

$$G_2 = \left[\alpha_1^{[2]} \ \alpha_2^{[2]} \ \cdots \ \alpha_n^{[2]} \right] \tag{2}$$

$$\vdots \qquad\qquad\qquad\qquad\qquad \vdots$$

$$\text{and} \quad G_n = \left[\alpha_1^{[n]} \ \alpha_2^{[n]} \ \cdots \ \alpha_n^{[n]} \right] \tag{n}$$

where $\{\alpha_1, \alpha_2, \ldots, \alpha_n\}$ is a *trace-orthogonal basis* in $GF(2^n)$. For each $j = 1, 2, \ldots, n$, using G_j as the *distance* matrix, define j generator matrices as follows:

$$\mathscr{G}_{1j} = G_j \otimes G_1$$
$$\mathscr{G}_{2j} = G_j \otimes G_2$$
$$\vdots$$
$$\text{and} \quad \mathscr{G}_{jj} = G_j \otimes G_j.$$

For each $j = 1, 2, \ldots, n$, let Γ_{ij} denote the $(n^2, 1, n^2)$ $GF(2^n)$-ary code obtained by $\mathscr{G}_{ij} = G_j \otimes G_i$ for $i = 1, 2, \ldots, j$. It is already shown that these $\frac{n(n+1)}{2}$ codes constitute an $(\{n^2\}, \{1\}, \{n^2\})$ $\left[\frac{n(n+1)}{2}\right]$-*Direct* code [8], increasing the *constituent* codes from n to $\frac{n(n+1)}{2}$. However, one can further increase the number of *constituent* codes. Using the fact that Kronecker product is not commutative, by changing the order of operands in the Kronecker products considered, one can obtain $\frac{n(n-1)}{2}$ more generator matrices leading to the total of n^2 distinct codes. For each $i = 2, 3, \ldots, n$, consider the following $i - 1$ generator matrices:

$$\mathscr{G}_{i1} = G_1 \otimes G_i$$
$$\mathscr{G}_{i2} = G_2 \otimes G_i$$
$$\vdots$$
$$\text{and} \quad \mathscr{G}_{i(i-1)} = G_{i-1} \otimes G_i.$$

Similarly, for each $i = 2, 3, \ldots, n$, $\mathscr{G}_{ij} = G_j \otimes G_i$ defines an $(n^2, 1, n^2)$ code, denote it by Γ_{ij} for $j = 1, 2, \ldots, i$. It remains now to verify that the newly obtained n^2 codes $(\Gamma_{ij}, i, j = 1, 2, \ldots, n)$ constitute an $(\{n^2\}, \{1\}, \{n^2\})$ n^2-*Direct* code.

Theorem 2.1. The set $\{\Gamma_{ij} \mid i, j = 1, 2, \ldots, n\}$ of codes constitute an n^2-*Direct* code.

Proof: For every (i, j) and (r, s):

$$\begin{aligned}
\mathscr{G}_{ij}\mathscr{G}_{rs}^{\mathrm{T}} &= (G_i \otimes G_j)(G_r \otimes G_s)^{\mathrm{T}} \\
&= (G_i \otimes G_j)(G_r^{\mathrm{T}} \otimes G_s^{\mathrm{T}}) \\
&= (G_i G_r^{\mathrm{T}}) \otimes (G_j G_s^{\mathrm{T}}) \\
&= \begin{cases} 1, & (i, j) = (r, s) \\ 0, & (i, j) \neq (r, s) \end{cases}.
\end{aligned}$$

Thus, $\{\Gamma_{ij} \mid i, j = 1, 2, \ldots, n\}$ form an $(\{n^2\}, \{1\}, \{n^2\})$ n^2-*Direct* code. $\square$

The n^2 codes thus obtained constitute an n^2-*Direct* code. This construction of n^2-*Direct* codes from n-*Direct* codes (over the same underlying field $GF(2^n)$) enables more users to participate simultaneously in a multi-user environment. However, even further increase in the number of *constituent* codes for an n^2-*Direct* code under the same field $GF(2^n)$ is possible, which is described in the next section.

3 Construction of $n^{[m]}$-*Direct* Codes

This section describes a generalized construction of $(\{n^{[m]}\}, \{1\}, \{n^{[m]}\})$ $n^{[m]}$-*Direct* codes for $m \geq 0$, where here and after $[m] = 2^m$. As the construction requires the use of tensors to represent the quantities such as *constituent* codes and generator matrices, we briefly outlines the notion of tensors.

An $n^{\times\Delta}$-tensor (Δ-order tensor) is an $n^{\times\Delta}$-array $\left[a_{i_1 i_2 \ldots i_\Delta}\right]^n_{i_1, i_2, \ldots, i_\Delta = 1}$, whose entries $a_{i_1 i_2 \ldots i_\Delta}$ are thought of as mathematical objects. A tensor is a natural generalization of a vector (first order tensor) and a matrix (second order tensor). A line in an $n^{\times\Delta}$-tensor Γ is a set of n entries in Γ which are indexed by Δ-tuples $(i_1, i_2, \ldots, i_\Delta)$, in which $\Delta - 1$ (arbitrary indices) out of Δ indices $i_1, i_2, \ldots, i_\Delta$ are fixed, whereas the Δth index (remaining index) ranges over the integers between 1 and n. In fact, lines in tensors are generalizations of rows and columns in matrices. In this paper, an $n^{\times\Delta}$-tensor $\left[a_{i_1 i_2 \ldots i_\Delta}\right]^n_{i_1, i_2, \ldots, i_\Delta = 1}$ is sometimes conveniently represented as an n^Δ-tuple vector $\left(a_{i_1 i_2 \ldots i_\Delta}\right)^n_{i_1, i_2, \ldots, i_\Delta = 1}$.

Consider an $(\{n^{[m]}\}, \{1\}, \{n^{[m]}\})$ $n^{[m]}$-*Direct* code $\left(\Gamma^{(m)}_{i_1 i_2 \ldots i_{[m]}}\right)^n_{i_1 i_2 \ldots i_{[m]} = 1}$, where the $(i_1, i_2, \ldots, i_{[m]})^{th}$ *constituent* code $\Gamma^{(m)}_{i_1 i_2 \ldots i_{[m]}}$ is defined by the $1 \times n^{[m]}$ generator matrix $\mathscr{G}^{(m)}_{i_1 i_2 \ldots i_{[m]}}$ for $m \geq 0$. These $n^{[m]}$ generator matrices can be conveniently represented as an $n^{\times[m]}$-array $\left[\mathscr{G}^{(m)}_{i_1 i_2 \ldots i_{[m]}}\right]^n_{i_1 i_2 \ldots i_{[m]} = 1}$. In particular, for $m = 0$, these $n^{[m]}$ generator matrices are precisely the generator matrices as in equations $(1) - (n)$.

Define $\mathscr{G}^{(m+1)}_{i_1 \ldots i_{[m]} j_1 \ldots j_{[m]}} = \mathscr{G}^{(m)}_{i_1 \ldots i_{[m]}} \otimes \mathscr{G}^{(m)}_{j_1 \ldots j_{[m]}}$ for $1 \leq i_1, \ldots, i_{[m]}, j_1, \ldots, j_{[m]} \leq n$.

Let $\Gamma^{(m+1)}_{i_1 \ldots i_{[m]} j_1 \ldots j_{[m]}}$ denote the resultant $(n^{[m+1]}, 1, n^{[m+1]})$ $GF(2^n)$-ary code obtained from $\mathscr{G}^{(m+1)}_{i_1 \ldots i_{[m]} j_1 \ldots j_{[m]}}$. The following theorem affirms the fact that the set of codes constructed indeed constitute an $n^{[m+1]}$-*Direct* code.

Theorem 3.1. The set $\left\{\Gamma^{(m+1)}_{i_1 i_2 \ldots i_{[m+1]}}\right\}^n_{i_1 i_2 \ldots i_{[m+1]} = 1}$ constitutes an $(\{n^{[m+1]}\}, \{1\}, \{n^{[m+1]}\})$ $n^{[m+1]}$-*Direct* code $\left(\Gamma^{(m+1)}_{i_1 i_2 \ldots i_{[m+1]}}\right)^n_{i_1 i_2 \ldots i_{[m+1]} = 1}$.

Proof: For any $(i_1, \ldots, i_{[m]}, j_1, \ldots, j_{[m]})$ and $(r_1, \ldots, r_{[m]}, s_1, \ldots, s_{[m]})$:

$$
\begin{aligned}
\mathscr{G}^{(m+1)}_{i_1 \ldots i_{[m]} j_1 \ldots j_{[m]}} \mathscr{G}^{(m+1)\mathbf{T}}_{r_1 \ldots r_{[m]} s_1 \ldots s_{[m]}} &= (\mathscr{G}^{(m)}_{i_1 \ldots i_{[m]}} \otimes \mathscr{G}^{(m)}_{j_1 \ldots j_{[m]}})(\mathscr{G}^{(m)}_{r_1 \ldots r_{[m]}} \otimes \mathscr{G}^{(m)}_{s_1 \ldots s_{[m]}})^{\mathbf{T}} \\
&= (\mathscr{G}^{(m)}_{i_1 \ldots i_{[m]}} \otimes \mathscr{G}^{(m)}_{j_1 \ldots j_{[m]}})(\mathscr{G}^{(m)\mathbf{T}}_{r_1 \ldots r_{[m]}} \otimes \mathscr{G}^{(m)\mathbf{T}}_{s_1 \ldots s_{[m]}}) \\
&= (\mathscr{G}^{(m)}_{i_1 \ldots i_{[m]}} \mathscr{G}^{(m)\mathbf{T}}_{r_1 \ldots r_{[m]}}) \otimes (\mathscr{G}^{(m)}_{j_1 \ldots j_{[m]}} \mathscr{G}^{(m)\mathbf{T}}_{s_1 \ldots s_{[m]}}) \\
&= \begin{cases} 1, & (i_1, \ldots i_{[m]}, j_1, \ldots, j_{[m]}) = (r_1, \ldots, r_{[m]}, s_1, \ldots, s_{[m]}) \\ 0, & (i_1, \ldots i_{[m]}, j_1, \ldots, j_{[m]}) \neq (r_1, \ldots, r_{[m]}, s_1, \ldots, s_{[m]}) \end{cases}
\end{aligned}
$$

where we have used the following fact for the $(\{n^{[m]}\}, \{1\}, \{n^{[m]}\})$ $n^{[m]}$-Direct code:

$$\mathscr{G}_{i_1 \ldots i_{[m]}}^{(m)} \mathscr{G}_{j_1 \ldots j_{[m]}}^{(m)\,\mathbf{T}} = \begin{cases} 1 \, , \, (i_1, \ldots i_{[m]}) = (j_1, \ldots, j_{[m]}) \\ 0 \, , \, (i_1, \ldots i_{[m]}) \neq (j_1, \ldots, j_{[m]}) \end{cases}.$$

Hence, $\left\{ \Gamma_{i_1 i_2 \ldots i_{[m+1]}}^{(m+1)} \mid 1 \leq i_1, i_2, \ldots, i_{[m+1]} \leq n \right\}$ constitute an $n^{[m+1]}$-Direct code. $\square$

As detailed above, the *distance* construction allows recursive construction of $(\{n^{[m]}\}, \{1\}, \{n^{[m]}\})$ $n^{[m]}$-*Direct* codes for the values of $m \geq 0$. In the above recursive construction, the dimensions of *constituent* codes were one only, allowing maximum possible error correcting capabilities of individual *constituent* code. Due to this restriction on the dimension, the *constituent* codes were to support same code rate. But, this is not the situation in practice; the real-time users require variable-rate (*constituent*) codes to be assigned. Keeping this in mind, in the next section, we constitute a $\mathscr{T}^2$-*Direct* code from a $\mathscr{T}$-*Direct* code with variable rates for *constituent* codes.

4 Construction of $\mathscr{T}^2$-Direct Codes

Let $\{\alpha_1, \alpha_2, \ldots, \alpha_n\}$ be a *trace-orthogonal basis* in $GF(2^n)$. Let $k_1, k_2, \ldots, k_{\mathscr{T}} > 0$ be a set of positive integers such that $k_1 + k_2 + \cdots + k_{\mathscr{T}} \leq n$. Consider an $(\{n\}, \{k_i\}, \{d_i\})$ $\mathscr{T}$-*Direct* code $(\Gamma_1, \Gamma_2, \ldots, \Gamma_{\mathscr{T}})$ such that the ith *constituent* code Γ_i is an (n, k_i, d_i) $GF(2^n)$-ary MRD code generated by $k_i \times n$ generator matrix G_i:

$$G_i = \left[\alpha_s^{[k_o + k_1 + \cdots + k_{i-1} + r]} \right]_{r,s=1}^{k_i, n} \quad (i = 1, 2, \ldots, \mathscr{T})$$

where $k_o = 0$ and it is so used to write G_i for brevity. Consider taking the Kronecker products between the generator matrices of the *constituent* codes as follows: Define $\mathscr{G}_{ij} = G_i \otimes G_j$ for $i, j = 1, 2, \ldots, \mathscr{T}$. Clearly, each $\mathscr{G}_{ij}$ defines an $(n^2, k_i k_j, d_i d_j)$ $GF(2^n)$-ary code Γ_{ij} (say). The $\mathscr{T}^2$ variable-rate codes $(\Gamma_{ij}, i, j = 1, 2, \ldots, \mathscr{T})$ indeed constitute an $(\{n^2\}, \{k_i k_j\}, \{d_i d_j\})$ $\mathscr{T}^2$-*Direct* code $(\Gamma_{11}, \ldots, \Gamma_{1\mathscr{T}}, \Gamma_{21}, \ldots, \Gamma_{2\mathscr{T}}, \ldots, \Gamma_{\mathscr{T}1}, \ldots, \Gamma_{\mathscr{T}\mathscr{T}})$ as proved in the following theorem.

Theorem 4.1 The set $\left\{ \Gamma_{ij} \mid i, j = 1, 2, \ldots, \mathscr{T} \right\}$ of codes constitute a $\mathscr{T}^2$-*Direct* code.

Proof: For every (i, j) and (r, s) pairs:

$$\begin{aligned} \mathscr{G}_{ij} \mathscr{G}_{rs}^{\mathbf{T}} &= (G_i \otimes G_j)(G_r \otimes G_s)^{\mathbf{T}} \\ &= (G_i \otimes G_j)(G_r^{\mathbf{T}} \otimes G_s^{\mathbf{T}}) \\ &= (G_i G_r^{\mathbf{T}}) \otimes (G_j G_s^{\mathbf{T}}) \\ &= \begin{cases} \mathbf{I} \, , \, (i, j) = (r, s) \\ (\mathbf{0}) \, , \, (i, j) \neq (r, s) \end{cases} \end{aligned}$$

where $\mathbf{I}$ is the $k_i k_j \times k_r k_s$ identity matrix and $(\mathbf{0})$ is the $k_i k_j \times k_r k_s$ zero-matrix. $\square$

Thus, the *distance* construction method increases the number of *constituent* codes (of variable rates) from $\mathscr{T}$ to $\mathscr{T}^2$ and thereby supporting more users in a multi-user environment for the channel under consideration. The following example facilitates the construction of $(\{16\},\{1,3,3,9\},\{16,8,8,4\})$ 4-*Direct* code from $(\{4\},\{1,3\},\{4,2\})$ 2-*Direct* code.

Example 1. Consider the $(\{4\},\{1,3\},\{4,2\})$ 2-*Direct* code (Γ_1,Γ_2) defined over $GF(2^4)$ along with the generator matrices of the two variable-rate *constituent* codes:

$$G_1 = \begin{bmatrix} \alpha^6 & \alpha^{11} & \alpha^9 & \alpha^{14} \end{bmatrix} \quad \text{and} \quad G_2 = \begin{bmatrix} \alpha^{12} & \alpha^7 & \alpha^3 & \alpha^{13} \\ \alpha^9 & \alpha^{14} & \alpha^6 & \alpha^{11} \\ \alpha^3 & \alpha^{13} & \alpha^{12} & \alpha^7 \end{bmatrix}$$

Then the generator matrices $\mathscr{G}_{11},\mathscr{G}_{12},\mathscr{G}_{21},\mathscr{G}_{22}$ can be obtained by taking Kronecker product of G_1, G_2 with each of G_1 and G_2 as follows:

$$\mathscr{G}_{11} = G_1 \otimes G_1$$
$$= \begin{bmatrix} \alpha^{12} & \alpha^2 & 1 & \alpha^5 & \alpha^2 & \alpha^7 & \alpha^5 & \alpha^{10} & 1 & \alpha^5 & \alpha^3 & \alpha^8 & \alpha^5 & \alpha^{10} & \alpha^8 & \alpha^{13} \end{bmatrix}$$

$$\mathscr{G}_{12} = G_1 \otimes G_2$$
$$= \begin{bmatrix} \alpha^3 & \alpha^{13} & \alpha^9 & \alpha^4 & \alpha^8 & \alpha^3 & \alpha^{14} & \alpha^9 & \alpha^6 & \alpha & \alpha^{12} & \alpha^7 & \alpha^{11} & \alpha^6 & \alpha^2 & \alpha^{12} \\ 1 & \alpha^5 & \alpha^{12} & \alpha^2 & \alpha^5 & \alpha^{10} & \alpha^2 & \alpha^7 & \alpha^3 & \alpha^8 & 1 & \alpha^5 & \alpha^8 & \alpha^{13} & \alpha^5 & \alpha^{10} \\ \alpha^9 & \alpha^4 & \alpha^3 & \alpha^{13} & \alpha^{14} & \alpha^9 & \alpha^8 & \alpha^3 & \alpha^{12} & \alpha^7 & \alpha^6 & \alpha & \alpha^2 & \alpha^{12} & \alpha^{11} & \alpha^6 \end{bmatrix}$$

$$\mathscr{G}_{21} = G_2 \otimes G_1$$
$$= \begin{bmatrix} \alpha^3 & \alpha^9 & \alpha^6 & \alpha^{11} & \alpha^{13} & \alpha^4 & \alpha & \alpha^6 & \alpha^9 & 1 & \alpha^{12} & \alpha^2 & \alpha^4 & \alpha^{10} & \alpha^7 & \alpha^{12} \\ 1 & \alpha^6 & \alpha^3 & \alpha^8 & \alpha^5 & \alpha^{11} & \alpha^8 & \alpha^{13} & \alpha^{12} & \alpha^3 & 1 & \alpha^5 & \alpha^2 & \alpha^8 & \alpha^5 & \alpha^{10} \\ \alpha^9 & 1 & \alpha^{12} & \alpha^2 & \alpha^4 & \alpha^{10} & \alpha^7 & \alpha^{12} & \alpha^3 & \alpha^9 & \alpha^6 & \alpha^{11} & \alpha^{13} & \alpha^4 & \alpha & \alpha^6 \end{bmatrix}$$

$$\mathscr{G}_{22} = G_2 \otimes G_2$$
$$= \begin{bmatrix} \alpha^9 & \alpha^4 & 1 & \alpha^{10} & \alpha^4 & \alpha^{14} & \alpha^{10} & \alpha^5 & 1 & \alpha^{10} & \alpha^6 & \alpha & \alpha^{10} & \alpha^5 & \alpha & \alpha^{11} \\ \alpha^6 & \alpha^{11} & \alpha^3 & \alpha^8 & \alpha & \alpha^6 & \alpha^{13} & \alpha^3 & \alpha^{12} & \alpha^2 & \alpha^9 & \alpha^{14} & \alpha^7 & \alpha^{12} & \alpha^4 & \alpha^9 \\ 1 & \alpha^{10} & \alpha^9 & \alpha^4 & \alpha^{10} & \alpha^5 & \alpha^4 & \alpha^{14} & \alpha^6 & \alpha & 1 & \alpha^{10} & \alpha & \alpha^{11} & \alpha^{10} & \alpha^5 \\ \alpha^6 & \alpha & \alpha^{12} & \alpha^7 & \alpha^{11} & \alpha^6 & \alpha^2 & \alpha^{12} & \alpha^3 & \alpha^{13} & \alpha^9 & \alpha^4 & \alpha^8 & \alpha^3 & \alpha^{14} & \alpha^9 \\ \alpha^3 & \alpha^8 & 1 & \alpha^5 & \alpha^8 & \alpha^{13} & \alpha^5 & \alpha^{10} & 1 & \alpha^5 & \alpha^{12} & \alpha^2 & \alpha^5 & \alpha^{10} & \alpha^2 & \alpha^7 \\ \alpha^{12} & \alpha^7 & \alpha^6 & \alpha & \alpha^2 & \alpha^{12} & \alpha^{11} & \alpha^6 & \alpha^9 & \alpha^4 & \alpha^3 & \alpha^{13} & \alpha^{14} & \alpha^9 & \alpha^8 & \alpha^3 \\ 1 & \alpha^{10} & \alpha^6 & \alpha^{16} & \alpha^{10} & \alpha^5 & \alpha & \alpha^{11} & \alpha^9 & \alpha^4 & 1 & \alpha^{10} & \alpha^4 & \alpha^{14} & \alpha^{10} & \alpha^5 \\ \alpha^{12} & \alpha^2 & \alpha^9 & \alpha^{14} & \alpha^7 & \alpha^{12} & \alpha^4 & \alpha^9 & \alpha^6 & \alpha & \alpha^3 & \alpha^8 & \alpha & \alpha^6 & \alpha^{13} & \alpha^3 \\ \alpha^6 & \alpha & 1 & \alpha^{10} & \alpha & \alpha^{11} & \alpha^{10} & \alpha^5 & 1 & \alpha^{10} & \alpha^9 & \alpha^4 & \alpha^{10} & \alpha^5 & \alpha^4 & \alpha^9 \end{bmatrix}$$

Clearly, $\mathscr{G}_{11},\mathscr{G}_{12},\mathscr{G}_{21},\mathscr{G}_{22}$ define $(16,1,16)$, $(16,3,8)$, $(16,3,8)$ and $(16,9,4)$ codes, respectively. It is easy to verify that the codes $\Gamma_{11},\Gamma_{12},\Gamma_{21},\Gamma_{22}$ obtained indeed constitute a $(\{16\},\{1,3,3,9\},\{16,8,8,4\})$ 4-*Direct* code $(\Gamma_{11},\Gamma_{12},\Gamma_{21},\Gamma_{22})$.

5 Conclusion

Employing *distance* construction approach, several constructions of the class of $\mathscr{T}$-*Direct* codes are presented. Firstly, a class of n^2-*Direct* codes are constructed from n-*Direct* codes, by making use of non-commutativity of Kronecker product. Then, a generalized construction of $(\{n^{[m]}\}, \{1\}, \{n^{[m]}\})$ $n^{[m]}$-*Direct* codes are given for $m = 0, 1, 2, \ldots$. In these constructions, the *constituents* are of constant-rate codes. To support wide range of users of a multi-user environment, variable-rate *constituent* codes are desirable. To address this issue, the paper also obtains $(\{n^2\}, \{k_i k_j\}, \{d_i d_j\})$ $\mathscr{T}^2$-*Direct* codes from an arbitrary class of $(\{n\}, \{k_i\}, \{d_i\})$ $\mathscr{T}$-*Direct* codes. The *distance* construction when employed not only increases the minimum distance of the *constituent* codes, but also enables the multi-user code to support more users with flexible code-rate options.

References

1. Gabidulin, E.M.: Theory of codes with maximum rank distance. Problems in Information Transmission 21, 1–12 (1985)
2. Massey, J.L.: Linear codes with complementary duals. Discrete Mathematics 106-107, 337–342 (1992)
3. Urbanke, R., Rimoldi, B.: Coding for the F-Adder Channel: Two applications for Reed-Solomon codes. In: Proceedings of IEEE International Symposium on Information Theory, San Antonio, p. 85 (1993)
4. Rimoldi, B.: Coding for the Guassian multiple access channel: An algebraic approach. In: Proceedings of International Symposium on Information Theory, Austin, TX (1993)
5. Vasantha, W.B., Raja Durai, R.S.: $\mathscr{T}$-*Direct* codes: An application to $\mathscr{T}$-user BAC. In: Proceedings of the IEEE Information Theory Workshop, Bangalore, India, p. 214 (2002)
6. Vasantha, W.B., Raja Durai, R.S.: Some results on $\mathscr{T}$-*Direct* codes. In: Proceedings of the 3rd Asia-Europe Workshop on Information Theory, Kamogawa, Chiba, Japan, pp. 43–44 (2003)
7. Raja Durai, R.S., Devi, M.: Construction of $(\mathscr{N} + \mathscr{M})$-*Direct* codes in $GF(2^{\mathscr{N}})$. In: Proceedings of World Congress on Information and Communication Technologies, Mumbai, India, pp. 770–775 (2011)
8. Raja Durai, R.S., Devi, M.: On the class of $\mathscr{T}$-*Direct* codes over $GF(2^N)$. International Journal of Computer Information Systems and Industrial Management Applications 5, 589–596 (2013)

Weighted Optimization of Various Parameters for Droplet Routing in Digital Microfluidic Biochips

Indrajit Pan and Tuhina Samanta

Abstract. Digital microfluidic biochip is often deployed for multiplexing several bioassays under optimized space and time constraints. Different parametric optimizations lead to performance enhancement and reusability of a biochip. In this paper, we propose an optimization algorithm to minimize three parameters simultaneously, named (i) Electrode usage, (ii) Latest arrival time, and (iii) Actuation pin assignment. A composite objective function has been designed and deployed at different instances with adjustable probabilistic weight factors (λ), to study the characteristic of routing schedules. Chosen parameters or objectives are minimized, leveraging the choice of weight factor at various runs, and dependency between the parameters are also studied. Shortest path based routing approach is adopted for optimal electrode usage and minimized routing completion time, and a graph coloring based on a dependency graph construction is navigated for optimal pin assignment. Experimental study of the proposed technique shows better result over some standard algorithms.

Keywords: Digital microfluidic biochip, Droplet routing, Multiple objectives, Weighted objective function, Algorithm.

1 Introduction

Digital microfluidic biochip (DMFB) is a recent technology that enables efficient on-chip fluid management. This technology is being used in large scale for

Indrajit Pan
RCC Institute of Information Technology, Kolkata, West Bengal, India
e-mail: p.indrajit@gmail.com
Tuhina Samanta
Bengal Engineering & Science University, Shibpur, Howrah, India
e-mail: t_samanta@it.becs.ac.in

S.M. Thampi et al. (eds.), *Recent Advances in Intelligent Informatics*,
Advances in Intelligent Systems and Computing 235,
DOI: 10.1007/978-3-319-01778-5_14, © Springer International Publishing Switzerland 2014

developing full custom chip design. DMFB is a portable and cost effective device. In recent days, DMFB has a vast area of medical application, starting from clinical diagnosis to drug discovery [1]. In DMFB, droplet movements are controlled by actuation of control pins along the desired path on electrode array. Electrodes are activated and deactivated in sequence through a time varying supply voltage [9]. Various algorithms were proposed to optimize resource utilization during droplet routing. A bypassibility and concession based approach was proposed in [1]. ILP based approach, and clique partitioning method for droplet routing appeared in [2] and [4] respectively. Recently, some meta-heuristic approaches are adopted in the work [9, 11] for resource optimization during droplet routing. Design of fully customized biochip by proper calibration of associated objectives is a major challenge. Prime objectives being used in this purpose are, (i) total number of electrode usage during routing completion of an assay (f_1), (ii) latest arrival time (f_2) and (iii) number of control pins required (f_3). In recent times various electrode control mechanisms have also been proposed to address the issues with growing number of control pins [3, 6, 7 and 8]. However, not much work has been done to find dependency between multiple objective functions, and their simultaneous optimization during droplet routing in DMFB.

In this paper, we propose a multi objective optimization technique which addresses the aforesaid three objective parameters. Here, different probabilistic weights have been assigned to control three simultaneous runs by putting customized emphasis on each objective to achieve three different routing schedules. In order to study the dependency between the objective functions, a weighted objective function has been derived. An efficient routing method has been used to study the droplet trace during optimization of f_1 and f_2, while to optimize the pin count function f_3, we have used a method of minimal coloring of dependency graph.

The rest of the paper is organized as follows; Section 2 describes problem formulation and section 3 elaborates the proposed method. Section 4 discusses the experimental results. Finally, Section 5 concludes with possible future directives.

2   Problem Formulations

2.1   Droplet Routing under Different Fluidic Constraints

Let d_i at location (x^t_i, y^t_i) and d_j located at (x^t_j, y^t_j) denote two independent droplets at time t. Then, the following constraints, generally called *Fluidic Constraint*, should be satisfied over every timestamp t during routing;

- Static Fluidic Constraint: $|x^t_i - x^t_j| > 1$ or $|y^t_i - y^t_j| > 1$
- Dynamic Fluidic Constraint: $|x^{t+1}_i - x^t_j| > 1$ or $|y^{t+1}_i - y^t_j| > 1$ or $|x^t_i - x^{t+1}_j| > 1$
  or $|y^t_i - y^{t+1}_j| > 1$

2.2   Multi Objective Optimization

2.2.1   Minimization of Total Electrode Usage

A set of nets $[N_i \mid i = 1 \text{ to } k]$ are given along with their individual pair of source and target location $[S_i - T_i]$. The main goal is to derive a routing path from S_i to T_i for all i. The number of unique electrode used by all the nets over the period to complete their routing is called total electrode usage. The objective is to find a schedule, where all nets will find their target by minimum electrode usage under static and fluidic constraints. Here, we consider the function f_1 to represent the objective constraint on total electrode usage. If N_i (E) be the set of electrodes on the path between S_i and T_i then objective function,

$$f_1 = \mathcal{M}inimize\,[\{N_1\,(E)\,\cup\,N_2\,(E)\,\cup\,...\,\cup\,N_k\,(E)\} - \{N_1\,(E)\,\cap\,N_2\,(E)\,\cap\,...\,\cap\,N_k\,(E)\}] \qquad (1)$$

2.2.2   Minimization of Latest Arrival Time (Routing Completion Time)

Second aspect is routing completion time of each net N_i. This is to calculate the latest arrival time among all droplets. The latest arrival time is the completion time of the net finishing last among all. The objective of achieving better routing completion time is functionally represented by f_2. If L^{AT} (N_i) represents the latest arrival time of net N_i then objective function,

$$f_2 = \mathcal{M}inimize\,[\mathcal{M}ax\,\{L^{AT}\,(N_i)\}] \quad \text{where, i = 1 to k.} \qquad (2)$$

2.2.3   Minimization of Total Pin Count

Third aspect covers the target of optimal pin assignment through compound use of same control pin to perform external biasing. If N_i (E) represents a set of electrodes on the routing path of net N_i, then let us assume N_i (P) is set of control pin required to maintain actuation sequence of those electrodes, where N_i (P) $\ll N_i$ (E). During this pin assignment, $[N_i$ (P) $\cap N_j$ (P) $= \Phi]$ if N_i (E) and N_j (E) are diagonally or non-diagonally adjacent and $[N_i$ (P) $\cap N_j$ (P) $\neq \Phi]$ if N_i (E) and N_j (E) are non-adjacent. Here we consider the function f_3 to represent the objective function on total pin count. If N_i (P) be the set of pins required to maintain the actuation sequence of N_i (E) on the path between S_i and T_i, then objective function,

$$f_3 = \mathcal{M}inimize\,[\{N_1\,(P)\,\cup\,N_2\,(P)\,\cup\,...\,\cup\,N_k\,(P)\} - \{N_1\,(P)\,\cap\,N_2\,(P)\,\cap\,...\,\cap\,N_k\,(P)\}] \qquad (3)$$

2.2.4   Composite Objective Function

The objectivities mentioned in sub section 2.2.1 to 2.2.3 are partially dependent and partially conflicting. So a calibration is highly required among these three parameters to avail optimized schedule. Initially, it seems like if the total number of electrode usage gets minimized then routing completion time will be least and number of distinct pins required to monitor the actuation sequence will also be minimum. But in order to minimize the electrode usage, if the same electrode is shared among multiple routes, then the routing may get delayed due to stall at multiple intersecting locations. Sometimes, over optimization leads to deadlock situations, which in turn may increase the number of failed nets. Apart from these, extensive use of neighboring electrodes may increase the difficulty in assigning activation pins, which will lead to increase in the number of assigned pin.

Thus a composite objective function has been designed where all the above mentioned objectivities (f_1, f_2 and f_3) have been combined together along with an individual probabilistic weight factor λ_k, where k = 1, 2 and 3. The objective function is,

$$\mathcal{F} = \lambda_1.f_1 + \lambda_2.f_2 + \lambda_3.f_3 \tag{4}$$

Here in equation 4, λ_k represent probabilistic weight factors which are basically deployed to perform calibration. According to our assumption ($\lambda_1 + \lambda_2 + \lambda_3 = 1$). λ takes fractional values either 0.6 or 0.2. At any particular run any one of the three functions is assigned with 0.6 and rest of two are assigned with 0.2. Here 0.6 signifies higher emphasis and maximum optimization of associated objective function where 0.2 ensures restricted optimization of associated objective functions.

3   Proposed Method

3.1   Droplet Routing

Droplet routing path for a net N_i aims to find a path from S_i to T_i through a shortest possible distance. Optimal routing path is the Manhattan distance between S_i and T_i. During droplet movement, every time a list of probable locations is marked out. Then the Manhattan distance between T_i and these listed locations are computed. Then the new location is chosen by selecting the location having minimum Manhattan distance. In this process, the distance of target location T_i gradually decreases. If a net N_i moves from (E^S, x^s, y^s) to ($E^{S'}$, $x^{s'}$, $y^{s'}$), then

$$|E^S - E^{S'}| = |x^{s'} - x^s| + |y^{s'} - y^s| \tag{5}$$

Here current droplet location is E^S and moving to $E^{S'}$, and their respective coordinates are (x^s, y^s) and ($x^{s'}$, $y^{s'}$).

During routing all static and fluidic constraints are observed accordingly. Stalling and Detour are incorporated during routing to find the best possible solution. In figure 1, we have depicted a scenario for this routing using a [6 × 6] two dimensional board. Here in figure 1, a net initiates from (5, 1) as its source (S) and intends to reach at T (2, 5). The dotted circles indicate a probable path, where values within the circle indicate remaining Manhattan distance.

However being at (5, 5), we see that (5, 6) or (6, 5) locations are away from the goal, as Manhattan value is higher, but (4, 5) is suitable for next movement and accordingly the movement is assigned. A logical description of the method is given in Table 1.

Fig. 1 Droplet Routing through Manhattan Distance Analysis

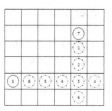

Table 1 Algorithm for Droplet Routing

Input: a) List of nets	
b) Blocked coordinates	
Step 1:	Select each of the nets and choose a move (it may be either horizontal or vertical)
Step 2:	According to the move calculate a probable destination electrode for next timestamp
Step 3:	Check the probable location with blocked set of electrodes if matches then select alternative move and recalculate a probable destination
Step 4:	Check the next destination for fluidic constraints with other concurrent droplets and make necessary action if any violation noticed
Repeat Step 1 to Step 4 until all nets find target	

3.2 Pin Count Aware Routing

Optimal number of control pins is assigned on the routed path so that each of the electrodes can be activated properly and unambiguously. Here the basic objective is to reassign the same pins in to some locations which are non-conflicting. Hence to determine the non-conflicting electrode locations, we have proposed a dependency graph model. A dependency or adjacency graph is constructed first on the basis of initial routing trace. Any droplet path is normally represented in a dependent sequence of three consecutive locations. In figure 2, we assume a portion of path starting from E^1 up to E^6 along with curvy connection among mutually dependent electrodes. Here, as shown in the diagram, the dependent locations cannot be assigned with same pin as it may result in improper and conflicting actuations. Here in the diagram E^1 is dependent with E^2 and E^3. Hence two separate connecting edges are there between E^1 and E^2, E^1 and E^3. Similarly E^2 is dependent with E^3 and E^4, E^3 is with E^4 and E^5, E^4 is with E^5 and E^6 and E^5 with E^6.

Fig. 2 Dependency of Electrodes

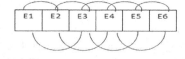

Fig. 3 Adjacency Graph from Two Concurrent Droplet Routes

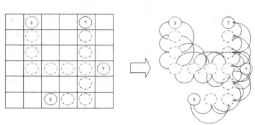

Dependency among these electrodes helps in deriving an adjacency graph model as shown in figure 3. Graph coloring concept for pin assignment is adopted after deriving this adjacency graph. In this graph coloring approach, no two dependent nodes will have same color, yet the overall requirement of different colors should be minimal. The requirement for total number of different colors is equated with minimal number of pins that may be required for controlling the actuation sequences. The logical description of the method is discussed in Table 2.

Table 2 Algorithm for Minimal Pin Assignment using Graph Coloring Approach

Input:	a) Adjacency graph with vertex set, V and edge set, E
Step 1:	Compute degree of all vertices in V and sort the vertices in descending order of degrees
Step 2:	Tag all vertices with unassigned status and repeat from Step 3 and Step 4 until there are un assigned vertices
Step 3:	Pick a vertex from unassigned set an assign with a Pin number.
Step 4:	Check the remaining unassigned set, if it is not adjacent to node of Step 3 then assign the current pin to it

3.3 *Optimizing Weighted Objective Function during Routing*

Proposed multi-objective optimization is based on three objective parameters f_1, f_2 and f_3 as mentioned in section 2.2.4. The optimization function $\mathcal{F}$ is minimizing in nature, which tries to generate an optimal tradeoff between f_1, f_2 and f_3. Now the probabilistic weight factors λ_1, λ_2 and λ_3 mostly operate on two values [0.6 and 0.2]. As they are assumed to be probabilistic in nature, sum of these three λ is always 1. Thus it ensures that at any time domain only one weight is chosen to be 0.6 and rest two will be 0.2. As per the assumption, 0.6 is of higher importance than 0.2. Now during iterative execution, proposed method updates the locations for concurrent droplets routing, with an objective to minimize the composite objective function. If there arise any conflict with the aspect of this optimization, then priority is given to the objective with associated weight $\lambda = 0.6$. As all the three objectives are minimizing in nature, overall functional yield ($\mathcal{F}$) will be gradually descending in nature.

4   Experimental Results

Proposed method has been experimented on some real life benchmarks like in-vitro, protein and a few hard test sets as used in [1]. A brief description of these standard benchmarks is given in Table 3.

The algorithm is implemented in Linux platform on a dual core T4440 Intel processor with a speed of 2.20 GHz and 2 GB memory. The results recorded during simulation are summarized in Table 4. We have executed the method under three different sets of values for λ_1, λ_2 and λ_3 . The values are [$\lambda_1 = 0.6$, $\lambda_2 = 0.2$ and $\lambda_3 = 0.2$], [$\lambda_1 = 0.2$, $\lambda_2 = 0.6$ and $\lambda_3 = 0.2$] and [$\lambda_1 = 0.2$, $\lambda_2 = 0.2$ and $\lambda_3 = 0.6$].

Table 3 Brief Description of Benchmarks

Benchmark Name	Grid Size	Droplet #
invitro_1	16 × 16	26
invitro_2	14 × 14	34
protein_1	21 × 21	179
protein_2	13 × 13	170
test_1	12 × 12	12
test_6	16 × 16	16

Table 4 Experimental Results

	$\lambda_1 = 0.6, \lambda_2 = 0.2, \lambda_3 = 0.2$			$\lambda_1 = 0.2, \lambda_2 = 0.6, \lambda_3 = 0.2$			$\lambda_1 = 0.2, \lambda_2 = 0.2, \lambda_3 = 0.6$		
Test Suite	Total Electrode	Latest Arrival Time	Total Pin Required	Total Electrode	Latest Arrival Time	Total Pin Required	Total Electrode	Latest Arrival Time	Total Pin Required
invitro_1	195	15	32	207	12	32	207	15	30
invitro_2	201	13	34	213	13	34	213	13	31
protein_1	1288	18	23	1321	18	23	1321	18	10
protein_2	847	15	21	847	13	21	847	15	17
test_1	56	53	36	77	48	36	77	53	33
test_6	95	51	41	95	51	41	95	51	41

Table 5 Comparison on Total Electrode Usage

Benchmark	[1]	[10]	Proposed Method
invitro_1	258	200	195
protein_1	1688	1546	1288
test_1	67	94	56
test_6	119	177	95

Table 6 Comparison on Routing Completion Time (Latest Arrival Time)

Benchmark	[1]	[10]	Proposed Method
invitro_1	-	15	12
protein_1	-	24	18
test_1	100	58	48
test_6	55	64	51
- Data Unavailable			

Table 7 Comparison on Pin Count

Benchmark	[5]	[10]	Proposed Method
invitro_1	-	30	30
protein_1	12	19	10
test_1	-	48	33
test_6	-	48	41
- Data Unavailable			

The results of Table 4 establish that during optimization of three different parameters, the parameters having greater emphasis is showing better yield. We have also compared the results with the standard results of [1, 5 and 10]. The comparison for total electrode usage is shown in Table 5, Table 6 compares the latest arrival time and Table 7 summarizes the pin count comparison. This comparative analysis establishes the efficacy of our proposed method. We have taken the best results that we achieved against $\lambda = 0.6$.

5   Conclusions

This paper deals with the multi objective optimization of various design metrics for droplet routing in digital microfluidic biochip. It has simultaneously attempted to optimize and calibrate three possible and desired outcomes, namely the total electrode usage, latest arrival time or routing completion time and total number of pins required. A composite objective function has been used for resource optimization during routing, subject to fluidic and electrode actuation constraints. The strategy used for calibration of optimizing function is heuristic in nature and is tuned with a predefined probabilistic weight adjustment factor λ. Experimental results show consistent improvement in all simulations.

Design automation for digital microfluidic biochips is becoming an essential part for fully customized chip design. The advent of nano-scale design and fabrication has led challenges towards some more stringent restrictions. Hence reconfigurability and reusability test can be considered as a new avenue of research. Dynamic reconfigurable placement design may also be studied as a possible future directive.

References

1. Cho, M., Pan, D.: A high-performance droplet routing algorithm for digital microfluidic bio-chips. IEEE Transaction on Computer-Aided Design of Integrated Circuits and Systems 27(10), 1714–1724 (2008)
2. Yang, C.L., Yuh, P.H., Sapatnekar, S., Chang, Y.W.: A progressive ILP based routing algo-rithm for cross-referencing biochips. In: Proc. of Design Automation Conference, pp. 284–289 (2008)

3. Xu, T., Chakrabarty, K.: Broadcast electrode-addressing for pin-constrained multi-functional digital microfluidic biochips. In: Proc. of Design Automation Conference, California, USA, pp. 173–178 (2008)
4. Xu, T., Chakrabarty, K.: A droplet manipulation method for achieving high throughput in cross referencing based digital microfluidic biochips. IEEE Transaction on Computer Aided Design of Integrated Circuits and Systems 27(11), 1905–1917 (2008)
5. Lin, C.C.Y., Chang, Y.W.: ILP based pin-count aware design methodology for microfluidic biochips. In: Proc. of ACM Design Automation Conference 2009, San Francisco, pp. 258–263 (2009)
6. Zhao, Y., Chakraborty, K.: Co-optimization of droplet routing and pin assignment in dispos-able digital microfluidic biochips. In: Proc. of ACM/ IEEE International Symposium on Physical Design, California, pp. 69–76 (2011)
7. Huang, T.W., Ho, T.Y.: A two-stage integer linear programming-based droplet routing algo-rithm for pin-constrained digital microfluidic biochips. IEEE Transaction on Computer Aided Design of Integrated Circuits and Systems 30(2), 215–228 (2011)
8. Lin, C.C.Y., Chang, Y.W.: Cross-contamination aware design methodology for pin-constrained digital microfluidic biochips. Transaction on Computer Aided Design of Integrated Circuits and Systems 30(6), 817–828 (2011)
9. Pan, I., Samanta, T., Rahaman, H., Dasgupta, P.: Ant colony optimization based droplet routing technique in digital microfluidic biochip. In: Proc. of IEEE International Symposium on Electronic System Design, pp. 223–229 (2011)
10. Mukherjee, R., Rahaman, H., Banerjee, I., Samanta, T., Dasgupta, P.: A heuristic method for co-optimization of pin assignment and droplet routing in digital microfluidic biochip. In: Proc. of IEEE International Conference on VLSI Design, pp. 227–232 (2012)
11. Pan, I., Samanta, T.: Efficient droplet router for digital microfluidic biochip using particle swarm optimizer. In: Proc. of SPIE International Conference on Communication and Electronics System Design, vol. 8760, pp. 87601Z-1 – 87601Z-10 (2013)

Phoneme-Based Recognizer to Assist Reading the Holy Quran

Yahya Ould Mohamed Elhadj, Mansour Alghamdi, and Mohammad Alkanhal

Abstract. This paper presents a new phase of our ongoing efforts for building a high performance speaker independent recognizer for Quran recitation. An in-house developed and annotated sound database of about eight hours is used for this purpose. Since this sound database is segmented and annotated on both allophone and phoneme levels, we are developing two separate baseline recognizers for respectively allophones and phonemes. We employed the same approach for developing both phoneme and allophone recognizers to be able to make some kind of comparison between them. The Cambridge HTK tools are used for the development of these recognizers. We present in this paper the development of the phoneme-based recognizer to measure its appropriateness for the sake of our ultimate goal of building a high performance speaker independent recognizer to assist reading and memorizing the Holy Quran; the details of the allophonic recognizer is being published separately. Each Quarnic phoneme is modeled by an acoustic Hidden Markov Model (HMM) with 3-emitting states. A continues probability distribution using 16 Gaussian mixture distributions is used for each emitting state. Results give 92% of average recognition rate, which is very promising, compared to 88% for the allophonic recognizer.

Keywords: Automatic Speech Recognition, Hidden Markov Models, Speech Corpus, Sound Corpus, Phonemes, Allophones, Phontetic Transcription, Speech Segmentation, Speech Annotation, Holy Quran Recitation, Quran Learning, Quran Sound Pronunciations, Tajweed Rules.

Yahya Ould Mohamed Elhadj
Center for Islamic and Arabic Computing
Al-Imam Mohammad Ibn Saud Islamic University, Riyadh, Kingdom of Saudi Arabia,
P.O. Box 5701, Riyadh 11432, KSA
e-mail: yelhadj@ariscom.org

Mansour Alghamdi · Mohammad Alkanhal
Computers and Electronics Research Institute, KACST, Riyadh, Saudi Arabia
e-mail: {mgamdi,mkanhal}@kacst.edu.sa

S.M. Thampi et al. (eds.), *Recent Advances in Intelligent Informatics*,
Advances in Intelligent Systems and Computing 235,
DOI: 10.1007/978-3-319-01778-5_15, © Springer International Publishing Switzerland 2014

1   Introduction

The use of statistical approach based on Hidden Markov Models (HMM), in the domain of Automatic Speech Recognition (ASR), was led to an enormous progress [1-5]. A typical HHM-based ASR approach consists of mapping each word of the application's vocabulary to a sequence of small linguistic units that are then modeled separately by a HMM with emitting states. The speech signal is segmented at small interval of times producing a sequence of contiguous frames that are then recognized by computing their likelihoods or emission probabilities. The emission probability is computed by a Gaussian distribution that is associated with each state of the model.

In general, the development of HMM-based ASR systems, especially speaker independent Large Vocabulary Continuous Speech Recognizers, requires many efforts and represent a difficult and long process. To facilitate and accelerate this process, different ASR engines have been developed by the scientific community. Hidden Markov Tool Kit (HTK)[1] [6-7] and Sphinx[2] [8], developed respectively at Cambridge and Carnegie Melon Universities, are the most known ones. While no benchmark tests are available to indicate which one of these two engines is better than the other at the best of our knowledge, HTK remains the most popular one due to its modularity and portability

In this paper we present the development of a phoneme based recognizer for Quranic Recitation using HTK. It is a part of our ongoing efforts to build a high performance speaker independent recognizer to assist reading and memorizing the Holy Quran [9]. We started this ambitious work by developing a specific sound database for Quranic recitations composed of ten reciters reading the last part of the Holy Quran (the 30th part); a manual segmentation and annotation of speech signals related to the Quranic Ayahs (parts of Quran chapters) were performed on three levels: word, phoneme, and allophone [10-11]. To exploit these levels of segmentation, we particularly interested to develop two baseline recognizers for allophones and phonemes. The idea behind these works is to compare two alternatives to select an appropriate approach for our ultimate goal indicated above: one trying to recognize all the Quranic sounds directly based on an appropriate modeling of these sounds; the other one aims at firstly recognizing the basic sounds of the Holy Quran (phonemes) and then use specific acoustic features to identify allophonic variations such as duration of sounds, which are very important and crucial in the Quran recitation. Here, we will just present the first stage of the second alternative related to the phoneme recognizer development as the second one is still under investigation. Works related the allophonic recognizer is being considered for a separate publication.

---

[1] http://htk.eng.cam.ac.uk
[2] http://www.speech.cs.cmu.edu/sphinx/index.html

2   Short Presentation of the Quran Sound Database

To build a high performance speech recognition system, it is necessary to have accurate acoustic models. Such kind of models cannot be obtained without having a well designed speech corpus. Since an appropriate speech corpus of the Quranic sounds was not available, at the best of our knowledge, it was primordial to firstly collect speech data from reciters memorizing the Quran and then focusing on their segmentation and labeling. To this end, we selected a part of the Holy Quran (the last one from 30 parts as it was very difficult to consider all the Holy Quran at the same time) and then chosen ten speakers to record their recitations in an appropriate environment. We used Audacity[3] as tool for recording and manipulating sounds (see figure 1) and the sounds' center at the King Abdulaziz City for Science and Technology (KACST)[4] as a place and environment of recording.

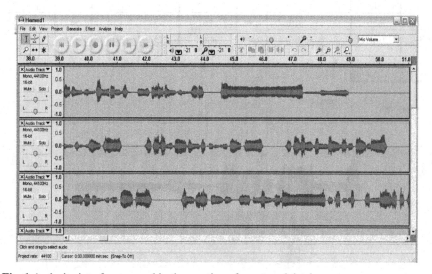

Fig. 1 Audacity interface as used in the creation of our sound database

Recordings were subdivided to small chunks depending on the pauses made by reciters during ayahs (part of the Quran chapters) of each recited surah (chapter of the Quran). Thus, sound files are coded on the following formats: SSS_XXX_YYY_ZZ, where SSS represents initial letters of the reciter's name, XXX represents sourah's number (1 to 114), YYY represents ayah's number, and ZZ represents pauses number made by the reciter inside ayahs. Text files of the spoken counterparts were created and named in a similar manner. We have a total of 5935 sound files with an average of about 594 files and 50 minutes of duration per reciter [9-10].

---

[3] http://audacity.sourceforge.net
[4] www.kacst.edu.sa

A suitable labeling scheme covering all Quarnic sounds was necessary to be proposed as the available coding systems (e.g. IPA[5], SAMPA[6], BEEP[7], etc.) were not appropriate for Classical Arabic (language of the Holy Quran) and also the Modern one. Thus, consistent labels composed of four characters are proposed for each sound at both phoneme and allophone levels [12]: The first two are letters that represent the Arabic phonemes which are taken from KACST Arabic Phonetic Database [13]. The third character is a number which symbolizes sound

Table 1 Labeling Scheme for Phonemes (A "Arabic Letter" and L "Label")

A	L	A	L	A	L	A	L	A	L
ĺ	as10	ج	Jb10	ز	Zs10	ظ	zb10	ل	ls10
ĺ	us10	ح	hb10	س	Ss10	ع	cs10	م	ms10
ĺ	is10	خ	xs10	ش	js10	غ	gs10	ن	ns10
ء	hz10	د	ds10	ص	sb10	ف	fs10	ه	hs10
ب	bs10	ذ	vb10	ض	db10	ق	qs10	و	ws10
ت	ts10	ر	rs10	ط	tb10	ك	ks10	ي	ys10
ث	vs10								

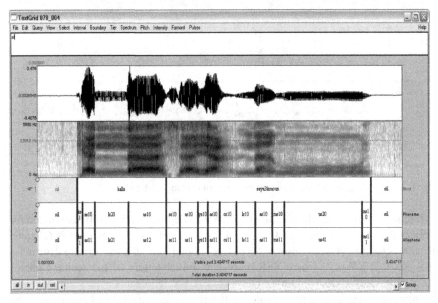

Fig. 2 A screenshot of the customized Praat interface: 1) wave, 2) spectrogram, 3) word-level transcription, 4) phoneme-level transcription, 5) allophone-level transcription

---

[5] http://www.arts.gla.ac.uk/ipa/ipa.html

[6] http://coral.lili.uni-bielefeld.de/Documents/sampa.html

[7] http://www.ldc.upenn.edu/Catalog/readme_files/wsjcam0/abstract.html

duration including geminates. The fourth character is another number that represent the allophonic variations. Table 1 presents the complete set of the sound system of the Classical Arabic at the phoneme level. The set consists of 31 phonemes that represent the single vowels and consonants. As it can be seen, the first number is always "1" which means that the sound is single, and the second number is always "0" which means that the sound is a phoneme. To represent the geminate counterparts of these phonemes, the first number must be "2".

A segmentation and labeling phase was conducted manually on three levels: word, phoneme, and allophone. The labels of the single and geminate phonemes were used to transcribe the Holy Quran recitation at the phoneme level. We used Praat tools to perform the segmentation and labeling at the aforementioned levels. Figure 2 gives an example.

3 Phoneme-Based Recognizer Development

3.1 Pronunciations Lexicon and List of Phonemes

As the transcription was performed using praat and the results were saved in textgrid files, which have well defined structures, we developed an appropriate program to extract Quranic words and their pronunciations across all reciters. Table 2 gives

Table 2 Extract of phoneme-based pronunciations lexicon

أَبَدًا	hz10 as10 hz10 is10 vb10 as20
أَئِنَّا	hz10 as10 hz10 is10 ns20 as20
أَأَنْتُمْ	hz10 as10 hz10 as10 ns10 ts10 us10 ms10
أَبَابِيلَ	hz10 as10 bs10 as20 bs10 is20 ls10
ابْتِغَاءَ	bs10 ts10 is10 gs10 as20 hz10 as10
ابْتَلَاهُ	bs10 ts10 as10 ls10 as20 hs10 us10
أَبَدَا	hz10 as10 bs10 as10 ds10 as20
إِبْرَاهِيمَ	hz10 is10 bs10 rs10 as20 hs10 is20 ms10 as10
أَبْصَارُهَا	hz10 as10 bs10 sb10 as10 rs10 us10 hs10 as20
أَبْوَابًا	as10 bs10 ws10 as20 bs10 as20
أَبِي	hz10 as10 bs10 is20
أَتَاكَ	hz10 as10 ts10 as20 ks10 as10
سُمَّ	is10 ss10 ms10 as10
	ss10 ms10 as10

an extract of the phoneme-based pronunciations lexicon for a proof of concept. As we can see from the table, almost all words have been pronounced in the same manner by reciters, which is logic as we are considering the basic sounds at the phoneme level. Moreover, this also indicates that reciters have applied correct rules of pronunciations during their recitations (what is called "tajweed" rules). Indeed, even obeying to tajweed rules, slight variations are still possible but might be distinguished as allophonic variation on the low level of segmentation (allophonic level). Precisely, statistics show that the average of pronunciation per word was 1.3 at the phoneme level while it was 1.7 at the allophone level, which can be seen as one average pronunciation for phonemes against two for allophones.

We also extracted the list of unique phonemes from the pronunciations lexicon (see table 3). A total number of 60 phonemes were identified, from which a specific phoneme called "sil" representing the silence at the beginning, at the end of ayahs, and the short pauses for breathing during recitations. The number of allophones was 110, which is almost the double of the phonemes.

Table 3 List of phonemes

as10	fs10	jb20	ns20	tb10	ws10
as20	fs20	js10	qs10	tb20	ws20
bs10	gs10	js20	qs20	ts10	xs10
bs20	hb10	ks10	rs10	ts20	xs20
cs10	hs10	ks20	rs20	us10	ys10
cs20	hs20	ls10	sb10	us20	ys20
db10	hz10	ls20	sb20	vb10	zb10
db20	is10	ms10	sil	vb20	zb20
ds10	is20	ms20	ss10	vs10	zs10
ds20	jb10	ns10	ss20	vs20	zs20

3.2 Acoustic Models for Phonemes

Figure 3 shows a standard 3-emitting state Hidden Markov Model (HMM) architecture (taken from [5]), which we used for Quran phonemes. It was also the same model used for Quran allophones.

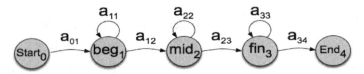

Fig. 3 HMM model for Quran phonemes

A Gaussian Mixture Models (GMM) is associated with each state of the indi-cated HMM acoustic model to identify the characteristics of the sound portion at this state.

Once the architecture of acoustic model is defined, the structure of feature vec-tors has to be specified. This means that an input wave-form is transformed to a sequence of feature vectors, each of them representing the signal in a small time window; usually Hamming window with size 10 to 30 ms. Among the different possible parameterizations of signal, MFCC (Mel Frequency Cepstral Coeffi-cients) is the most known and popularly used for speech processing. Current ASR systems limit the number of MFCC coefficients considered to the first twelve ones representing the static features of the signal portion, for which the first and second derivatives (velocity and acceleration) are added to capture the dynamic features. Energy of the signal portion and its derivatives are also considered, and thus a feature vector of 39 coefficients is obtained. Figure 4 shows the standard MFCC extraction steps taken from [5].

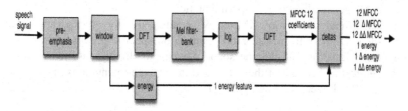

Fig. 4 Feature extraction process

3.3   *Language Model*

As we used in the allophone recognizer, we are simply employing a flat language model to allow different possibilities of sounds combination, and thus to avoid altering sound pronunciations. Table 4 gives the language model used in the pho-neme-based recognizer.

Table 4 Language Model for phonemes

$Phon = as10 | as20 | us10 | us20 | is10 | is20 | hz10 | bs10 | bs20 | ts10 | ts20 | xs10 | xs20 | vs10 | vs20 | ds10 | ds20 | jb10 | jb20 | vb10 | vb20 | hb10 | rs10 | rs20 | sb10 | sb20 | db10 | db20 | tb10 | tb20 | zb10 | zb20 | cs10 | cs20 | gs10 | ls10 | ls20 | fs10 | fs20 | ms10 | ms20 | qs10 | qs20 | ns10 | ns20 | ks10 | ks20 | hs10 | hs20 | zs10 | zs20 | ss10 | ss20 | js10 | js20 | ws10 | ws20 | ys10 | ys20 | sil ; (< $Phon >)

4 Experimentations

In this work, we conducted similar types of experimentations as we did in the allophone recognizer for comparison purposes. Thus, two main series of experimentations were conducted: the first one aims to select the best combination of HTK training tools to ensure a good training of the models, while the second one is related to the combination of GMMs to determine the optimal number appropriate for our sound database. For the training, we start using HInit + HRest and then HCompV + HERest; the best results were obtained from the first combination as we noticed in the allophone recognizer. This is somehow expected as our manual segmentation was done with an acceptable level of accuracy. So, guided by this inspiration that HInit gives a good initialization and thus could be considered as appropriate starting point, we conducted two other experimentations, one using HInit + HERest, and another one using HInit + HRest + HERest. Also, we got the best results from the last combination exactly as it was in the allophone recognizer. So, the combination (HInit + HRest + HERest) will be considered each time the training is needed in a given experimentation.

For the optimal number of GMMs, we conducted a lot of experimentations varying their numbers from 1 to more than 16 on specific groups of training & testing sets defined as follows: the sound database is divided into ten groups of training and testing sets; all of them are used in the experimentations one at a time and then a global average is computed. For each group, we consider a particular

Table 5 Percentage of Average Recognition Rates for 1 to 10 and 16 GMMs

GMMs	AAH	AAS	AMS	ANS	BAN	FFA	HSS	MAS	MAZ	SKG
1	81.16	80.91	75.48	83.56	82.78	78.65	80.59	79.11	79.20	82.1
2	85.27	82.97	79.82	87.21	85.86	84.36	87.1	83.22	84.91	88.14
3	87.10	86.4	83.81	89.95	90.19	85.16	88.01	85.5	87.20	89.05
4	88.36	90.74	85.29	91.21	89.17	87.33	90.07	86.19	89.26	89.85
5	88.93	91.54	85.06	91.78	91.56	87.56	91.21	89.38	88.34	90.08
6	90.53	92.34	85.86	91.1	93.5	88.58	91.21	89.84	90.97	91.33
7	90.53	92.00	87.57	92.12	93.73	90.41	92.24	90.41	90.29	91.11
8	90.75	91.89	86.66	92.47	92.7	88.13	92.24	89.84	89.03	90.42
9	91.78	92.80	86.66	92.12	92.7	89.04	92.35	89.61	91.20	91.9
10	92.58	93.26	86.32	93.04	93.27	90.30	92.35	90.41	90.97	92.13
16	92.12	93.37	88.03	90.87	94.18	89.16	92.81	91.10	92.69	92.47

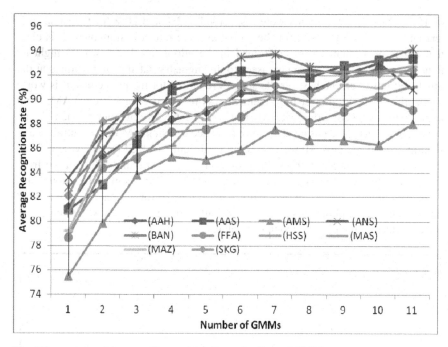

Fig. 5 Percentage of Average Recognition Rates for Several GMMs

reciter to construct the testing set by extracting the first ayah of each sourah from it; the remaining ayahs of this reciter as well as all ayahs of the other reciters are used for training. Since our sound database contains 38 sourahs and 572 ayahs, so in each group the training and testing sets are respectively composed of 534 and 38 ayahs. This means that in each group, about 93% of the corpus is used for training and 7% is used for testing.

We report in the table 5, the results from 1 to 10 and 16 GMMs as the other numbers did not show more significant recognition rates. These results are plotted in the figure 5.

5 Discussing and Commenting the Results

From the results showed above, we notice good recognition rates as the lowest one is beyond 75% (for the reciter AMS) and a global average of 80% using one GMM. Compared to the results obtained from allophones, the lowest recognition rate was 67% for the same reciter and 73% as global average using one GMM; this represents a difference of about 12% and 10% respectively in favor of phonemes.

When increasing the number of GMMs, we obtained huge improvements varying from 11% to 17% respectively for reciters ANS and AMS (also MAZ). The global average rate was increased from 80% to 92% representing an enhancement of 15%. These improvements were respectively 12% and 20% for the same reciters and 21% (73% to 88%) for the global average rate for allophones. It is clear that big improvements were performed in the allophonic level, which might be logic as several GMMs were necessary to distinguish related sound features at the this level; however, the best results are still recorded for the phonemes. It is worth to mention that the highest improvement was recorded for the reciter AMS who had the lowest recognition rate when using one GMM for both phonemes and allophones. This indicates clearly the importance of using GMMs to neutralize and separate sound features. However, the optimal number of GMMs may depend on the model parameters and the volume of data used for training. We can see from the above reported results that between 6 and 7 GMMs seem to be enough for phonemes in this application, while this was around 10 for allophones. We remind that the number of phonemes is almost the half of the allophones, which may have a great effect on the global recognition rates.

```
Aligned transcription: Phoneme_Level/mfcc/MAS_000_001_00.lab vs Phoneme_Level/mfcc/MAS_000_001_00.rec
LAB: sil bs10 is10 ss10 ms10 is10 ls20 as20 hs10 is10 rs20 as10 hb10 ms10 as20   ns10 is10 rs20      as10 hb10      is20 ms10      sil
REC: sil bs10 is10 ss10 ms10 is10 ls20 as20 hs10 is10 rs20 as10 hb10 ms10 as20 ms10 ys10 zb10 rs20 us10 rs10 us10 hb10 is20 ys20 is20 ms10 zb10 jb20 ts20
Aligned transcription: Phoneme_Level/mfcc/MAS_001_001_00.lab vs Phoneme_Level/mfcc/MAS_001_001_00.rec
LAB: sil hz10 as10 ls10 hb10 as10 ms10 ds10 us10 ls10 is10 ls20 as20   hs10 is10   rs10 as10 bs20 is10 ls10 cs10 as20 ls10   as10 ms10   is20   ns10 sil
REC: sil hz10   ls10 hb10   ms10 ds10 us20 ls10 is10 ls20 as20 as10 hs10 is10 us10 rs10 as10 bs20 is10 ls10 cs10 as20 ls10 hs10 as10 ms10 ys20 is20 is10
ns10 sil
Aligned transcription: Phoneme_Level/mfcc/MAS_078_001_00.lab vs Phoneme_Level/mfcc/MAS_078_001_00.rec
LAB:    sil cs10 as10            ms20 as10 ys10 as10 ls10 as20 hz10 as10 ls10      us10 ns10 sil
REC: ts20 qs10 cs10 as10 ms10 ms20 ms20 ms20 as10   as10 ls10 bs10 ss10 as20 hz10 as10 ls10 us20 ws20 us20 ns10 ts20
Aligned transcription: Phoneme_Level/mfcc/MAS_079_001_00.lab vs Phoneme_Level/mfcc/MAS_079_001_00.rec
LAB: sil ws10 as10 ns20    as20 zs10 is10 cs10 as20 ls10 is10    gs10 as10 rs10 qs10 as20   sil
REC: sil ws10 as10 ns20 as10 as20 zs10 ys10 cs10 as20 ts10 is10 hs10 gs10 as10 rs20 qs10 as20 zb10 ts20

Aligned transcription: Phoneme_Level/mfcc/MAS_103_001_00.lab vs Phoneme_Level/mfcc/MAS_103_001_00.rec
LAB: sil ws10 as10 ls10 cs10 as10 sb10 rs10 sil
REC: sil ws10 as10 ls10 cs10 as10 sb10 qs10 sil

Aligned transcription: Phoneme_Level/mfcc/MAS_114_001_00.lab vs Phoneme_Level/mfcc/MAS_114_001_00.rec
LAB: sil   qs10 us10 ls10 hz10   as10 cs10   us20   vb10 us10 bs10 is10 rs10 as10 bs20 is10   ns20 as20   ss10            sil
REC: sil ts20 qs10 us10 ls10 hz10 as10 as10 cs10 hs10 us20 us10 zs10 us10 bs10 is10 rs10 as10 bs20 is10 ns10 ns20 as20 ns10 ss10 sil sil sil
================= HTK Results Analysis =======================
Date: Thu Sep 11 02:26:46 2009
Ref : Phoneme_Level/mfcc/MLF_Ref_test.mlf
Rec : Phoneme_Level/recout.mlf
----------------- Overall Results -----------------
WORD: %Corr=90.87, Acc=74.77 [H=796, D=17, S=63, I=141, N=876]
```

Fig. 6 Alignment of transcription between a recognition output and its reference

For a deep analysis of the results, we considered the correlation matrices and the alignment of outputs and references (see figure 6 for some examples) for both phonemes and allophones to determine which kind of errors are present. We remarked that many confusions between related sounds (such as degrees of vowel prolongation or lengthening, heaviness and lightness which means making some sounds emphatic or non-emphatic, degrees of vibrations or unrest which means producing the voiced stop consonants with a schwa-like sound at the end, etc) performed at allophone level have been largely reduced at the phoneme level. The variations of these sounds are very close and thus were difficult to be distinguished by a standard HTK approach considering all allophones. This problem somehow persists for some sounds at the phoneme level and was particularly seen for geminated phonemes. So, it might be worth to limit also the phonemes to their basic ones -single phonemes- (about 30) without considering any variation neither geminating nor others and to deal with all these variations on a separate level taking in consideration specific cues such as sound durations and other relevant characteristics.

6 Conclusion and Future Works

In this paper, we presented the development of a phoneme-based recognizer for Quran sounds. It has been preceded by the development of an allophone-based recognizer for Quarnic sound variations. From the development of these two similar recognizers, we were interested to compare two approaches based on these baseline recognizers to determine which one will seem to be appropriate for building an accurate speaker independent recognizer to assist Quran reading and memorization and its sounds' pronunciation. While the first approach is trying to directly recognize all the Quranic sounds based on an appropriate modeling of these sounds, the second one is looking firstly to identify the basic sounds of the Holy Quran (single phonemes) and then employing specific acoustic features to identify allophonic variations in a separate level.

We followed the same methodology for building the two aforementioned baseline recognizers to allow a direct comparison of their results. A lot of experimentations were conducted and the results showed a neat improvement of global recognition rates for phonemes, which is very logic due to the difference of models' number.

An in-depth analysis of results put in evidence a good improvement in terms of confusion of several sounds' variations. The number of insertion and deletion was also reduced in the case of phonemes. However, we perceived that the geminate counterparts of phonemes were confused with their single ones. This shed the light on the importance of considering just the basic sounds at the phoneme level and thus to consider the geminates as also a variation of the underlying sound. The number of basic phonemes is very limited, around 30th, and is very appropriate to be considered and well studied and modeled.

Based on the above indications, we are planning to work on two directions, for which we think would be appropriate: 1) build a recognizer for the basic sound phonemes and tune it as much as possible to be able to identify these basic sounds with high accuracy, 2) conduct a study on the basic phonemes to determines their different allophonic variations and their characteristics, 3) find an appropriate way to exploit these sounds properties to identify which allophonic variation is most likely to be appeared in a given context.

Acknowledgements. This paper is part of the "Computerized Teaching of the Holy Quran" Project funded by King Abdulaziz City for Science and Technology under the grant number AT-25-113, Riyadh, Saudi Arabia. We thank all the project members.

Some of the experimentations presented here had been conducted during a research stay within the SAMOVA team, IRIT, Toulouse.

References

[1] Rabiner, L.: A Tutorial on Hidden Markov Models and Selected Applications in Speech Recognition. Proceedings of the IEEE 77(2) (1989)

[2] Rabiner, L., Juang, B.H.: Fundamentals of Speech Recognition. Prentice Hall (1993)

[3] Jelinek, F.: Statistical Methods for Speech Recognition. MIT Press, Cambridge (1998)

[4] Huang, X., Acero, A., Hon, H.: Spoken Language Processing. Prentice Hall (2001)

[5] Jurafsky, D., Martin, J.H.: Speech and Language Processing: An introduction to natural language processing, computational linguistics, and speech recognition, 2nd edn. Prentice Hall (2008)

[6] Young, S.: The HTK Hidden Markov model toolkit: design and philosophy (Tech. Rep. CUED/FINFENG/TR152). Cambridge University Engineering Dept, UK (1994)

[7] Young, S., et al.: HTK Book (V.3.4). Cambridge University Engineering Dept, UK (2009)

[8] Huang, X., Alleva, F., Hon, H.W., Hwang, M.Y., Rosenfeld, R.: The SPHINX-II speech recognition system: an overview. Computer Speech and Language 7(2), 137–148 (1993)

[9] Elhadj, Y.O.M., Alsughayeir, I.A., Alghamdi, M., Alkanhal, M., Ohali, Y.M., Alansari, A.M.: Computerized teaching of the Holy Quran. Final Technical Report, King Abdulaziz City for Sciences and Technology (KACST), Riyadh, KSA (2012) (in Arabic)

[10] Elhadj, Y.O.M., AlGhamdi, M., AlKanhal, M., Alansari, A.M.: Sound Corpus of a part of the noble Quran. In: Proc. of the International Conference on the Glorious Quran and Contemporary Technologies, King Fahd Complex for the Printing of the Holy Quran, Almadinah, Saudi Arabia, October 13-15 (2009) (in Arabic)

[11] Elhadj, Y.O.M.: Preparation of speech database with perfect reading of the last part of the Holly Quran. In: Proc. of the 3rd IEEE International Conference on Arabic Language Processing (CITAL 2009), Rabat, Morocco, May 4-5, pp. 5–8 (2009) (in Arabic)

[12] AlGhamdi, M., Elhadj, Y.O.M., AlKanhal, M.: A manual system to segment and transcribe Arabic Speech. In: Proceedings of IEEE ICSPC 2007, Dubai, UAE, pp. 233–236 (2007) ISBN 1-4244-1236-6

[13] Alghamdi, M.: KACST Arabic Phonetics Database. In: The Fifteenth International Congress of Phonetics Science, Barcelona, pp. 3109–3112 (2003)

A Secure Two Party Hierarchical Clustering Approach for Vertically Partitioned Data Set with Accuracy Measure

Ipsa De and Animesh Tripathy

Abstract. Data mining has been a popular research area for more than a decade because of its ability of efficiently extracting statistics and trends from large sets of data. However, there are many applications where the data set are distributed among different parties. This makes the privacy an issue of concern for each individual/organization. This paper makes an approach towards privacy preserving clustering problem for vertically partitioned data set(VPD). We propose a secure hierarchical clustering algorithm for two parties over vertically partitioned data set with accuracy measure. Each site only learns the final results about the clusters, but nothing about the individual's data.

1 Introduction

Data Mining has been a popular research area for more than a decade because of its ability of efficiently extracting statistics and trends from large sets of data. Basically data mining refers to extraction or mining knowledge from large amounts of data.[1] Basically data mining means processing large volumes of data which may be stored in a database, searching for statistics and knowledge about the data and relationships between them. The overall goal of the data mining process is to extract information from a data set and transform it into an understandable structure for further use.

A well known approach to extract desired information from a data set is clustering of related data. However, when dealing with such sensitive information, the privacy issues become major concerns, as any leakage or compromise of data may result in potential harm to individuals or financial losses to the corporate. It is widely used in the applications of financial affairs, marketing, insurance, medicine, chemistry, machine learning, data mining, etc.[3,4,6,7,8]

Ipsa De · Animesh Tripathy
School of Computer Engineering, KIIT University, Bhubaneswar, Odisha, India
e-mail: 1155007@kiit.ac.in, animeshtripathy@gmail.com

S.M. Thampi et al. (eds.), *Recent Advances in Intelligent Informatics*,   153
Advances in Intelligent Systems and Computing 235,
DOI: 10.1007/978-3-319-01778-5_16, © Springer International Publishing Switzerland 2014

So, the main privacy preserving clustering problem is that there are 'n' data objects distributed over physically apart partitioned databases and those objects should be clustered in 'k' number of clusters based on their similarity without compromising about the privacy of the databases. The existing clustering algorithms [1,4,6], are quite straightforward. But when the data is distributed among many parties, [9] who want to preserve their private data, then solution becomes quite challenging. Depending on the problems there can be various approaches to solve the problem. These kinds of problems have been studied in the field of privacy-preserving data mining or PPDM for short [1,8,9,10].

But most of the clustering problem solutions for vertically partitioned data set are based on k-means algorithm. Only in [13] a secure version of BIRCH is given over vertically partitioned data. In this paper a secure hierarchical clustering approach over vertically partitioned data is provided which increases the accuracy of the clusters over the existing approaches. The proposed approach can be found to be more efficient in terms of running time and cluster accuracy than usual k-means clustering algorithm in identifying cluster centers.

In the next section preliminaries are given. A secure algorithm for computing clusters over vertically partitioned data set is given in section 3. Efficiency and privacy of the algorithm is discussed in section 4. Experimental results are shown in section 5. We conclude in section 6 with some discussions and future research work direction.

2   Preliminaries

In this section, some preliminaries are presented.This includes different types of partitioning of data sets and some basic privacy preserving techniques which can be used in privacy consideration.

2.1   Distributed Data Mining

In this model, data sources are assumed to be distributed across multiple sites. A simple approach to mine data over multiple sites can be done by combining data into a single site securely. Such data set can be partitioned in two ways horizontally or vertically. A brief overview of horizontally and vertically partitioned data set is given below:

Horizontal Partitioning: In horizontal partitioning different sites collect the same set of information, but about different entities. Credit card databases of two different (local) credit unions can be an example of horizontal partitioning. Taken together, one may find that fraudulent customers often have similar transaction histories, etc.[2]

Vertical Partitioning: Vertical partitioning of data implies that though different sites gather information about the same set of entities, they collect different feature sets. An illustrative example of vertical partitioning is given below. Let there are two databases; one contains medical records of people while another contains cell phone information for the same set of people. Mining the joint global database might reveal information like Cell phones with Li/Ion batteries lead to brain tumors in diabetics.[2]

2.2 Cluster Analysis

Clustering finds sub-families among a large collection of data. More formally given a set of data points, each having a set of attributes, and a similarity measure among them, find clusters such that (1) Data points in one cluster are more similar to one another and (2) Data points in separate clusters are less similar to one another.[1] Many clustering algorithms exist in literature. Among them hierarchical clustering algorithm is one of the most important clustering algorithm.

Hierarchical clustering algorithms: It creates a hierarchical decomposition of the given set of data objects. It can be either agglomerative or divisive, based on how the hierarchical decomposition is formed. The agglomerative approach (bottom-up approach) starts with each object forming a separate group. It successively merges the objects or groups that are close to one another, until all of the groups are merged into one or until a termination condition holds. The divisive approach (top-down approach) starts with all of the objects in the same cluster. In each successive iteration, a cluster is split up into smaller clusters, until eventually each object is in one cluster, or until a termination condition holds. In this paper the work is mainly based on agglomerative approach. This approach is chosen as it is better than k-means clustering algorithm in terms of number of iteration and number of database scanning.[1]

2.3 Some Basic Privacy Preserving Techniques

The main motto for privacy preserving data mining is to reduce the risk of data misuse. With this end in view, in this section we introduce some existing cryptographic protocols. A number of effective methods for data mining have been proposed considering the privacy factor. Most of these methods are devised with slight modification to the main mining algorithm to achieve security without hampering the flavor of mining. Some basic concepts that are used to achieve privacy are given below:

Random Share: It is a popular method in current privacy preserving data mining studies. Each computed intermediate values are designated as uniformly distributed random values, with each party holding one of these values. Their sum is the actual intermediate value. It basically adds noise to the individual data values so that they can't be recovered and aggregation of those randomized values should provide the final output and the final result should be available to all the parties.[5,7]

Homomorphic Encryption: Encryption method mainly resolves the problems that people jointly conduct mining tasks based on the private inputs they provide. The encryption method can ensure that the transformed data is exact and secure. Homomorphic encryption schemes allow certain computations on encrypted values. In particular, an encryption scheme is additively homomorphic if there is some operation $\oplus$ (XOR operation) on encryption such that for all clear-text values a and b, $E(a) \oplus E(b) = E(a + b)$.[4,7,12]

Secure Scalar Product Protocol: In this scheme the scalar product or dot product of two vectors is computed securely. It means that the actual values of the vectors are not revealed only the result is revealed. Let a vector $X = (x_1, \ldots, x_n)$, held by Alice, and $Y = (y_1, \ldots, y_n)$, held by Bob. They need to securely compute the scalar product $X \times Y = \sum_{i=(1-n)} (x_i \times y_i)$. So, at the end of the protocol Alice knows only $X.Y$ not Y and so for Bob.[5,7,11]

Secure Add and Compare: It is based on the homomorphic encryption. It builds a circuit that has two inputs from each party, sums the first input of both parties and the second input of both parties, and returns the result of comparing the two sums. This is a semantically secure approach.[7]

3   Proposed Work

Problem definition of the proposed work in relation to privacy preserving clustering algorithm and finding solution thereof are discussed below:

3.1   Privacy Preserving Clustering Problem

Assume that a data set D consists of n instances with m attributes and it is vertically partitioned between Alice and Bob. It is assumed that both parties contain the same number of data records and Alice contains some attributes for all records while Bob contains the other attributes. So, Alice contains her own database with first m_1 attributes and Bob contains his own database with remaining m_2 attributes, where $m_1+m_2=m$. Alice and Bob want to find the final k-clusters over the total partitioned data but trying to protect their own privacy.

3.2   Secure Hierarchical Clustering on Vertically Partitioned Data Set

The used strategy behind this algorithm is 'divide-conquer-merge'. Firstly, each party computes k number of clusters on their own private data set. Secondly, both parties compute the distance between each instance and each of the k cluster centers. Thus, each party gets a n×k matrix. All these computations are done by the concerned parties on their private databases without loss of privacy. Thirdly, both parties exchange their distance matrices along with their k cluster centers in randomized form. Fourthly, each party computes the all possible combinations of cluster centers from the total 2k clusters. In the process, k^2 cluster centers are formed and each party computes the distance between each record and the k^2 cluster centers. Minimum closest cluster is chosen by each party for n data points and finally they are merged into k-clusters on the basis of best pair of clusters[15]. A best pair of clusters is one with least error. If C_i and C_j are best pair, then they are replaced by $C_i \cup C_j$.

The error between the clusters can be computed by using the following formula.Let C_1 and C_2 be two clusters being considered for a merge and $C.weight$ denote the number of objects in cluster C and $dist^2(C_1, C_2)$ is the distance between the clusters C_1 and C_2 [6], then

$$Error\,(C_1 \cup C_2) = C_1.weight \times C_2.weight \times dist^2(C_1, C_2) \tag{1}$$

This computed error is basically the distance between two cluster centers which is calculated by computing the difference for each attribute of one cluster with the corresponding attribute of the other clusters. So, minimum error leads to closest clusters which are merged.

3.3 Flow Diagram and Algorithm for Secure Hierarchical Clustering

In the last section we have presented the proposed approach in brief. In this section, we present the flow diagram and algorithm of the approach in detail. After each of Alice and Bob locally computes k-cluster centers from their own part of the data, each party computes the following n×k matrix. Alice computes her corresponding distance matrix M_{Alice}

$$\begin{pmatrix} a_{1,1} & a_{1,2} & \cdots & a_{1,k} \\ a_{2,1} & a_{2,2} & \cdots & a_{2,k} \\ \cdot & \cdot & \cdots & \cdot \\ \cdot & \cdot & \cdots & \cdot \\ a_{n,1} & a_{n,2} & \cdots & a_{n,k} \end{pmatrix}$$

In the above matrix each row represents the distance between an instance to all the k cluster center i.e $a_{i,j}$ represents the distance between i^{th} instance and j^{th} cluster center. Similarly Bob computes the same matrix based on his data set i.e M_{Bob}.

Flow Diagram for the Proposed Work

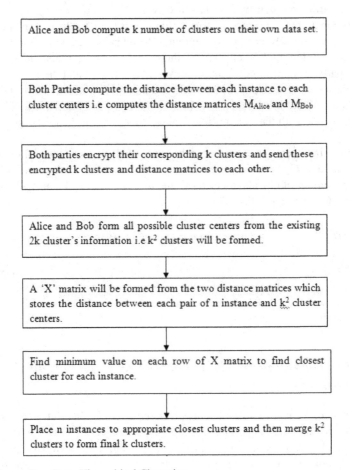

Alice and Bob compute k number of clusters on their own data set.

Both Parties compute the distance between each instance to each cluster centers i.e computes the distance matrices M_{Alice} and M_{Bob}

Both parties encrypt their corresponding k clusters and send these encrypted k clusters and distance matrices to each other.

Alice and Bob form all possible cluster centers from the existing 2k cluster's information i.e k^2 clusters will be formed.

A 'X' matrix will be formed from the two distance matrices which stores the distance between each pair of n instance and k^2 cluster centers.

Find minimum value on each row of X matrix to find closest cluster for each instance.

Place n instances to appropriate closest clusters and then merge k^2 clusters to form final k clusters.

Fig. 1 Secure Two Party Hierarchical Clustering

Thus each of them possesses k-cluster centers and distance matrix which they randomly share after encrypting their corresponding vector C and sending it to the other party. Each party forms the all possible combination of cluster centers from the received encrypted cluster center information and his/her cluster center information. Finally both parties possess same k^2 number of cluster center. Then each party computes the distance between each n instances and all the k^2 cluster centers by using the two distance matrices and then a cluster is chosen which has the minimum distance to an instance. Both parties then place all records to closest clusters and merge k^2 clusters to form final k clusters.

Algorithm: A Secure Two Party Hierarchical Clustering Algorithm

Input: A vector of cluster centers C of the form $(C_1, C_2, ..., C_k)$ for Alice and a vector of cluster centers C of the form $(C_{k+1}, C_{k+2}, ..., C_{2k})$ for Bob.

Output: Assignment of cluster number to each object by both Alice and Bob.

1. Alice and Bob compute the distance matrices M_{Alice} $(n \times k)$ and M_{Bob} $(n \times k)$ respectively.

2. Alice computes encryption of vector C as $(E(C_1), ..., E(C_k))$ and sends to Bob along with M_{Alice} while Bob computes encryption of vector C as $(E(C_{k+1}), ..., E(C_{2k}))$ and sends to Alice along with M_{Bob}.

3. Alice gets random share $(C_1, C_2, ..., C_k, E(C_{k+1}), ..., E(C_{2k}))$ and Bob gets random share $(E(C_1), ..., E(C_k), C_{k+1}, ..., C_{2k})$.

4. Alice and Bob form all possible cluster centers from the existing 2k cluster's information i.e k^2 cluster will be formed.

5. Both Alice and Bob compute a matrix X $(n \times k^2)$, which holds the distance between each pair of n points and k^2 cluster centers, using Compute-X sub-procedure as given below:

    5.a. for p= 1 to n
    5.b.   l=0
    5.c.   for q= 1 to k
    5.d.   for r=1 to k
    5.e.   l=l+1
    5.f. $X_{pl} = a_{pq} + b_{pr}$
    5.g.   end for
    5.h.   end for
    5.i. end for
    5.j. Return X

6. Both Alice and Bob find minimum value on each row of X matrix to get closest cluster for each instance i.e if i^{th} column has minimum value in j^{th} row then j^{th} instance will be closest to i^{th} cluster.

7. Both Alice and Bob place n instances to appropriate closest clusters and then merge k^2 clusters to form final k clusters.

4 Efficiency and Privacy Analysis

Alice and Bob compute k clusters on their own data set and then they share it to each other after encrypting the values. The encryption takes O(k) time for each party. Next, the computational complexity for computing the distance matrix by each party is O(nk). Step 4 of the proposed algorithm takes $O(k^2)$ time. Again the computational complexity for Compute-X sub-procedure is $O(nk^2)$. Step 6 runs n times for each instance and for each instance it takes $O(k^2)$ time. So, total complexity for step 6 is of $O(nk^2)$. Total computational complexity for this algorithm is of $O(nk^2)$.

For communication complexity, Alice and Bob send k encrypted values to each other. If we assume that it takes c bits to represent each encryption then the total communication complexity is O(kc). Again to send the distance matrix O(nk)time is taken. Thus it totally takes O(nk) communication time.

Both parties compute k clusters independently from the data objects they have. Then these information are shared between the two but in the encrypted form. So, it does not reveal the intermediate cluster center. Next the distance matrices are shared. But, it only holds the distance between cluster center and the instances, no private information. So, it does not leak any private information. After merging, final k cluster centers are revealed to each party according to their share. Hence the privacy-preserving hierarchical k clustering algorithm is secure and does not leak any information.

5 Experimental Results

All of our experiments have been conducted on a computer with 2.60 GHz Intel CORE i5 CPU and 4 GB main memory. The operating system of the computer is Microsoft Windows 7. The algorithms are all implemented in NetBeansIDE 7.1.

We performed our experiments on the UCI data-sets:WBC, Ionosphere, iris, weather[14]. We assume that the data-sets are vertically partitioned between Alice and Bob. The description of the data-sets is presented in Table 1. n denotes the number of instance, m is the number of attributes, Alice holds the first m_1 attributes, Bob holds the last m_2 attributes. k is the number of clusters. We report the running time in Table 2.

Additionally we have implemented an existing work for privacy preserving clustering process where the k-means clustering algorithm is implemented over VPD. Table 3 shows the experimental results for both running time and cluster accuracy

Table 1 Descriptions of Data-sets

Data-set	n	m	m_1	m_2	k
WBC	699	9	5	4	2
IONO	351	34	17	17	2
Iris	150	5	2	3	3
Weather	14	5	2	3	2

Table 2 Running Time on WBC, IONO, Iris, Weather

Data-set	Run time(Sec)
WBC	1.145
IONO	0.541
Iris	0.450
Weather	0.326

for different clusters which can be visualized through Fig.2. In view of using only one iteration, this approach reduces the running time as compared to the existing work. Cluster accuracy has been calculated using Davis-Bouldin Index[16] and Dunn Index for both the ideas. Clusters with low intra-cluster distances (high intra-cluster similarity) and high inter-cluster distances (low inter-cluster similarity) will have a low DaviesBouldin index. The Dunn index[17] aims to identify dense and well-separated clusters. It is defined as the ratio between the minimal inter-cluster distances to maximal intra-cluster distance. So, a low value of DaviesBouldin index and high value of Dunn index is desired. The Table 3 indicates the results as desired. It therefore reveals that the proposed approach is superior to the existing work.

Table 3 Comparative Study among Clusterers on Weather Data Set

Parameter	VPD with k-means clustering	VPD with hierarchical clustering
Run Time(Sec)	2.579	0.326
Davis Bouldin Index	1.41	1.36
Dunn Index	1.29	1.34

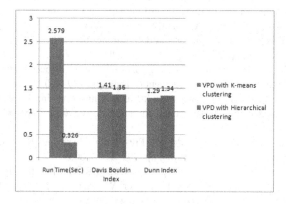

Fig. 2 Comparative Study among Clusterers on Weather Data Set

6 Conclusion and Future Research Direction

Here the privacy preserving clustering problem is analyzed for vertically partitioned data set. There are various approaches to solve this problem like adding noise or encrypting data values. In this paper a novel secure hierarchical clustering approach is given to cluster the data objects which are vertically partitioned among two parties. The future research work can be to find a solution for hierarchical clustering for data set which is vertically partitioned among multiple parties without giving up the efficiency.

References

1. Han, J., Kamber, M.: Data Mining: Concepts and Techniques, 2nd edn., pp. 383–460. China Machine Press, Beijing (2006)
2. Vaidya, J.: Privacy Preserving Data Mining Over Vertically Partitioned Data. Ph.D Thesis, Purdue University, pp. 1–149 (2004)
3. Vaidya, J., Clifton, C.: Privacy Preserving K-Means Clustering over Vertically Partitioned Data. In: Proceedings of the Ninth ACM SIGKDD International Conference on Knowledge Discovery and Data Mining, pp. 206–215. ACM, Washington, DC (2003)
4. Yu, T.-K., Lee, D.T., Chang, S.-M., Zhan, J.: Multi-Party k-Means Clustering with Privacy Consideration. In: International Symposium on Parallel and Distributed Processing with Applications, pp. 200–207. IEEE Computer Society (2010)
5. Jagannathan, G., Wright, R.: Privacy Preserving Distributed k-Means Clustering over Arbitrarily Partitioned Data. In: Proceedings of the 11th ACM, SIGKDD International Conference on Knowledge Discovery and Data Mining, pp. 1–7. ACM, USA (2005)
6. Jagannathan, G., Pillaipakkamnatt, K., Wright, R.: A New Privacy Preserving Distributed k-Clustering Algorithm. In: Proc. of the 6th SIAM International Conference on Data Mining, pp. 492–496. SIAM (2006)
7. Jagannathan, G., Pillaipakkamnatt, K., Wright, R., Umano, D.: Communication-Efficient Privacy-Preserving Clustering. Transactions on Data Privacy 3(1), 1–25 (2010)
8. Estivill-Castro, V.: Private Representative-Based Clustering for Vertically Partitioned Data. In: Proceedings of the Fifth Mexican International Conference in Computer Science (ENC 2004), pp. 1–8. IEEE Computer Society (2004)
9. Lindell, Y., Pinkas, B.: Privacy preserving data mining. In: Bellare, M. (ed.) CRYPTO 2000. LNCS, vol. 1880, pp. 36–54. Springer, Heidelberg (2000)
10. Agrawal, R., Srikant, R.: Privacy-preserving data mining. In: Proceedings of the 2000 ACM SIGMOD Conference on Management of Data, pp. 439–450. ACM (2000)
11. Bunn, P., Ostrovsky, R.: Secure Two-Party k-Means Clustering. In: Proceedings of the 14th ACM Conference on Computer and Communications Security, pp. 486–497. ACM (2007)
12. Jha, S., Kruger, L., McDaniel, P.: Privacy Preserving Clustering. In: 10th European Symp. on Research in Computer Security, pp. 397–417 (2005)
13. Prasad, P.K., Pandu Rangan, C.: Privacy Preserving BIRCH Algorithm for Clustering over Vertically Partitioned Databases. In: Jonker, W., Petković, M. (eds.) SDM 2006. LNCS, vol. 4165, pp. 84–99. Springer, Heidelberg (2006)
14. Asuncion, A., Newman, D.J.: UCI Machine Learning Repository (2007), http://www.ics.uci.edurmlearnIMLRepository.html
15. Tripathy, A., De, I.: Privacy Preserving Two-Party Hierarchical Clustering Over Vertically Partitioned Dataset. A Journal of Software Engineering and Applications 6, 26–31 (2013)
16. Davies David, L., Bouldin Donald, W.: A Cluster Separation Measure. IEEE Transactions on Pattern Analysis and Machine Intelligence PAMI-1(2), 224–227 (1979)
17. Dunn, J.: Well separated clusters and optimal fuzzy partitions. Journal of Cybernetics 4(1), 95–104 (1974)

Document Classification: An Approach Using Feature Clustering

B.S. Harish and B. Udayasri

Abstract. In this paper, we propose a new method of representing text documents based on feature clustering approach. The proposed representation method is very powerful in reducing the dimensionality of feature vectors for text classification. Further, the proposed method is used to form a symbolic representation (interval valued representation) for text documents. To corroborate the efficacy of the proposed model, we conducted extensive experimentation on standard text datasets. We have compared our classification accuracy achieved by the symbolic classifier with the other existing classifiers like: Naïve Bayes, k-NN, Centroid based and SVM classifiers. The experimental results reveal that the achieved classification accuracy is better than that of the existing methods. In addition our method is based on a simple matching scheme; it requires negligible time for classification.

Keywords: Documents, Symbolic, Representation, Features, Clustering, Classification.

1 Introduction

Text classification is one of the important research issues in the field of text mining, where the documents are classified with a supervised knowledge. Based on a likelihood of the training set, a new document is classified. The task of text classification is to assign a boolean value to each pair $(d_j, k_i) \in D \times K$, where $'D'$ is

B.S. Harish
Department of Information Science & Engineering, S.J. College of Engineering,
Mysore 570 006, Karnataka, India
e-mail: bsharish@ymail.com

B. Udayasri
Department of Computer Science and Engineering, Vidyavardhaka College of Engineering,
Mysore 570 002, Karnataka, India
e-mail: udaya.sri84@gmail.com

S.M. Thampi et al. (eds.), *Recent Advances in Intelligent Informatics*,
Advances in Intelligent Systems and Computing 235,
DOI: 10.1007/978-3-319-01778-5_17, © Springer International Publishing Switzerland 2014

the domain of documents and $'K'$ is a set of predefined categories. The task is to approximate the true function $\phi : D \times K \rightarrow \{1,0\}$ by means of a function $\hat{\phi} : D \times K \rightarrow \{1,0\}$ such that ϕ and $\hat{\phi}$ coincide as much as possible. The function $\hat{\phi}$ is called a classifier. A classifier can be built by training it systematically using a set of training documents D, where all of the documents belonging to D are labeled according to K [1], [2]. The major challenges and difficulties that arise in the problem of text classification are: High dimensionality (thousands of features), variable length, content and quality of the documents, sparse distribution of terms in documents, loss of correlation between adjacent words and understanding complex semantics of terms in a document [3]. To tackle these problems a number of methods have been reported in literature for effective classification of text documents. Many representation schemes like binary representation [4], ontology [5], N-Grams [6], multiword terms as vector [7], UNL [8], symbolic representation for web documents [9], [10], Bayes formula to vectorize documents [11], [12], Vector Space Model [13], Latent Semantic Indexing [14], Locality Preserving Indexing [15], Regularized Locality Preserving Indexing [16] are proposed as text representation schemes for effective text classification.

Also, in conventional supervised classification an inductive learner is first trained on a training set, and then it is used to classify a testing set, about which it has no prior knowledge. However, for the classifier it would be ideal to have the information about the distribution of the testing samples before it classifies them. Thus in this paper, to deal with the problem of learning from training sets of different sizes, we exploited the information derived from clusters of the term frequency vectors of documents.

Clustering has been used in the literature of text classification as an alternative representation scheme for text documents. Several approaches of clustering have been proposed. Given a classification problem, the training and testing documents are both clustered before the classification step. Further, these clusters are used to exploit the association between index terms and documents [17]. In [18], words are clustered into groups based on distribution of class labels associated with each word. Information bottle neck method is used to find a word cluster that preserves the information about the categories. These clusters are used to represent the documents in a lower dimensional feature space and naïve bayes classifiers are applied [19]. Also, in [20] information bottleneck is used to generate a document representation in a word cluster space instead of word space, where words are viewed as distributions over document categories. [20] proposed an information theoretic divisive algorithm for word clustering and applied it to text classification. Classification is done using word clusters instead of simple words for document representation. Two dimensional clustering algorithms are used to classify text documents in [21]. In this method, words/terms are clustered in order to avoid the data sparseness problem. In [22] clustering algorithm is applied on labeled and unlabeled data, and introduces new features extracted from those clusters to the patterns in the labeled and unlabeled data. The clustering based text classification approach in [23] first clusters the labeled and unlabeled data. Some of the unlabeled data are then labeled based on the clusters obtained. Feature clustering is a powerful method to reduce the dimensionality of feature vector of text classification [24]. The words in the feature vector of document set are grouped into clusters based on similarity test. Words that are similar to each other are grouped into the same cluster.

All in all, the above mentioned clustering based classification algorithms work on conventional word frequency vector. Conventionally the feature vectors of term document matrix (very sparse and very high dimensional feature vector describing a document) are used to represent the class. Later, this matrix is used to train the system using different classifiers for classification. Generally, the term document matrix contains the frequency of occurrences of terms and the values of the term frequency vary from document to document in the same class. Hence to preserve these variations, we propose a new interval representation for each document. An initial attempt is made in [3] by giving an interval representation by using maximum and minimum values of term frequency vectors for the documents. However, in this paper we are using mean and standard deviations to give the interval valued representation for documents. Thus, the variations of term frequencies of document within the class are assimilated in the form of interval representation. Moreover conventional data analysis may not be able to preserve intraclass variations but unconventional data analysis such as symbolic data analysis will provide methods for effective representations by preserving intraclass variations. The recent developments in the area of symbolic data analysis have proven that the real life objects can be better described by the use of symbolic data, which are extensions of classical crisp data. The aim of the symbolic data analysis is to provide suitable methods for managing aggregated data described by multi valued variables, where the cells of the data contain sets of categories, intervals, or weight distributions. Symbolic data analysis provides a number of clustering methods for symbolic data. These methods differ in the type of considered symbolic data, in their cluster structures and/or in the considered clustering criterion [27], [28]. The previous issues motivated us to use symbolic data rather than using a conventional classical crisp data to represent a document. To preserve the intraclass variations we create multiple clusters for each class. Term frequency vectors of documents of each cluster are used to form an interval valued feature vector. With this backdrop, the work presented in [3] is extended towards creating multiple representatives per class using clustering after symbolic representation.

The rest of the paper is organized as follows: A detailed literature survey and the limitations of the existing models are presented in section 1. The working principle of the proposed method is presented in section 2. Details of dataset used, experimental settings and results are presented in section 3. Along with the future work, the paper is concluded in section 4.

2   Proposed Method

The proposed method has 2 stages: 1) Representation stage and 2) Classification stage.

2.1   Representation Stage

In the proposed method, initially documents are represented by term document matrix. The obtained term document matrix is of very high dimension and huge sparse matrix. To eliminate the dimensionality problem we employed feature clustering method over the term document matrix. The main idea of feature clustering is to group the original features into cluster with a high degree of pairwise semantic relatedness [24]. To employ feature clustering technique, we should reduce the higher dimensional matrix D to its lower dimensional matrix D'. Such reduction is possible by employing effective feature selection approaches. Let W be the wordset $W = \{w_1, w_2, ..., w_m\}$ represents the feature vector of the document set. The feature reduction is to reduce a wordset W to a new wordset $W' = \{w_1', w_2', ..., w_k'\}$, $k \ll m$, such that W and W' equally obeys the properties of D. In this work, we used Information Gain [25] to measure the weight of a word W_j. The weight of a word W_j can be calculated as

$$Information\ Gain\ IG\left(w_j\right) = -\sum_{l=1}^{p} P(c_l) \log P(c_l)$$
$$+ P\left(w_j\right) \sum_{l=1}^{p} P\left(c_l \middle| w_j\right) \log P\left(c_l \middle| w_j\right) \tag{1}$$
$$+ P\left(\bar{w}_j\right) \sum_{l=1}^{p} P\left(c_l \middle| \bar{w}_j\right) \log P\left(c_l \middle| \bar{w}_j\right)$$

Where, $P(c_l)$ denotes prori probability for class c_l, $P(w_j)$ denotes prori probability for feature w_j, $P\left(\bar{w}_j\right)$ is identical to $1 - P(w_j)$, and $P(c_l \mid w_j)$ and $P\left(c_l \mid \bar{w}_j\right)$ denotes the probability for class c_l with the presence or absence, of w_j, respectively. The words of top k weights in w are selected as the features in w'.

Further, it is recommended to employ feature clustering methods over the wordset w'. Unfortunately, the existing clustering methods poses number of challenges like: 1. Deciding the parameter k, 2. Calculating the similarities when the variances among the clusters are not considered, and 3. All words in a cluster have the same degree of contribution to the resulting extracted features [24]. To overcome these limitations we have used feature clustering technique proposed by [24]. Let G be a cluster containing q word patterns $x_1, x_2, ..., x_q$. Let $X_j = \langle x_{j1}, x_{j2}, ..., x_{jp} \rangle$, $1 \le j \le q$. Then the mean $\mu = \langle \mu_1, \mu_{2,...,}\mu_p \rangle$ and the deviation $\sigma = \langle \sigma_1, \sigma_2, ..., \sigma_p \rangle$ of G are defined as:

$$\mu_i = \frac{\sum_{j=1}^{q} x_{ji}}{|G|} \qquad (2)$$

$$\sigma_i = \sqrt{\frac{\sum_{j=1}^{q} \left(x_{ji} - \mu_{ji}\right)^2}{|G|}} \qquad (3)$$

For $1 \le i \le p$, where $|G|$ denotes the size of G, i.e., the number of word patterns contained in G. The fuzzy similarity of a word pattern $X = \langle x_1, x_2, ..., x_p \rangle$ to cluster G is defined by

$$\mu_G(X) = \prod_{i=1}^{p} \exp\left[-\left(\frac{x_i - \mu_i}{\sigma_i}\right)^2\right] \qquad (4)$$

Where, $0 \le \mu_G(X) \le 1$.

$$if \begin{cases} \mu_G(X) \approx 1; word\ pattern\ is\ similar\ to\ cluster \\ \mu_G(X) \approx 0; word\ pattern\ is\ far\ from\ the\ cluster \end{cases}$$

Now, we recommend capturing intraclass variations in each ith feature values of the gth cluster in the form of an interval valued features $\left[f_{g_i}^-, f_{g_i}^+\right]$ where $f_{g_i}^- = \mu_{gi} - \sigma_{gi}$; $f_{g_i}^+ = \mu_{gi} + \sigma_{gi}$. Thus, each interval $\left[f_{g_i}^-, f_{g_i}^+\right]$ representation depends up on mean and its standard deviation of the respective individual features if a cluster. The interval $\left[f_{g_i}^-, f_{g_i}^+\right]$ represents the upper and lower limits of the feature value of a document cluster in the knowledge base. Now, the reference document for a cluster G_g, is formed by representing each feature in the form of an interval and it is given by

$RF_g = \left[f_{g_1}^-, f_{g_1}^+\right], \left[f_{g_2}^-, f_{g_2}^+\right], ..., \left[f_{g_i}^-, f_{g_i}^+\right]$, where $g = 1, 2, ..., G$ represents the number of clusters of documents samples of a class. Is shall be noticed that unlike conventional feature vector, this is a interval valued features and this symbolic feature vector is stored in the knowledge base as a representative of the ith cluster.

2.2 Document Classification

The document classification proposed in this work considers a test document, which is described by a set of w feature values of type crisp and compares it with the corresponding interval type feature values of the respective cluster stored in the knowledge base. Let, $F_i = [f_{i1}, f_{i2}, ..., f_{iw}]$ be a w dimensional feature vector describing a test document.

Let RF_g be the interval valued symbolic feature vector of gth cluster of lth class. Now, each wth feature value of the test document is compared with the corresponding interval in RF_j to examine whether the feature value of the test document lies within the corresponding interval. The number of features of a test document, which fall inside the corresponding interval, is defined to be the degree of belongingness. We make use of Belongingness Count B_c as a measure of degree of belongingness for the test document to decide its class label.

$$B_c = \sum_{k=1}^{w} C\left(f_{ik}, \left[f_{jk}^{-}, f_{jk}^{+}\right]\right) \text{ and} \tag{5}$$

$$C\left(f_{ik}, \left[f_{jk}^{-}, f_{jk}^{+}\right]\right) = \begin{cases} 1; & if \left(f_{ik} \geq f_{jk}^{-} \text{ and } f_{ik} \leq f_{jk}^{+}\right) \\ 0; & Otherwise \end{cases}$$

The crisp value of a test document falling into its respective feature interval of the reference class contributes a value 1 towards B_c and there will be no contribution from other features which fall outside the interval. Similarly, we compute the B_c value for all clusters of remaining classes and the class label of the cluster which has highest B_c value will be assigned to the test document as its label. Further, there may be chances of having the same B_c value for two or more classes. Under such circumstances we recommend to resolve the conflict by making use of the similarity measure proposed by [26], to measure the similarity between the test feature vector R_q and the gth class representative R_g as given by,

$$Total\_Sim(R_q, R_g) = \sum_{l=1}^{m} Sim(f_{ql}, [f_{gl}^{-}, f_{gl}^{+}]) \tag{6}$$

Here, $[f_{gl}^{-}, f_{gl}^{+}]$ represents the lth feature interval of the gth class, and

$$Sim(f_{ql}, [f_{jl}^{-}, f_{jl}^{+}]) = \begin{cases} 1 & if \left(f_{ql} \geq f_{jl}^{-} \text{ and } f_{ql} \leq f_{jl}^{+}\right) \\ \max\left(\dfrac{1}{1+\left|f_{ql} - f_{jl}^{-}\right|}, \dfrac{1}{1+\left|f_{ql} - f_{jl}^{+}\right|}\right) & Otherwise \end{cases} \tag{7}$$

Similarity values of the test document with all k classes are computed and then the test document is classified to be a member of the class for which it has a highest similarity value.

3 Experimental Setup

3.1 Datasets

To test the efficacy of the proposed model, we have used the following datasets.

For experimentation we have used the classic Reuters 21578 collection has the benchmark dataset. Originally Reuters 21578 contains 21578 documents in 135 categories. However, in our experiment, we discarded those documents with multiple category labels, and selected the largest ten categories. For the smooth conduction of experiments we used ten largest classes in the Reuters 21578 collection with number of documents in the training and test sets as follows: earn (2877 vs 1087), trade (369 vs 119), acquisitions (1650 vs 179), interest (347 vs 131), money-fx (538 vs 179), ship (197 vs 89), grain (433 vs 149), wheat (212 vs 71), crude (389 vs 189), corn (182 vs 56). The second dataset is standard 20 Newsgroup Large. It is one of the standard benchmark dataset used by many text classification research groups. It contains 20000 documents categorized into 20 classes. For our experimentation, we have considered the term document matrix constructed for 20 Newsgroup [16]. The third data set contains WWW-pages collected from computer science departments of various universities in January 1997 by the World Wide Knowledge Base (Web-Kb) project of the CMU text learning group. The 8,282 pages were manually classified into the following categories: student (1641), faculty (1124), staff (137), department (182), course (930), project (504), other (3764). The fourth dataset [11] consists of vehicle characteristics extracted from wikipedia pages (Vehicles - Wikipedia). The dataset contains 4 categories that have low degrees of similarity. The dataset contains four categories of vehicles: Aircraft, Boats, Cars and Trains. All the four categories are easily differentiated and every category has a set of unique key words.

3.2 Experimentation

In this section, we present the results of the experiments conducted to demonstrate the effectiveness of the proposed method on all the four datasets viz., Reuters 21578, 20 Newsgroup, 4 University Dataset and Vehicles Wikipedia. During experimentation, we conducted two different sets of experiments. In the first set of experiments, we used 50% of the documents of each class of a dataset to create training set and the remaining 50% of the documents for testing purpose. On the other hand, in the second set of experiments, the numbers of training and testing documents are in the ratio 60:40. Both experiments are repeated 5 times by choosing the training samples randomly. As measures of goodness of the proposed method, we computed percentage of classification accuracy. The minimum, maximum and the average value of the classification accuracy of all the 5 trials are presented in Table 1 to Table 4.

For both the experiments, we have randomly selected the training documents to create the symbolic feature vectors for each class. To do the task of classification, we have used four well known classifier viz., Naïve Bayes, k-NN, Centroid classifier and Support Vector Machine (SVM). Further, we have also used symbolic classifier [3] to classify the text documents. From Table 1 to Table 4 it is analyzed that the proposed method achieves better classification accuracy.

Table 1 Classification Accuracy of the proposed model for dataset: Reuters 21578

Dataset	Classifier Name	Training Vs Testing	Minimum Accuracy	Maximum Accuracy	Average Accuracy
Reuters 21578	Naïve Bayes	50 vs 50	76.50	78.25	77.40
		60 vs 40	81.00	83.45	82.95
	K-NN	50 vs 50	77.10	79.50	78.65
		60 vs 40	86.45	87.00	86.80
	Centroid Based	50 vs 50	84.50	86.00	85.75
		60 vs 40	86.70	88.50	87.35
	SVM	50 vs 50	84.50	85.50	84.70
		60 vs 40	88.10	89.90	88.85
	Symbolic Classifer	50 vs 50	86.50	87.45	87.10
		60 vs 40	91.50	92.35	92.10

Table 2 Classification Accuracy of the proposed model for dataset: 20 Newsgroup

Dataset	Classifier Name	Training Vs Testing	Minimum Accuracy	Maximum Accuracy	Average Accuracy
20Newsgroup	Naïve Bayes	50 vs 50	84.35	86.90	85.40
		60 vs 40	87.60	88.40	87.95
	K-NN	50 vs 50	84.20	85.15	84.60
		60 vs 40	85.50	86.90	86.20
	Centroid Based	50 vs 50	83.40	85.25	84.60
		60 vs 40	85.20	87.10	86.40
	SVM	50 vs 50	86.70	87.70	86.90
		60 vs 40	89.10	90.10	89.40
	Symbolic Classifer	50 vs 50	87.40	89.60	88.95
		60 vs 40	91.10	92.65	91.95

Table 3 Classification Accuracy of the proposed model for dataset: 4 University Dataset

Dataset	Classifier Name	Training Vs Testing	Minimum Accuracy	Maximum Accuracy	Average Accuracy
4 University Dataset	Naïve Bayes	50 vs 50	71.50	73.25	72.85
		60 vs 40	76.40	77.90	76.95
	K-NN	50 vs 50	75.40	76.10	75.60
		60 vs 40	77.80	78.60	78.15
	Centroid Based	50 vs 50	72.40	74.35	73.45
		60 vs 40	78.90	80.10	79.10
	SVM	50 vs 50	84.35	85.40	84.90
		60 vs 40	86.70	88.90	87.45
	Symbolic Classifer	50 vs 50	87.40	88.90	88.15
		60 vs 40	89.90	91.10	90.60

Table 4 Classification Accuracy of the proposed model for dataset: Vehicles Wikipedia

Dataset	Classifier Name	Training Vs Testing	Minimum Accuracy	Maximum Accuracy	Average Accuracy
Vehicles Wikipedia	Naïve Bayes	50 vs 50	84.30	86.50	85.90
		60 vs 40	86.70	88.90	87.60
	K-NN	50 vs 50	82.40	83.65	82.90
		60 vs 40	83.50	84.50	84.10
	Centroid Based	50 vs 50	86.70	87.20	86.95
		60 vs 40	89.90	91.20	90.55
	SVM	50 vs 50	88.45	90.60	89.95
		60 vs 40	90.40	92.10	91.85
	Symbolic Classifer	50 vs 50	90.05	93.45	92.15
		60 vs 40	91.25	94.50	93.60

4   Conclusions

The paper presents a new method of representing the text documents based on feature clustering. The proposed method uses the symbolic interval valued representation for text documents. The enhancement in terms of classification accuracy and time complexity can be achieved to the greater extent. To corroborate the efficacy of the proposed method, we conduct extensive experimentation on standard datasets like 20 Newsgroup, Reuters 21578, 4 University Dataset and vehicles Wikipedia. The experimental results reveal that the proposed method outperforms the other existing methods.

In the future, our research will emphasize in enhancing the ability and performance of our model by considering other parameters to effectively capture the variations between the classes, which in turn improves the classification accuracy. Along with this, we have a plan of exploiting other similarity/dissimilarity measures, selection of dynamic threshold value and studying the classification accuracy for the varying dimensions. Besides this we are also targeting towards the study of complexity issues of the proposed model with the existing representation models.

References

[1]   Seabastiani, F.: Machine Learning in Automated Text Categorization. ACM Computing Surveys 34, 1–47 (2002)
[2]   Jiang, S., Pang, G., Wu, M., Kuang, L.: An improved K-nearest-neighbor algorithm for text categorization. Journal of Expert Systems with Applications 39, 1503–1509 (2012)
[3]   Guru, D.S., Harish, B.S., Manjunath, S.: Symbolic representation of text documents. In: Proceedings of Third Annual ACM Compute, Bangalore (2010)
[4]   Li, Y.H., Jain, A.K.: Classification of Text Documents. The Computer Journal 41, 537–546 (1998)
[5]   Hotho, A., Nürnberger, A., Paaß, G.: A Brief Survey of Text Mining. Journal for Computational Linguistics and Language Technology 20, 19–62 (2005)
[6]   Cavnar, W.B.: Using an N-Gram based document representation with a vector processing retrieval model. In: Third Text Retrieval Conference (TREC-3), pp. 269–278 (1994)
[7]   Milios, E., Zhang, Y., He, B., Dong, L.: Automatic term extraction and document similarity in special text corpora. In: Sixth Conference of the Pacific Association for Computational Linguistics (PACLing 2003), Canada, pp. 275–284 (2003)
[8]   Choudhary, B., Bhattacharyya, P.: Text clustering using Universal Networking Language representation. In: Proceedings of Eleventh International World Wide Web Conference (2002)
[9]   Craven, M., DiPasquo, D., Freitag, D., McCallum, A., Mitchell, T.M., Nigam, K., Slattery, S.: Learning to Extract Symbolic Knowledge from the World Wide Web. In: Proceedings of AAAI/IAAI, pp. 509–516 (1998)
[10] Esteban, M., Rodrıguez, O.R.: A Symbolic Representation for Distributed Web Document Clustering. In: Proceedings of Fourth Latin American Web Congress, Cholula, Mexico (2006)
[11] Isa, D., Lee, L.H., Kallimani, V.P., Rajkumar, R.: Text document preprocessing with the Bayes formula for classification using the support vector machine. IEEE Transactions on Knowledge and Data Engineering 20, 23–31 (2008)
[12] Wan, C.H., Lee, L.H., Rajkumar, R., Isa, D.: A hybrid text classification approach with low dependency on parameter by integrating K-nearest neighbor and support vector machine. Journal of American Society of Information Science 41(16), 391–407 (1990)
[13] Salton, G., Wang, A., Yang, C.S.: A Vector Space Model for Automatic Indexing. Communications of the ACM 18, 613–620 (1975)

[14] Deerwester, S.C., Dumais, S.T., Landauer, T.K., Furnas, G.W., Harshman, R.A.: Indexing by Latent Semantic Analysis. Journal of the Expert Systems with Applications 39(15), 11880–11888 (2012)

[15] He, X., Cai, D., Liu, H., Ma, W.Y.: Locality Preserving Indexing for document representation. In: Proceedings of International Conference on Research and Development I Information Retrieval (SIGIR 2004), UK, pp. 96–103 (2004)

[16] Cai, D., He, X., Zhang, W.V., Han, J.: Regularized Locality Preserving Indexing via Spectral Regression. In: Proceedings of Conference on Information and Knowledge Management (CIKM 2007), pp. 741–750 (2007)

[17] Kyriakopoulou, A., Kalamboukis, T.: Text classification using clustering. In: Proceedings of ECML-PKDD Discovery Challenge Workshop (2006)

[18] Pereira, F., Tishby, N., Lee, L.: Distributional clustering of English words. In: Proceedings of the 31st Annual Meeting of the Association for Computational Linguistics, pp. 183–190 (1993)

[19] Slonim, N., Tishby: The power of word clustering for text classification. In: Proceedings of the European Colloquium on IR Research, ECIR 2001 (2001)

[20] Dhillon, I., Mallela, S., Kumar, R.: Enhanced word clustering for hierarchical text classification. In: Proceedings of the 8th ACM SIGKDD International Conference on Knowledge Discovery and Data Mining, Canada, pp. 191–200 (2002)

[21] Takamura, H., Matsumoto, Y.: Two-dimensional clustering for text categorization. In: Proceedings of the Sixth Conference on Natural Language Learning (CoNLL 2002), Taiwan, pp. 29–35 (2002)

[22] Raskutti, B., Ferr, H., Kowalczyk, A.: Using unlabeled data for text classification through addition of cluster parameters. In: Proceedings of the 19th International Conference on Machine Learning ICML, Australia, pp. 514–521 (2002)

[23] Zeng, H.J., Wang, X.H., Chen, Z., Lu, H., Ma, W.Y.: CBC: Clustering based text classification requiring minimal labeled data. In: Proceedings of the 3rd IEEE International Conference on Data Mining, USA, pp. 443–450 (2003)

[24] Jiang, J.Y., Liou, R.J., Lee, S.J.: A Fuzzy Self-Constructing Feature Clustering Algorithm for Text Classification. IEEE Transactions on Knowledge and Data Engineering 23, 335–349 (2011)

[25] Yang, Y., Pedersen, J.P.: A Comparative Study on Feature Selection in Text Categorization. In: Proceedings of the Fourteenth International Conference on Machine Learning, pp. 412–420 (1997)

[26] Guru, D.S., Nagendraswamy, H.S.: Symbolic Representation of Two-Dimensional Shapes. Pattern Recognition Letters 28, 144–155 (2006)

[27] Bock, H.H., Diday, E.: Analysis of symbolic Data. Springer (1999)

[28] Billard, L., Diday, E.: From the statistics of data to the statistics of knowledge: Symbolic data analysis. J. American Statistics Association 98(462), 470–487 (2003)

Classification Approach Based on Rough Mereology

Mahmood A. Mahmood, Nashwa El-Bendary, Aboul Ella Hassanien, and Hesham A. Hefny

Abstract. This article presents a classification approach based on granular computing combined with rough set. The proposed classification approach used the theory of rough mereology and fuzzification in order to classify input datasets into sets of optimized granules. The proposed approach was applied to five datasets of the UC Irvine Machine Learning Repository. The Abalone dataset that consists of 4177 objects and eight attributes was selected as an illustrative example. Empirically obtained experimental results demonstrated that the proposed rough mereology based classification approach obtained better performance compared to other experienced proposed classification approaches.

Mahmood A. Mahmood
ISSR, Computer Sciences and Information Dept., Cairo University, Cairo - Egypt,
Scientific Research Group in Egypt (SRGE)
e-mail: mahmood.moneim@egyptscience.net
http://www.egyptscience.net

Nashwa El-Bendary
Arab Academy for Science,Technology, and Maritime Transport, Cairo - Egypt,
Scientific Research Group in Egypt (SRGE)
e-mail: nashwa.elbendary@ieee.org
http://www.egyptscience.net

Aboul Ella Hassanien
Information Technology Dept., Faculty of Computers and Information, Cairo University,
Cairo - Egypt,
Scientific Research Group in Egypt (SRGE)
e-mail: aboitcairo@gmail.com
http://www.egyptscience.net

Hesham A. Hefny
ISSR, Computer Sciences and Information Dept., Cairo University, Cairo - Egypt
e-mail: h.hefny@ieee.org

S.M. Thampi et al. (eds.), *Recent Advances in Intelligent Informatics,*
Advances in Intelligent Systems and Computing 235,
DOI: 10.1007/978-3-319-01778-5_18, © Springer International Publishing Switzerland 2014

1 Introduction

Granular computing is a methodology that emulates human brain for solving different problems in the real world. One main classifier that is defined under granular computing is the theory of rough mereology, which is a paradigm for reasoning under uncertainty with primitive notion as a part to a degree. Complexity time of indiscernibility of rough mereology is non-linear, so rough inclusion technique is an efficient way to compute the indiscernibility of rough mereology via linear complex time. Some recent research work that applied granular computing, rule generation, and rough mereology approaches are done by T.Y.Lin et al. in [1] and Hong Zhen Zheng et al. in [2] presented bitmap based association rule algorithms using granular computing technique. Authors in both researches indicated that bitmap and granular computing techniques can greatly enhance the performance for finding association rule. Moreover, Denise and Jean in [3] proposed an extension for the conventional upper approximation, termed $upper^{\alpha}$ approximation, that restricts the elements of the region of transition between a given class and its complement with respect to an $\alpha - cut$. Authors used that new concept for utilizing fuzzy rules to compose a fuzzy classifier. Furthermore, Jiujiang et al. in [4] presented a rule generation algorithm based on granular computing (RGAGC) that generates rule from the granule space without considering the selection of attributes like many classic decision tree methods. Hence, RGAGC approach used uncertainty support method to build granules, while the previously presented approaches used the ID3 (Iterative Dichotomiser 3) algorithm to divide the information into granules.

This article presents a classification approach based on granular computing combined with rough set. The proposed classification approach used the theory of rough mereology and fuzzification in order to classify input datasets into sets of generated rules in three phases; namely pre-processing, clustering, and voting by objects. The proposed approach was applied to five datasets of the UC Irvine Machine Learning Repository [5]. The Abalone dataset that consists of 4177 objects and eight attributes was selected as an illustrative example. The rest of this article is organized as follows. Section 2 presents an overview about rough mereology approach. Section 3 describes the proposed rough mereology based classification approach. Section 4 introduces and discusses experimental results. Finally, section 5 addresses conclusions and discusses future work.

2 Rough Mereology: An Overview

Rough mereology can be classified according to the measurement of similarity. The similarity functions [6] that satisfy certain similarity properties, namely Monotonic (MON), Identity (ID), Extreme, or proportionality (EXT), can be defined as follows:

- (MON) if similarity (x, y, 1) then for each z, where $d(z,x) > d(z,y)$, from similarity (z, x, r) it follows that similarity (z, y, r).
- (ID) similarity (x, x, 1) for each x.
- (EXT) if similarity (x, y, r) and $s \leq r$ then similarity (x, y, s).

Rough mereology proposed by Lesniewski in [7] as the theory of concept, where the relation of mereology is a part of relation, e.g. x mereology y means x is a part of y, according to Polkowski [8], the mereology relation described in equation (1), where $\pi(u,w)$ is a partial relation (proper part) and $ing(u,w)$ is ingredient relation means an improper part.

$$ing(u,w) \Leftrightarrow \pi(u,w) \; or \; u = w \tag{1}$$

$\mu(x,y,r)$ means rough mereology relation x is part of y at least degree r, also described as shown in equation (2).

$$\mu(x,y,r) = sim_\delta(x,y,r) \Leftrightarrow \rho(x,y) \le (1 - -r) \tag{2}$$

Computing the indiscernibility relation to get the object can be achieved by using rough inclusion, which is of less complexity time than the indiscernibility relation computed by rough set technique. Rough inclusion from metric, according to Polkowski in [8], can be computed by the Euclidean metric space or Manhattan space, where

$$\mu_h(x,y,r) \Leftrightarrow \rho(x,y) \le 1 - -r$$

Rough inclusion technique satisfies the similarity properties. Datasets are formalized as decision systems of the form of triple (U, A, d) or information system of the form (U, A), where U is a finite set of objects, A is a finite set of attributes, each attribute $a \in A$ described as mapping $a : U \to V_a$ of objects in U into the value set of a, and $d \notin A$ is the decision. The indiscernibility relation Ind can be computed as shown in equation (3).

$$Ind(x,y) = \frac{|IND(x,y)|}{|A|} \tag{3}$$

Then equation (2) becomes:

$$\mu_h(x,y,r) \Leftrightarrow Ind(x,y) \ge r \tag{4}$$

and

$$IND(x,y) = a \in A : a(x) = a(y) \tag{5}$$

where a is an attribute(s) in an information system A, $a(x)$ is the value of tuple x in attribute a, $a(y)$ is the value of tuple y in attribute a, and $|A|$ is the cardinality of a set A.

3 Classification Approach Based on Rough Mereology

The proposed classification approach applied Rough mereology and rough inclusion approaches in order to classify input datasets into sets of generated rules in three phases; namely pre-processing, clustering, and voting by objects phases, as shown in figure 1.

1. **Pre-processing phase:** maps the experienced input dataset into a normalized dataset of the range [0,1], where the values 0 and 1 represent the smallest the largest values in each dataset's attribute, respectively.
2. **Clustering phase:** uses theory of rough mereology and rough inclusion technique to classify the resulted dataset into sets of granules with different radius and produce rough inclusion table that reflects similarity degree among parameters.
3. **Voting by objects phase:** applies voting by training objects algorithm to select the optimized granules of dataset and to produce the optimal similarity measurement.

Algorithm 3 shows the detailed steps for the proposed rough mereology based classification approach.

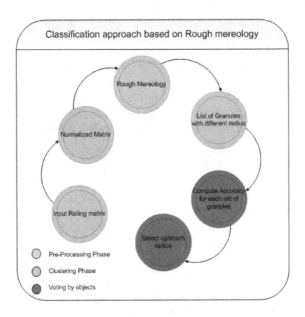

Fig. 1 Phases of the proposed classification approach

To describe the proposed rough mereology based classification approach, we considered the Abalone dataset in UCI [5] that consists of 4177 objects as an illustrative example. Table 1, clarifies a sample part of the Abalone dataset, which consists of eight attributes: Sex (S), Length (L), Diameter (D), Height (H), Whole weight (W), Shucked weight (SW), Viscera weight (V), and Shell weight (SH).

Algorithm 1. Classification Approach based on Rough Mereology

---

1: Input: An information System table (*IST*)
2: Output: Rule granule set (*RGS*)
3: Initialize set of Rule sets *SRS* = φ
4: **for** Each radius from 0 to 1 step 0.1 **do**
5:    Initialize Rough Inclusion set *RIS* = φ
6:    **for** Each column in (*IST*) **do**
7:       Compute rough inclusion and put the values in *RIS*
8:    **end for**
9:    Initialize Rule generation set *RGS* = φ
10:   **for** Each row in *RIS* **do**
11:      Get the rule according to definition of rough inclusion, and put the rule in *RGS*
12:      Remove duplicates rules from *RGS*
13:      Output rule generation set *RGS*
14:   **end for**
15:   Store *RGS* in *SRS*
16: **end for**
17: Initialize *ACCRateset* = φ
18: **for** Each rules in *SRS* **do**
19:    compute Accuracy measure rate and put in *ACCRateset* as shown in equation (7)
20: **end for**
21: Initialize *bestACC* as the first element in set *ACCRateset*
22: **for** Each element in *ACCRateset* **do**
23:    **if** element ¿ *bestACC* **then**
24:       *bestACCset*=element
25:    **end if**
26: **end for**
27: **for** Each element in *ACCRateset* **do**
28:    **if** element = *bestACC* **then**
29:       Store element in *bestACCset*
30:    **end if**
31: **end for**
32: get the corresponding set of rules of *bestACCset* and put in *RGS*
33: apply voting by object equations as shown in to refine *RGS*
34: Exit

Table 1 Sample part of the Abalone Dataset

S	L	D	H	W	SW	V	SH
M	0.455	0.365	0.095	0.514	0.2245	0.101	0.15
M	0.35	0.265	0.09	0.2255	0.0995	0.0485	0.07
F	0.53	0.42	0.135	0.677	0.2565	0.1415	0.21
M	0.44	0.365	0.125	0.516	0.2155	0.114	0.155
I	0.33	0.255	0.08	0.205	0.0895	0.0395	0.055
I	0.425	0.3	0.095	0.3515	0.141	0.0775	0.12
F	0.53	0.415	0.15	0.7775	0.237	0.1415	0.33
F	0.545	0.425	0.125	0.768	0.294	0.1495	0.26
M	0.475	0.37	0.125	0.5095	0.2165	0.1125	0.165
F	0.55	0.44	0.15	0.8945	0.3145	0.151	0.32

3.1 Pre-processing Phase

Pre-processing phase takes dataset as a data input in the matrix form, called *rating matrix*, so that the proposed approach normalizes this matrix by finding the largest value and the smallest value in each attribute, which are called max_a and min_a, respectively. The produced new matrix, *Normalized Matrix*, contains values in the range $[0, 1]$ and is calculated as shown in equation (6). Table 2 shows the normalized matrix of Abalone Dataset presented in table 1.

$$NormalizedV_a = \frac{V_a - min_a}{max_a - min_a} \qquad (6)$$

Table 2 Normalized matrix of Abalone Dataset

L	D	H	W	SW	V	SH
0.80	0.82	0.18	0.52	0.43	0.43	0.41
0.81	0.81	0.16	0.55	0.47	0.53	0.40
0.81	0.77	0.15	0.54	0.54	0.42	0.34
0.81	0.80	0.17	0.53	0.42	0.49	0.42
0.82	0.83	0.17	0.67	0.60	0.54	0.49
0.82	0.81	0.16	0.51	0.43	0.33	0.46
0.85	0.83	0.19	0.51	0.44	0.43	0.46
0.85	0.80	0.15	0.55	0.41	0.52	0.44
0.86	0.84	0.16	0.76	0.84	0.49	0.43
0.88	0.85	0.16	0.63	0.59	0.48	0.43

3.2 Clustering Phase

Clustering phase received the normalized rating matrix and produced list of set of granules that describes rule generation according to different radius r in granules. This phase consists of two stages; namely 1) rough mereology and 2) list of granules. For the rough mereology stage, the normalized rating matrix is considered as an input and the equations of rough mereology with radius r, previously described in section 2, were applied to that matrix. The output of this stage is rough inclusion table that represents similarity measure of each object in each attribute of the dataset. In the list of granules stage, a set of rough inclusion tables is computed by re-applying the first stage of rough mereology ten times with different radius of r from 0.0 to 1.0 with step 0.1, as shown in table 3 and table 4, where table 3 describes the similarity measure at r_0 and table 4 describes the similarity measure at r_2.

Table 3 Rough Inclusion of Abalone Dataset with $r = r_0$

L	D	H	W	SW	V	SH
1.000	1.000	1.000	1.000	1.000	1.000	1.000
1.000	1.000	1.000	1.000	1.000	1.000	1.000
1.000	1.000	1.000	1.000	1.000	1.000	1.000
1.000	1.000	1.000	1.000	1.000	1.000	1.000
1.000	1.000	1.000	1.000	1.000	1.000	1.000
1.000	1.000	1.000	1.000	1.000	1.000	1.000
1.000	1.000	1.000	1.000	1.000	1.000	1.000

Table 4 Rough Inclusion of Abalone Dataset with $r = r_2$

L	D	H	W	SW	V	SH
1.000	0.997	0.022	0.677	0.576	0.577	0.585
0.997	1.000	0.025	0.680	0.579	0.580	0.588
0.022	0.025	1.000	0.345	0.446	0.445	0.436
0.677	0.680	0.345	1.000	0.896	0.896	0.902
0.576	0.579	0.446	0.896	1.000	0.915	0.900
0.577	0.580	0.445	0.896	0.915	1.000	0.914
0.585	0.588	0.436	0.902	0.900	0.914	1.000

3.3 *Voting by Objects*

Voting by objects phase takes as an input the list of similarity measure tables for each radius resulted from the clustering phase and computes the accuracy rate for each radius as shown in equation (7). Then, the optimum radius r_{opt} that represents the largest accuracy measure is selected. If more than one largest accuracy measure exist, then the proposed approach will generate set of largest accuracy measure. This phase is divided into two stages; namely 1) accuracy measure computation, which uses equation (7), where T is the number of similar tuples and N is the total number of tuples in each row, to compute the accuracy for each table representing the similarity measure at radius r and 2) optimization stage, where the table containing the largest accuracy measure at radius r is selected, so the radius r called the *optimum radius* r_{opt}.

$$Accuracyrate = (\frac{T}{N}) * 100 \tag{7}$$

Results of the illustrative example tested in this article, considering the Abalone database, showing that the accuracy measure of table 3 is 87.5 and the accuracy measure of table 4 is 86.21. So, it could be concluded that when $r = r_0$ is the largest accuracy measure, r_{opt} is r_0.

4 Experimental Results and Discussion

Simulation experiments have been conducted on 5 datasets; namely, Abalone Database (A), Car Evaluation database (CE), Nursery database (N), Pima Indian Diabetes database (PID), and Wisconsin Prognostic Breast database (WPE), shown in table 5 [5]. Table 6 represents the accuracy measure rate of all datasets with different values of r in interval [0,1], where r_0 and r_{10} represent the values of r = 0.0 and r =1.0, respectively, with step 0.1 for each increment of r_i, where i = 0, 1, ..., 10. The values of each tuple represents the accuracy measure for each radius r in different datasets. For example, the accuracy measure for Abalone database at r_5 was shown to be less than the accuracy measure at r_4. So, as presented in table 6, the optimum radius in that case is the radius when $r = r_0$.

Table 5 The Datasets

Dataset	Number of records	Number of attributes
Abalone Data Database (A)	4177	8
Car Evaluation Database (CE)	1782	6
Nursery Database (N)	12960	8
Pima Indian Diabetes Database (PID)	768	8
Wisconsin Prognostic Breast Database (WPB)	569	31

Table 6 Accuracy measure rate with $r[0,1]$

Radius / Dataset	A	CE	N	PID	WPB
r_0	87.50	85.71	88.89	88.89	90.00
r_1	87.48	49.26	55.56	40.80	26.09
r_2	86.21	51.34	55.56	39.28	20.59
r_3	58.11	78.15	51.85	41.08	20.59
r_4	47.90	74.79	52.38	43.31	21.24
r_5	32.47	83.42	52.38	50.09	43.95
r_6	48.01	76.50	52.13	53.55	90.00
r_7	59.00	0.00	13.79	66.71	90.00
r_8	0.00	0.00	0.00	60.37	90.00
r_9	0.00	0.00	0.00	0.00	90.00
r_{10}	0.00	0.00	0.00	0.00	0.00

From table 3 and 4, we found out that the data of both tables differs when we increase the radius of granular r_i, where r_i represents the radius of granules and the step of radius is $\frac{i}{10}$, where i = 0, 1, ..., 10. In the classification phase, we computed the inclusion, ten times as shown in table 6, where the radius changes from minimum value of each attribute to minimum value increased by step=0.1 for each iteration. So, we have ten inclusion tables with different granule's radius and has different data. The voting by objects phase is responsible to find the optimum radius, which gives more accurate rate for the generated rules. Initially, a part of the dataset is accustomed to generate rules with the proposed approach. Then, that accustomed part of the dataset tests the average accurate rate of rules. Simulation experimental results are shown in tables 6 that represents the average accurate measure rate for different values of radius and optimum radius and table 7 that represents a comparison between the proposed rough mereology based classification approach on the same dataset that experimented by researchers in different classifiers like ID3 and RGAGC. So, from table 7 we conclude that the proposed approach outperformed the accuracy measure obtained by other approaches.

Table 7 Accuracy measure rates for the proposed classification approach against RGAGC and ID3 classifiers

Dataset	Accuracy measure rate		
	The proposed approach	RGAGC	ID3
Abalone Data Database	87.50	33.7	6
Car Evaluation Database	85.71	66	63.2
Nursery Database	88.89	91.98	72.7
Pima Indian Diabetes Database	88.89	80.7	20.6
Wisconsin Prognostic Breast Database	90.00	0	0

5 Conclusions and Future Works

Granular Computing (GrC) aims to find a way to acquire knowledge for huge orderless very high dimensional perception information. Theory of rough mereology is a main classifier that is defined under granular computing. The primary contribution of the proposed classification approach, presented in this article, is that it returns rules from the granule space without considering the selection of attributes as though a lot of classic decision tree techniques and the "false preserving" property of quotient space theory is applied. Experimental results depicted that the proposed approach is valid for many aspects as it has been tested for 5 different datasets. We observed that the optimal classifier obtained with the general radius $r = min$ gave better accuracy than the classifier on training objects with r near to 1, which means that weighting heuristics slightly improves the quality of classification. Generally, experimental results using Irvine ML Repository datasets showed that the proposed rough mereology based classification approach outperformed both RGAGC and ID3 previously proposed algorithms. The performance measures used for the proposed rough mereology based classification approach are the same measures of the other approaches applied to the same datasets. For future work, we will consider to apply the proposed classification approach on datasets with text reviews for rule generation using opinion mining approach.

References

1. Lin, T.Y., Liau, C.-J.: Granular Computing and Rough Sets, IVth edn. Data Mining and Knowledge Discovery Handbook, pp. 535–561 (2005)
2. Zheng, H.Z., Chu, D.-H., Zhan, D.C.: Association Rule Algorithm Based on Bitmap and Granular Computing. Artificial Intelligence and Machine Learning (AIML) Journal 5(3), 51–54 (2005)
3. Guliato, D., de Sousa Santos, J.C.: Granular Computing and Rough Sets to Generate Fuzzy Rules. In: Kamel, M., Campilho, A. (eds.) ICIAR 2009. LNCS, vol. 5627, pp. 317–326. Springer, Heidelberg (2009)
4. An, J., Wang, G., Wu, Y., Gan, Q.: A Rule Generation Algorithm based on Granular Computing. In: Proceedings of the IEEE International Conference on Granular Computing, Beijing, China, vol. 1, pp. 102–107 (2005)

5. UCI ML Repository Datasets,
   http://www.ics.uci.edu/~mlearn/databases/
6. Veltkamp, R.C., Hagedoorn, M.: Shape Similarity Measures, Properties, and Constructions. In: Laurini, R. (ed.) VISUAL 2000. LNCS, vol. 1929, pp. 467–476. Springer,
   Heidelberg (2000)
7. Lesniewski, S.: On the foundations of set theory. Topoi 2, 7–52 (1982)
8. Polkowski, L., Artiemjew, P.: Granular Computing in the Frame of Rough Mereology. A
   Case Study: Classification of Data into Decision Categories by Means of Granular Reflec-
   tions of Data. International Journal of Intelligent Systems 26(6), 555–571 (2011)

Boosting Text Classification through Stemming of Composite Words

Marenglen Biba and Eva Gjati

Abstract. Text mining is a knowledge intensive process with the main purpose of effectively and efficiently processing large amounts of unstructured data. Due to the rapidly growing amount of raw text available there is a strong need for methods that are capable of dealing with this in terms of automatic classification or indexing. In this context, an essential task is the semantic processing of natural language in order to provide a sound input to the text classification or categorization task. One of the important tasks is stemming which is the process of reducing a certain word to its root (or stem). When a text is pre-processed for mining purposes, stemming is applied in order to bring words from their current variation to their original root in order to better process the natural language with subsequent steps. A challenging task is that of stemming composite words which in many languages form a large part of the daily used vocabulary. In this paper we develop a novel rule-based algorithm for stemming composite words and we show through extensive experiments that the text classification accuracy greatly improves by stemming composite words.

1 Introduction

The growing amount of information which comes in different forms and from different topic areas is becoming available almost everywhere: on the web, in a variety of social networks, on corporate or enterprise Intranets and other information centric applications. Due to the proliferation of information in the form of raw

Marenglen Biba
University of New York in Tirana, Albania
e-mail: marenglenbiba@unyt.edu.al

Eva Gjati
University of New York in Tirana and University of Greenwich, UK
e-mail: evagjati@unyt.edu.al

S.M. Thampi et al. (eds.), *Recent Advances in Intelligent Informatics*,
Advances in Intelligent Systems and Computing 235,
DOI: 10.1007/978-3-319-01778-5_19, © Springer International Publishing Switzerland 2014

text, which needs to be stored and processed, text mining has gained an increasing attention in the recent years. Unstructured data is the easiest form of data that can be found in any application scenario. As a result, there has been a huge need and demand to design algorithms and methods capable of effectively processing and mining textual data (Aggarwal and Zhai, 2012).

By employing techniques from Data Mining, Natural Language Processing, Machine Learning, Information Extraction (IE), Knowledge Management and Information Retrieval (IR), text mining attempts to solve the problem of information overload. This involves the discovery, extraction and pre-processing of large collections of documents, the storage of the intermediate representations and the techniques to analyze these intermediate representations (Feldman & Sanger, 2007). In this context, text mining largely exploits methodologies and techniques from corpus-based computational linguistics which is responsible for transforming raw, unstructured data into more carefully structured intermediate format.

One of the processing steps in the transformation of raw text into a form acceptable by data mining algorithms, is stemming which is a computational linguistic and normalization procedure that attempts to reduce different grammatical forms of a word with the same root to a common form, by removing the words affixes (prefixes and suffixes) (Lovins, 1968). What is very often observed is that most of the times the morphological variants of words have semantically similar terms (roots) which may differ in their affixes. For example, the words: *computer, computing, computation, computational* have all the same root *comput*. The idea is that before retrieving information from documents, stemming techniques are applied on the target data with the aim of reducing the data set and increasing the IR performance and effectiveness. Therefore the total number of distinct terms in a query or a document is reduced, which in turn will also reduce the time of processing of the final output.

Considerable effort has been dedicated through the years to stemmers for different languages such as: English, German, Italian, Swedish, etc, with the main purpose on text pre-processing by stripping away the word affixes to its root (stem) form. Unfortunately little research has been done for other languages and for the task of composite words. Albanian is a language that has not been much researched from the point of view of computational linguistics. The first attempt has only been made in (Sadiku, 2011) where the author introduced the first stemmer for Albanian and performed some experiments on text classification showing good performance of several classifiers on the stemmed documents. However, this work does not handle composite words which is a limitation due to the high number of this kind of words in Albanian.

In this paper we develop a stemming algorithm for handling composite words in stemming. The algorithm is based on a set of rules that are generated through a thorough analysis of the Albanian language morphology and structure. The algorithm is used in text classification tasks and we show that the classifier performs much better after composite words are stemmed, compared with the case of not using stemming. We believe that the results obtained can be generalized to other languages where the composite words are common.

The paper is organized as follows: Section 2 presents stemming as a natural language processing task and some of the most important algorithms for rule-based stemming. Section 3 presents some features of Albanian and Section 4 presents the developed algorithm. Section 4 presents the experiments and we conclude in Section 5 with some conclusions and directions for future work.

2   Stemming Algorithms

The first studies concerning stemming date back to the 1960s when it was proposed the first stemming algorithm (Lovins, 1968). Some stemming algorithms are known as rule based affix removal language dependent algorithms. In addition, other techniques known as statistical and mixed algorithms are alternative methods to stemming, developed to obviate the performance problems and language dependency difficulties of rule based stemmers.

The simplest stemming algorithm of this group was the Truncate (n) stemmer that truncated words at the n-th symbol, keeping unaffected words with n letters and removing the rest. When the length of the word is small the chances of over-stemming errors are increased. Although, this technique is not used in real stemming systems, it can serve for evaluating other algorithms (Paice, 1994). S-stemmer is another approach proposed in (Harman, 1991) which is able to conflate English nouns of plural and singular forms. This algorithm is applied only for words longer than three letters as per below set of rules.

2.1   The Lovin's Stemmer

The Lovin's stemmer was the first developed and well known stemming algorithm (Lovins, 1968). It is a context sensitive algorithm that removes endings, based on the principle of the longest match, and applies a number of contextual rules for preventing the removal of endings that can lead to incorrectly produced stems. This algorithm utilizes an extensive list of 294 endings each associated to one of the 29 contextual conditions of the algorithm which prevent removing the endings in certain circumstances. In addition it utilizes 35 transformation rules, designed to deal with the most frequent exceptions. When a word is presented for stemming an ending that meets a certain condition is found and removed. Next, an appropriate transformation rule is implemented in order to handle the word double consonants and irregular plurals. For example, applying the Lovin's stemmer to the word *nationally* there are two endings that match: *ationally* and *ionally*. The first ending is rejected based on the rule that stem must be at least 3 or more letters long, while the second ending is removed with no restriction producing the stem *nat*.

The advantage of the Lovin's stemmer is that it is very fast, able to handle the removal of double letters of words and the many irregular plurals. The disadvantage of this algorithm is the large amount of time and data, due to the number of special cases and rules that should be developed for each ending, which however will be able to handle only a small number of errors. Furthermore, the Lovin's stemmer often missed to reduce certain endings, due to the technical vocabulary used by the author.

2.2   The Dawson's Stemmer

The Dawson's algorithm (Dawson, 1974) can be considered as an improvement and extension of the Lovin's algorithm. Dawson used the same longest-match and single pass nature of Lovin's approach but covers an extensive list of approximately 1200 suffixes. In addition, it utilizes the partial matching technique that match stems that are equal within certain limits, instead of the recoding technique used by Lovin's stemmer that involves a number of transformations based on the letters of a stem.

The aim of this stemmer was to elaborate the rules of the original Lovin's stemmer and to improve any basic error. To achieve this, the first step was to include all plurals and simple suffix combinations that increase the ending list size to five hundred. The second step was to use the completion principle to complete any suffix within the ending list of all variants, flexions and combinations that increase the ending list to 1200 terms (Dawson, 1974). However, this extended list of suffixes presented two main problems: the storage limitation and the time required to test all suffixes with the list of endings. To obviate these problems the suffixes were stored in reversed order indexed by length and by last letter. Due to its complexity and deficiency of a standard reusable implementation the Dawson stemmer did not gain popularity.

2.3   The Porter's Stemmer

The Porter's stemmer, firstly proposed in the 1980s, is as of now one of the most popular and widely used suffix removal algorithms (Porter, 1980). Quickly adopted and extended, the Porter's algorithm became the word conflation standard approach for IR and a great inspiration for many later stemmers for English and a broad range of other languages. This stemmer was initially developed with the purpose of stemming texts in English language. Later on, the increased importance of IR systems in the 1990s resulted in an intense interest in conflation methods development that would improve text retrieval and provide a natural model for text processing in different languages.

Porter's algorithm employs a simple design of 60 suffixes, 2 rules responsible for recoding and a single context sensitive rule which decides if a suffix should be cut off or not (Willett, 2006). Implemented in five concise steps, this algorithm proved to be efficient in terms of computation complexity. Despite the problems of mis-stemming and over-stemming errors, which can potentially be reduced by using a comprehensive well structured dictionary, this algorithm continues to have good practical results and great utility in IR systems.

The lack of readily available stemming algorithms for languages other than English and the many incorrect implementations and misinterpretations of the original Porter's stemmer in different studies forced Porter to develop a rigorous framework for defining stemming algorithms, known as Snowball (Porter, 2001). Snowball is a rigorous system able to define stemming algorithms, where rules are expressed in natural way allowing programmers to develop their own stemmers for a chosen language. This framework includes an improved version of English stemmer together with a series of stemmers for other languages.

There are two main reasons why Porter's stemmer is considered important. First, the utilization of a simple technique to stemming that obviously gives good results in practice and can be implemented in many languages. Second, the tremendous interest of researchers in stemming as a separate research IR topic that this algorithm induced.

3   Albanian Language and Composite Word Formation

The Albanian language is part of the extensive Indo-European language family and thus related to a certain extent to almost all the other European languages. Nevertheless, studies have shown no close historical affinity of Albanian to any of the other language of the Indo-European family forming a distinct and unique language branch, which is the Albanian branch (Agalliu, et al., 2002).

From the morphological and structural point of view, Albanian words are considered as diversified and can be broken down in different groups and subgroups. From the point of view of the number of rooting morphemes, the words can be divided in two large groups:

- Simple words, which are created by a single rooting morpheme such as: *breg, bregore, drejt, drejtoj, i drejtë, punë, etc.*

- Composite words, which are created by two or more rooting morphemes *atdhe,* armë*pushim, dyvjeçar, juglindor, zemërmirë, etc.*

3.1   Composite Formation of Nouns

Composite words according to the syntactic relation between the parts are divided in two groups: composites with conjunctive relation and composites with subordinate relation (Agalliu, et al., 2002). Most composite nouns are those with subordinate relation where we distinguish the following types of composites:

- Noun + adverbial noun of the actor, formed with the *-(ë)s* suffix
- Noun + adverbial noun of the action, formed with *-je* or *-im* suffixes
- Noun + any other noun
- Pronoun or Numeral + Noun prefix *vetë-*
- Adjective + Noun
- Verb + Noun
- Noun + verb formed with suffix *–je*
- Noun + adverb formed with *bashkë-, drejt-, keq-, mirë- kundër-* prefixes
- Composites formed with nouns + agglutinations

3.2   Composite Formation of Adjectives

The composite adjectives are created concatenating two or sometimes three themes in a single word (Agalliu, et al., 2002). Like in composite nouns the composite adjectives consist of the same groups of composites:

    1.   Composites with conjunctive relation

This first group contains those adjectives which are formed without nodes and are equivalent from syntactic and semantic perspectives. The adjectives of this group neither define each other nor are dependent on each other. However they are mutually complementary to a certain extent to each-other.

    2.   Composites with subordinate relation

A special feature of this group is the determinant character of one of the composite parts. According to the word formation themes that form the adjectives of this group we distinguish the below subgroups:

- Composite adjectives formed with two nominal themes
- Composite adjectives formed with a noun + adjective
- Composite adjectives formed with two adjective themes
- Composite adjectives formed with an adverbial theme + noun
- Composite adjectives formed with a numeral + adjective
- Composite adjectives formed with a participle + adverb

4   Language Analysis and Algorithm Design

Almost all composite splitting algorithms developed for other languages use huge knowledge resources: like monolingual lexicons, monolingual or bilingual corpora and parallel corpora. These approaches rely on the availability of lexicons with compound parts using sometimes lazy learning and mostly language specific techniques.

Taking into consideration the pros and cons of rule based and dictionary based algorithms, as well as available resources for Albanian we decided to develop a rule based algorithm. The algorithm contains a number of rules each of which comprises one or more 'if' conditions depending on the various cases of composite formations. The set of rules has been conducted analyzing the linguistic and general characteristics of composite words. Each of the rules is executed one by one in a specified order. If the specified conditions are met the algorithm returns the stem of the word. For developing the algorithm we analyzed some groups of composites with common features and developed rules for each of them.

In the first group there are composites that contain a separating dash between the two word parts such as: *ekonomiko-shoqëror, politiko-ekonomik,* etc. These types of composites are very frequent in Albanian.

The second group of composites handled by the algorithm consists in composites formed with prefixes. In Albanian, this group of composites is very productive. A list of all possible prefixes has been conducted and integrated in a set of 60 rules. In addition, in this group of composites we distinguish composites formed with international prefixes such as: *hidro, bio, aero, biblio, deci, kilo*, etc. These composites are also very frequent in Albanian. A set of 29 such rules has been developed to handle these cases.

Composites formed with numerals are the third group of composites handled by the algorithm. A set of 10 rules has been developed considering the most frequent numeral composites.

The fourth group of composites considers a number of productive nouns which associated with adjectives or verbs form a large number of composites. Such nouns are: *jetë, gojë, ndihmës, zëvëndës, zemër*, etc. 59 rules have been developed for these cases.

The fifth group of composites considers some of the most frequent encountered second composite parts. This rule is created for handling composites that are randomly created and no common rule exists for their formation such as: *bukëpjekës, samarpunues, orëndreqës, gurskalitës*, etc. In addition, this ensures us that the algorithm will handle as many composites as possible.

The sequences of steps that the stemmer follows are illustrated in Fig. 1. All the other steps after splitting are the same as in the JStem algorithm of (Sadiku, 2011). The full set of the above rules coded in Java is presented in (Gjati, 2013).

5   Experimental Evaluation

In this section we will evaluate the performance of the designed algorithm in text mining tasks with documents in Albanian. The effectiveness of the stemming algorithm will be measured in terms of classification accuracy in a rich collection of documents. We will use different algorithms in order to prove that the performance improvement is not just due to a random coincidence with one algorithm, but can be generalized to the whole learning and text classification problem.

5.1   Experimental Setting

In order to measure the performance of the stemmer under the text mining perspective we need to perform text mining experiments. The experiments have been performed in Weka[1], an open source program that is widely used in academia and industry. Weka offers a number of tools for text pre-processing, classification, clustering, regression, etc and provides an integrated framework for machine learning and mining (Witten et al., 2011).

In our experiments we used a corpus developed in (Taullaj, 2012), a previous work done for comparing text mining algorithms for Albanian. The corpora consist of eight different areas of study (Biology, Chemistry, Culture, Curiosities, History, Economy, Literary and Sport) containing 40 documents each. Each document of the corpora is first stemmed by our algorithm and then converted by

---

[1] Weka can downloaded from http://www.cs.waikato.ac.nz/ml/weka/

Weka with the StringToWordVector filter that transforms text files into vectors that are stored in input files in .arff format. These files are then used as training and testing input for the text classification task. The data mining algorithms chosen in these tests are the Nearest Neighbour (IBK in Weka), Support Vector Machines and Naïve Bayes.

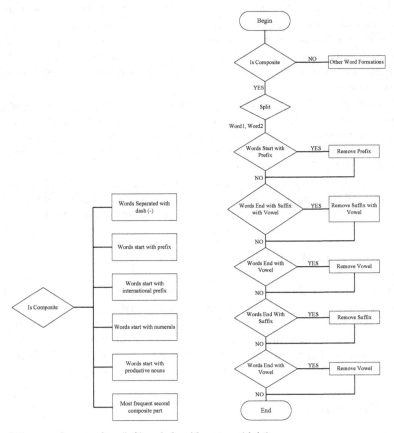

Fig. 1 Groups of composites (left) and algorithm steps (right)

5.2 Experimental Results

The experiments were performed between two or more datasets comparing the accuracy of the classifying algorithms in related and unrelated fields. We performed in total 10 experiments with stemmed and not-stemmed files in order to see if stemming will help in document classification. In each experiment, we used 10-fold cross-validation.

The experiments and the relative classes are as follows:

1   Biology and Chemistry
2   Biology and History

3	Chemistry and Literary
4	History and Literary
5	Sport and Culture
6	Biology, Chemistry and Economy
7	Biology, Chemistry, History and Literary
8	Culture, Curiosities, History and Literary
9	6 Classes: Biology, Chemistry, History, Literary, Economy and Culture
10	8 Classes: Biology, Chemistry, History, Literary, Economy, Culture, Curiosities, Sport

Table 1 Experimental results for the ten experiments

	Stemmed Documents			Terms	Not Stemmed Documents			Terms
Exp.	KNN	SVM	NB		KNN	SVM	NB	
1	85%	95%	96.25%	1897	57.5%	68.75%	78.75%	1638
2	90%	93.75%	91.25%	1744	88.75%	96.25%	91.25%	1770
3	96.25%	100%	98.75%	2004	80%	96.25%	95%	1811
4	93.75%	100%	96.25%	1804	87.50%	91.25%	88.75%	1604
5	55%	87.5%	91.25%	2587	55%	86.25%	91.25%	2579
6	81.66%	95%	97.50%	2569	44.16%	79.1667%	89.8333%	2344
7	75%	93.75%	91.87%	3185	55%	76.25%	76.875%	2791
8	73.12%	89.375%	91.87%	4179	55.62%	86.87%	88.75%	3964
9	69.58%	93.33%	94.58%	5201	51.66%	83.33%	84.58%	4835
10	52.50%	88.75%	91.25%	7117	39.37%	80.31%	83.43%	6739

As we can see from the Table 1, the classifier accuracy is significantly higher in the case of using stemmed documents. There is not even one case where the classifier performs better with the not stemmed documents. This is an indication that the stemming algorithm, including the composite word split, leads to significant improvements in the text classification task. In addition, the fact that all the algorithms show a growth of performance in the case of stemmed documents indicates that stemming of composite words is a task that helps the overall learning process and not just particular algorithms.

6 Conclusion

Text mining is a knowledge intensive process that requires techniques from natural language processing in order to provide an appropriate transformation of raw text into input for machine learning algorithms. In this paper we present a novel stemming algorithm that is able to split composite words and then stem these. We show through extensive experiments on text classification tasks with several machine learning algorithms that the classifiers perform significantly better in the case of stemmed documents compared with the not stemmed case. As future work we intend to develop machine learning models that can learn from corpora how to stem composite words. This however requires large labeled corpora in order to train the model.

References

1. Agalliu, F., Angoni, E., Demiraj, S., Dhrimo, A., Hysa, E., Lafe, E.: Gramatika e Gjuhes Shqipe, vol. 1. Akademia e Shkencave (2002) ISBN: 99927-761-6-1
2. Aggarwal, C.C., Zhai, C.X.: Mining Text Data. Springer, London (2012)
3. Dawson, J.: Suffix removal and word conflation. Bulletin of the Association for Literary and Linguistic Computing 2(3), 33–47 (1974)
4. Feldman, R., Sanger, J.: The Text Mining Handbook. Cambridge University Press, New York (2007)
5. Gjati, E.: Handling Composite Words in Stemming Albanian in a Text Classification Approach. Master Thesis. University of New York Tirana (2013)
6. Harman, D.: How effective is suffixing? Journal of the American Society for Information Science 42, 7–15 (1991)
7. Lovins, J.B.: Development of a stemming algorithm. Mechanical Translation and Computational Linguistics 1(1&2) (1968)
8. Paice, C.D.: Another stemmer. ACM SIGIR Forum 24(3), 56–61 (1990)
9. Paice, C.D.: An evaluation method for stemming algorithms. In: Annual International ACM SIGIR Conference on Research and Development in Information Retrieval, UK, vol. 17, pp. 42–50 (1994)
10. Porter, M.F.: An Algorithm for Suffix Stripping. Program Electronic Library and Information Systems 14(3), 130–137 (1980)
11. Porter, M.F.: Snowball: A language for stemming algorithms (October 2001), http://snowball.tartarus.org/texts/introduction.html (retrieved November 2012)
12. Porter, M.F.: Stemming algorithms for various European languages (2006), http://snowball.tartarus.org/texts/stemmersoverview.html (retrieved)
13. Sadiku, J.: A Novel Stemming Algorithm for Albanian text in a Data Mining Approach for Document Classification, Master Thesis. Tirana: University of New York Tirana (2011)
14. Taullaj, L.: Comparative Evaluation of Text Mining Algorithms for Albanian. Master Thesis. University of New York Tirana (2012)

Sentiment Analysis through Machine Learning: An Experimental Evaluation for Albanian

Marenglen Biba and Mersida Mane

Abstract. Opinions have always influenced our behaviours and they have a key role in human activities. Nowadays, online opinion resources such as newspapers, blogs, and reviews have enormously increased the amount of text data available for analysis. Sentiment Analysis (or Opinion Mining) is increasingly becoming an important tool for analysing text data in order to understand opinions correctly. In this context, machine learning methods have the potential to perform correct classification of texts as expressing positive or negative opinion for a certain topic. However, much research has been dedicated to languages such as English, Japanese, Chinese or German but no research has been made for other rare Indo-European languages such as the Albanian. In this paper, we present the first approach for Sentiment Analysis in Albanian. We show through extensive experiments with text data from political news consisting of five different topics, that the proposed approach is effective in classifying text documents as belonging to negative or positive opinion regarding the given topic.

1 Introduction

What other people think is increasingly becoming important for many decision-making processes [8]. Opinions have always influenced our behaviour and they have a key role in human activities. Long before, many of us looked for opinions by asking our friends and family. Consumers always asked other users of a

Marenglen Biba
University of New york in Tirana, Albania
email: marenglenbiba@unyt.edu.al

Mersida Mane
University of New York in Tirana and University of Greenwich, London, UK
email: mersidamane@unyt.edu.al

S.M. Thampi et al. (eds.), *Recent Advances in Intelligent Informatics*,
Advances in Intelligent Systems and Computing 235,
DOI: 10.1007/978-3-319-01778-5_20, © Springer International Publishing Switzerland 2014

product or service before doing a purchase. On the other hand, whenever an organization had to make decisions based on the opinions of the consumers about its products or services, it conducted surveys, focus groups or opinion polls.

With the explosive growth in the past few years of the social media content (e.g., blogs, forum discussions and posting in social network sites) on the Web, worldwide communication has changed [6]. We are not any more limited to ask our friends or family for their opinions. There are a great number of opinion–rich resources, such as user reviews on Web, forums, blogs and social sites, where individuals that are neither our acquaintances nor professional critics discuss with each other and express their opinions. There is no more need for the organizations to gather customer opinions by conducting surveys, focus groups or opinion polls, since there is plenty of information publicly available on the Web.

Sentiment analysis is an approach for extracting feelings from a given text. For this task to be effectively accomplished, it is required to process huge amounts of data available online and to retrieve the opinion hidden in it. However, in processing subjective information effectively by these systems, there are a number of challenges to overcome [8]:

First, applications integrated into a general-purpose search engine require determining if the user is looking for subjective material. This could not be a difficult problem if the queries contain indicator terms like "opinion", "review", or the application provides a checkbox to the users giving them the possibility to indicate that reviews are desired. However, query classification remains still a difficult problem.

Second, besides the still-open problem of determining which documents are topically relevant to an opinion-oriented query, an additional challenge is determining which documents or portions of documents contain review-like or opinionated material. For texts fetched from review aggregations sites like 'Amazon.com' or 'Epinions.com', this is a relatively easy problem because the review-oriented information has a stereotyped format. However, the material in other sites can vary widely in presentation, style, content and grammatical level.

Third, it is hard for computers to analyze free-form texts because of additional challenges such as quotations included in an article. The views expressed in each quotation must be attributed with the correct entity.

Finally, the sentiment information will be presented in a reasonably summarized fashion, for example, by highlighting some of the opinions, through representation of points of consensus and disagreement or by considering the level of authority of the opinion holders. Sometimes producing a visualization of sentiment data is more appropriate than a textual summary of it.

The last three challenges represent also the most active areas of research. In addition, these challenges highlight some concerns that are also faced during sentiment analysis in Albanian.

Significant research work on sentiment analysis has been done on languages such as English, Japanese, Chinese, German, and Romanian [2]. So far, to the best of the authors' knowledge, there have been no sentiment analysis approaches for Albanian. In this paper we build the first dataset tagged with sentiment information, and perform extensive experiments with several algorithms. We use in our approach a stemmer for Albanian in order to bring the words and their variations to the original form which is the stem or root. This greatly helps in reducing the number of terms that are used in the next stage for text classification. We show that by providing a sound and significant dataset for training, it is possible to achieve high results in the classification of texts expressing a certain opinion into the right category as being positive or negative.

The paper is organized as follows: Section 2 introduces Sentiment Analysis and related approaches. Section 3 contains a brief introduction to Albanian with some features that make it a complex language to deal with, Section 4 contains the experimental results and we conclude in Section 5.

2   Sentiment Analysis

In this section we introduce Sentiment Analysis and related issues.

2.1   The Problem of Sentiment Analysis

Everyone can express his own opinion about everything that might be for example a service, a product, an event or a topic, which is called target object or entity. The authors of an expressed opinion are known as opinion holders or opinion sources. Opinions fall into one of these two groups: regular opinions and comparative opinions. Regular opinions are known as opinions in the research literature [1]. Comparative opinions express preferences of the opinion holder in relation to two or more target objects based on their aspects. An opinion orientation can be positive, negative or neutral which are also known as polarities or semantic orientations [5].

A regular opinion is a *quintuple* (e_i, a_{ij}, oo_{ijkl}, h_k, t_l), where e_i is the name of an entity, a_{ij} is an aspect of e_i, oo_{ijkl} is the orientation of the opinion about aspect a_{ij} of entity e_i, h_k is the opinion holder, and t_l is the time when the opinion is expressed by h_k [1]. All the above components are important because if any of them is missing, it becomes complicated in general. For example, in the sentence 'The touch screen was perfect', it is impossible to decide to whom this touch screen belongs to, so this opinion is worthless. However, there are cases when not all the components are required. For example, when required to know the opinions of thousands of people, the opinion holders component in not needed. Also, there are cases when new components are needed to be added to the quintuple, such as the gender and the age of the opinion holders. This kind of definition for opinion plays a key role for the transformation of unstructured text to structured data. It helps with important information for both qualitative and quantitative analysis of opinions.

Two key concepts in opinion mining are subjectivity and emotion. As mentioned in [6], an objective sentence consists of facts, while a subjective sentence consists of personal feelings, attitudes or beliefs. Subjective expressions might be opinions, wishes, viewpoint, doubts, allegations or speculations and very often they do not offer opinions [10, 14]. Also, not all the objective sentences offer an opinion. For example, the objective sentence "*the battery of my laptop lasts 45 minutes*" does mean a negative opinion. The association of subjectivity with opinion is still confusing for the researchers. Even though subjective sentences are different from opinion sentences, there is great intersection between them.

Emotions are spontaneous feelings and thoughts. The six main emotions are love, happiness, surprise, anger, sorrow and fright which include a number of other secondary and tertiary ones [9]. The intensity of the emotion defines how strong an opinion is. There are a lot of cases when an opinion sentence has no emotion in it and when an emotion sentence contains no opinion.

2.2 Aspect-Based Opinion Summary

The number of opinions used for the majority of opinion mining application is significantly high. Opinions from one opinion holder are usually not enough. This means that in such cases it is preferable a summary of opinions. Opinion quintuples offer very useful resources of information for obtaining qualitative and quantitative summaries. Aspect-based opinion summary is an ordinary type of summary based on aspects [1].

Aspect-based opinion summary can be presented by using bar charts or text summary. Text summary is a brief overview of what the others think related to a product or service. The main disadvantage of text summary is that it is only qualitative where in cases such as analytical purposes it is not helpful. For example, a common text summary would be "Lots of people think that Apple computers are the best", while the quantitative summary for the same sentence would be"87% of people think that Apple computers are the best, while 13% prefer other brands". Quantitative presentation is as fundamental to many opinion mining applications as the traditional survey researches. In this context a lot of research is dedicated to changing the text summary into a more readable form.

2.3 Document-Level Sentiment Classification

In a document-level sentiment classification the whole document is considered as the fundamental information component for performing the classification and deciding if it expresses a positive opinion, negative opinion or sentiment [1].

Sentiment classification considers the document as expressing opinions for only one entity and the opinions are expressed from only one opinion holder. This is true for customer reviews because they are only for a product or service and are expressed by only one reviewer. In forums or blog postings this is not true because one can express his opinion for many products or services and can also compare these products or services by using comparative sentences.

2.4 Supervised and Unsupervised Learning for Sentiment Analysis

Sentiment classification can be described as a supervised learning problem which consists of these three classes: positive, negative and neutral. Product or service reviews are the common data used for training and testing. These rated reviews, for example from 1 to 5 stars, serve as ready data for further training and testing. For example, reviews assigned 1-2 stars are thought to be negative reviews, reviews assigned 3 stars are thought to be neutral ones and those reviews assigned 4-5 stars are thought to be positive reviews. Any of the current supervised learning methods can be used for sentiment classification.

The authors in [8] have shown that the usage of features such as unigrams in classification had the same performance by using both methods Naive Bayes and Support Vector Machines. Some of the most important features are: Terms and their frequency, Part of speech, Opinion words and phrases, Negations, Syntactic dependency. Different types of approaches are used to increase the classification accuracy, for example the use of a score function [4], feature weighting schemes [6], or the utilization of opinion words during the training procedure [12].

Opinion words and phrases are the key indicators for sentiment analysis, thus applying unsupervised learning methods based on these words and phrases is normal. An unsupervised approach is presented in [14], which has three main steps: in the first step adverbs and adjectives are extracted since they are helpful containers of opinions; in the second step, it is computed an equation for calculating the pointwise mutual information; in step three, for a given text, the algorithm calculates the average semantic orientation (SO) for all the phrases and based on the value of the average SO, the text is classified.

3 Albanian Language

Nowadays, Albanian is a language spoken in Albania, as well as in Kosovo, northwest of Macedonia, southwest of Montenegro and northwest of Greece. It is spoken even wider, in states such as Italy, Greece, USA and other countries where thousands of Albanians have emigrated and currently live [3]. During the years it has been influenced by other languages due to many invasions. However, Albanian as an Indo-European language has kept its originality with its special structure of phonetic, grammar and lexicon. Albanian alphabet is based on the Latin alphabet and consists of 36 letters, 7 vowels

Albanian is a very rich language in words with more than one meaning which are called polysemantic words. In the context of sentiment analysis these play an important role. For example, the phrase *'pranë zjarrit' (near the fire)* can mean really near the file, but also *'at home'*. In addition it is also a symbol of tranquillity, being safe and a symbol of family life [3]. These cases may mislead an automatic classifier since the opinion being expressed might be related to only one of the meanings of the word.

Lexical field consists of a set of words used to express almost the same idea. For example, the set of words used to express the notion of noises is called lexical field of noises. Words of a lexical field have different types of relations between them. Based on grammatical point of view, in the lexical field of noises the words could be nouns, for example *bubullim (thunder)*, *fëshfërim (whisper)*, *zhurmë (noise)*, and verbs, for example *kërcet, thërret (call)*, etc. Based on semantic point of view, the words could be classified as general, for example *zhurmë (noise), zëra (voices)*, and specific words, for example *kuit, cinzërit*, etc [3].

Another feature of Albanian is the figurative meaning of words. Many words have several meanings [3]. For example, the word *faqe (cheek)* means: the sides of the face (*Lotët i përshkuan faqet - Tears ran down her cheeks*); the front part of a hill (*Gjithë faqja e kodrës ishte e mbjellë me ullinj - The front of the hill was planted with olive trees.*). However, the word faqe can also have the meaning of honor: *si mbeti faqe, ska faqe ai*. In this last case, the word 'faqe' is used in figurative meaning.

Homonyms are words that are written or spelled the same, but they have different meanings, for example mbledhje which means sum or meeting, depending on the context used [3]. Antonyms are words with opposite meanings. Antonyms are classified as lexical antonyms and grammatical antonyms. Lexical antonyms just have opposite meanings, while grammatical antonyms have similar form, for example: *i ditur (cultured)– i paditur (uncultured), i armatosur (armed) – i çarmatosur (unarmed)*, etc. Antonyms have an important role in expressing opinions.

4 Experimental Evaluation

4.1 Experimental Setting

Experiments are performed in the Weka framework, a machine learning software written in Java that provides implementation of learning algorithms for different data mining tasks [15]. The workbench includes also many data processing tools. Weka can be used for applying a learning method on a given dataset and evaluate the results, for using learned models to make predictions for new instances and for evaluating the results and comparing the performance of learning algorithms.

Preprocessing tools are used for transforming the instances. They are known as "filters". Filters are grouped as supervised and unsupervised where supervised filters are used in case class information is available. Both supervised and unsupervised filters are further divided into attribute filters and instance filters.

In the pre-processing step, first, each document is stemmed by using the JStem algorithm [11]. The stemmed documents are then given in input to Weka where each document is transformed into a word vector with the filter StringToWordVector.

4.2 Corpus Building

For sentiment analysis in Albanian we used political news. They are classified as having positive or negative opinion based on the sentiment of the document. Our corpora consist of five different topics, each containing 40 documents with positive sentiment and 40 documents with negative sentiment. The five topics selected are:

Topic 1 - Opening of ballot boxes. An important long and complex debate between the two main parties, Democratic and Socialist in Albania, has been the opening of ballot boxes for the general elections of 2009. Many news articles can be found on web, where one party claims that the boxes should be opened and the other party insists for the opposite.

Topic 2 - Immunity reform: A current topic in Albania is the voting of immunity reform. The Democratic party recommends to the opposite party to vote the immunity reform on August 6th 2012, while the opposition leader strongly declares that his party will not vote it until all the proper Constitutional amends are discussed first.

Topic 3 - Negotiations between Kosovo – Serbia: Complex relations have been between the Republic of Kosovo and Republic of Serbia, since Kosovo declared its independence in 2008. During 2012 and 2013, several negotiation rounds have been hold between these two countries. However, in Kosovo the opposition is against the negotiations, while the government intends to go ahead.

Topic 4 - Economic Crisis in Albania: INSTAT (Institute of Statistics of Albania) and the opposition report about the inflation in different periods during the year expressing negative opinion for the economy. However, there are a few reports which show growth of economy or news where government states that economy is growing. These are used as documents expressing positive opinions.

Topic 5 - Electoral reform: No consensus between the two main parties and the smaller parties for voting the electoral reform. The Socialist and the smaller parties have several proposals to achieve the consensus. While the President and the Democratic party are recommending the voting of the electoral reform to be done as soon as possible.

The above topics were selected due to their complexity in terms of the complex language used and also due to the very rich technical vocabulary. In particular, the immunity reform and the electoral reform bear high complexity since in most of the texts of these categories, references to the constitution are made which could probably mislead the classifier. On the other side, most text documents that discuss the economic crisis, contain a high number of quantitative data in terms of percentages, rates etc, which coming in the form of numbers may again mislead the classifier.

4.3   Experimental Results

In this section we present the experimental results. Table 1 presents all the results obtained with the six different algorithms. We have chosen the most representative algorithms for mining that have been also shown previously to achieve high results in many classification tasks.

As we can see from the results by training on the corpus, we are able to produce classifiers with accuracy for the different problems between 86 and 92%. Even though the topics that we have chosen are not easy to deal with due to the complex language and vocabulary used, the approach by first stemming the texts with an Albanian stemmer and then classifying each text with a machine learning algorithm, succeeded in building effective classifiers.

Table 1 Results of mining political news

Algorithm	Topic 1	Topic 2	Topic 3	Topic 4	Topic 5	Average
BayesianLogistic Regression	77.5%	**88.75%**	86.25%	78.75%	78.75%	83.13%
Logistic Regression	**86.25%**	87.5%	81.25%	85%	75%	83.00%
SVM	76.25%	87.5%	86.25%	82.5%	82.5%	83.00%
Voted Perceptron	71.25%	82.5%	83.75%	72.5%	72.5%	76.50%
Naïve Bayes	70%	86.25%	**88.75%**	75%	75%	79.00%
Hyper Pipes	83.75%	87.5%	81.25%	**92.5%**	**90%**	87.00%

On average, the best performing classifier is HyperPipes. However, depending on the topic, the best performing algorithm for each topic varies. We have used 10-fold-cross-validation in each of the experiments in order to get reliable classification accuracy.

5 Conclusion

In this paper we have presented the first attempt for sentiment analysis in Albanian. We have built a corpus to train machine learning models and after stemming every text with an Albanian stemmer, we build classifiers that are able to classify at high accuracy the given texts. Possible future work includes improvements of the stemming algorithm which has a key role in producing the input for the classifier and therefore a direct effect on increasing the value of the performance obtained. In addition, the same experiments could also be performed with a larger dataset and the comparison with the current results would be important to know which should be the minimal or the most appropriate size of the dataset in order to achieve high classification accuracy.

References

1. Aggarwal, C.C., Zhai, C. (eds.): Mining Text Data. Springer (2012) ISBN 978-1-4614-3222-7
2. Banea, C., Mihalcea, R., Wiebe, J.: Multilingual Sentiment and Subjectivity Analysis. In: Multilingual Natural Language Processing. Prentice Hall (2012) ISBN 10: 0137151446
3. Beci, B.: Gramatika e gjuhes shqipe. Logos-A (2005) ISBN 9989580197
4. Dave, K., Lawrence, S., Pennock, D.: Mining the peanut gallery: Opinion extraction and semantic classification of product reviews. In: Proc. of Int'l Conf. on World Wide Web, WWW 2003 (2003)
5. Jindal, N., Liu, B.: Mining comparative sentences and relations. In: Proceedings of National Conf. on Artificial Intelligence, AAAI 2006 (2006)

6. Liu, B.: Sentiment Analysis: A Multi-Faceted Problem. IEEE Intelligent Systems 25(3), 76–80 (2010)
7. Paltoglou, G., Thelwall, M.: A study of information retrieval weighting schemes for sentiment analysis. In: Proceedings of Annual Meeting of the Association for Computational Linguistics, ACL 2010 (2010)
8. Pang, B., Lee, L.: Opinion Mining and Sentiment Analysis. Foundations and Trends in Information Retrieval Journal 2 (2008)
9. Parrott, W.G.: Emotions in social psychology: Essential readings. Psychology Press (2001) ISBN-10: 0863776833
10. Riloff, E., Patwardhan, S., Wiebe, J.: Feature subsumption for opinion analysis. In: Proceedings of the Conference on Empirical Methods in Natural Language Processing, EMNLP 2006 (2006)
11. Sadiku, J.: A Novel Stemming Algorithm for Albanian text in a Data Mining Approach for Document Classification, Master Thesis. Tirana: University of New York Tirana (2011)
12. Tan, S., Wang, Y., Cheng, X.: Combining learn-based and lexicon-based techniques for sentiment detection without using labeled examples. In: Proceedings of ACM SIGIR Conference on Research and Development in Information Retrieval, SIGIR 2008 (2008)
13. Turney, P.D.: Thumbs up or thumbs down? Semantic orientation applied to unsupervised classification of reviews. In: Proc. of the Association for Computational Linguistics, ACL 2002 (2002)
14. Wiebe, J.: Learning subjective adjectives from corpora. In: Proceedings of Nat'l Conf. on Artificial Intelligence, AAAI 2000 (2000)
15. Witten, I.H., Frank, E.: Data Mining: Practical machine learning tools and techniques, 2nd edn. Morgan Kaufmann Pub., San Francisco (2005)

Classification of Text Documents Using Adaptive Fuzzy C-Means Clustering

B.S. Harish, Bhanu Prasad, and B. Udayasri

Abstract. In this paper, we propose a new method of representing text documents based on clustering of term frequency vectors. Term frequency vectors of each cluster are used to form a symbolic representation (interval valued representation) by the use of mean and standard deviation. In order to cluster the term frequency vectors, we make use of fuzzy C-Means clustering method for interval type data based on adaptive squared Euclidean distance between vectors of intervals. Further, to corroborate the efficacy of the proposed model we conducted extensive experimentation on standard datasets like 20 Newsgroup Large, 20 Mini Newsgroup, Vehicles Wikipedia and our own created datasets like Google Newsgroup and Research Article Abstracts. We have compared our classification accuracy achieved by the Symbolic classifier with the other existing Naïve Bayes classifier, KNN classifier, Centroid based classifier and SVM classifiers. The experimental results reveal that the achieved classification accuracy is better than that of the existing methods. In addition, our method is based on a simple matching scheme; it requires negligible time for classification.

Keywords: Classification, Text Documents, Representation, Adaptive Fuzzy C-Means, Clustering Algorithms.

1 Introduction

Text classification is one of the important research issues in the field of text mining, where the documents are classified with a supervised knowledge. Based on a

B.S. Harish · B. Udayasri
Department of Information Science & Engineering, S.J. College of Engineering,
Mysore 570 006, Karnataka, India
e-mail: bsharish@ymail.com, udaya.sri84@gmail.com

Bhanu Prasad
Department of Computer and Information Sciences Florida A&M University, Tallahassee,
FL 32307, USA
e-mail: prasad@cis.famu.edu

S.M. Thampi et al. (eds.), *Recent Advances in Intelligent Informatics*,
Advances in Intelligent Systems and Computing 235,
DOI: 10.1007/978-3-319-01778-5_21, © Springer International Publishing Switzerland 2014

likelihood of the training set, a new document is classified. The task of text classification is to assign a boolean value to each pair $(d_j, k_i) \in D \times K$, where $'D'$ is the domain of documents and $'K'$ is a set of predefined categories. The task is to approximate the true function $\phi : D \times K \to \{1, 0\}$ by means of a function $\hat{\phi} : D \times K \to \{1, 0\}$ such that ϕ and $\hat{\phi}$ coincide as much as possible. The function $\hat{\phi}$ is called a classifier. A classifier can be built by training it systematically using a set of training documents D, where all of the documents belonging to D are labeled according to K [1], [2]. The major challenges and difficulties that arise in the problem of text classification are: High dimensionality (thousands of features), variable length, content and quality of the documents, sparse distribution of terms in documents, loss of correlation between adjacent words and understanding complex semantics of terms in a document [3]. To tackle these problems a number of methods have been reported in literature for effective classification of text documents. Many representation schemes like binary representation [4], ontology [5], N-Grams [6], multiword terms as vector [7], UNL [8] are proposed as text representation schemes for effective text classification. Also, in [9], [10] a new representation model for the web documents are proposed. Recently, [11] [12] used bayes formula to vectorize a document according to a probability distribution reflecting the probable categories that the document may belongs to. Although many representation models for the text document are available in literature, the Vector Space Model (VSM) is one of the most popular and widely used models for document representation [13]. Unfortunately, the major limitation of the VSM is the loss of correlation and context, of each term which are very important in understanding the document. To deal with these problems, Latent Semantic Indexing (LSI) was proposed by [14]. The LSI is optimal in the sense of preserving the global geometric structure of a document space. However, it might not be optimal in the sense of discrimination [14]. Thus, to discover the discriminating structure of a document space, a Locality Preserving Indexing (LPI) is proposed in [15]. An assumption behind LPI is that, close inputs should have similar documents. However, the computational complexity of LPI is very expensive. Thus it is almost infeasible to apply LPI over very large dataset. Hence to reduce the computational complexity of LPI; Regularized Locality Preserving Indexing (RLPI) has been proposed in [16]. The RLPI is being significantly faster obtains similar or better results when compared to LPI. This makes the RLPI an efficient and effective data preprocessing method for large scale text classification [16].

However the RLPI fails to preserve the intraclass variations among the documents of the different classes. Also in case of the RLPI, we need to select the number of dimensions m to be optimal. Unfortunately we cannot say that the selected m value is optimal. Also, in conventional supervised classification an inductive learner is first trained on a training set, and then it is used to classify a testing set, about which it has no prior knowledge. However, for the classifier it would be ideal to have the information about the distribution of the testing samples before it classifies them. Thus in this paper, to deal with the problem of learning from training sets of different sizes, we exploited the information derived from clusters of the term frequency vectors of documents.

Clustering has been used in the literature of text classification as an alternative representation scheme for text documents. Several approaches of clustering have been proposed. Given a classification problem, the training and testing documents are both clustered before the classification step. Further, these clusters are used to exploit the association between index terms and documents [17]. In [18], words are clustered into groups based on distribution of class labels associated with each word. Information bottle neck method is used to find a word cluster that preserves the information about the categories. These clusters are used to represent the documents in a lower dimensional feature space and naïve bayes classifiers are applied [19]. Also, in [20] information bottleneck is used to generate a document representation in a word cluster space instead of word space, where words are viewed as distributions over document categories. [20] proposed an information theoretic divisive algorithm for word clustering and applied it to text classification. Classification is done using word clusters instead of simple words for document representation. Two dimensional clustering algorithms are used to classify text documents in [21]. In this method, words/terms are clustered in order to avoid the data sparseness problem. In [22] clustering algorithm is applied on labeled and unlabeled data, and introduces new features extracted from those clusters to the patterns in the labeled and unlabeled data. The clustering based text classification approach in [23] first clusters the labeled and unlabeled data. Some of the unlabeled data are then labeled based on the clusters obtained.

All in all, the above mentioned clustering based classification algorithms work on conventional word frequency vector. Conventionally the feature vectors of term document matrix (very sparse and very high dimensional feature vector describing a document) are used to represent the class. Later, this matrix is used to train the system using different classifiers for classification. Generally, the term document matrix contains the frequency of occurrences of terms and the values of the term frequency vary from document to document in the same class. Hence to preserve these variations, we propose a new interval representation for each document. An initial attempt is made in [3] by giving an interval representation by using maximum and minimum values of term frequency vectors for the documents. However, in this paper we are using mean and standard deviations to give the interval valued representation for documents. Thus, the variations of term frequencies of document within the class are assimilated in the form of interval representation. Moreover conventional data analysis may not be able to preserve intraclass variations but unconventional data analysis such as symbolic data analysis will provide methods for effective representations by preserving intraclass variations. The recent developments in the area of symbolic data analysis have proven that the real life objects can be better described by the use of symbolic data, which are extensions of classical crisp data. The aim of the symbolic data analysis is to provide suitable methods for managing aggregated data described by multi valued variables, where the cells of the data contain sets of categories, intervals, or weight distributions. Symbolic data analysis provides a number of

clustering methods for symbolic data. These methods differ in the type of considered symbolic data, in their cluster structures and/or in the considered clustering
criterion [24]. The previous issues motivated us to use symbolic data rather than
using a conventional classical crisp data to represent a document. To preserve the
intraclass variations we create multiple clusters for each class. Term frequency
vectors of documents of each cluster are used to form an interval valued feature
vector. With this backdrop, the work presented in [3] is extended towards creating
multiple representatives per class using clustering after symbolic representation.

The rest of the paper is organized as follows: A detailed literature survey and
the limitations of the existing models are presented in section 1. The working
principle of the proposed method is presented in section 2. Details of dataset used,
experimental settings and results are presented in section 3. The paper is concluded along with future works in section 4.

2   Proposed Method

The proposed method has two stages: (i) Cluster based representation and (ii)
Document classification stage.

2.1   Cluster Based Representation

In the proposed method, initially documents are represented by term document
matrix. The obtained term document matrix is of very high dimension and huge
sparse matrix. We employ regularized locality preserving indexing (RLPI) dimensionality reduction technique, to reduce to a lower dimension. Unfortunately, the
RLPI features of documents of a class have considerable intraclass variations.
Thus, we propose to have an effective representation by capturing these variations
through clustering and representing each cluster by an interval valued feature vector called symbolic feature vector as follows.

The training documents of each class are first clustered on RLPI features. Let
$[D_1, D_2, D_3, ..., D_n]$ be a set of n documents of a document cluster say C_j;
$j = 1, 2, 3, ..., N$ (N denotes the number of clusters) and let M be the number of
classes and let $F_i = [f_{i1}, f_{i2}, f_{i3}, ..., f_{im}]$ be the set of m features characterizing the
document D_i of a cluster C_j. Let μ_{jk}; $k = 1, 2, ..., m$ be the mean of the kth feature
values obtained from all n documents of a cluster C_j.

i.e.,

$$\mu_{jk} = \frac{1}{n} \sum_{i=1}^{n} f_{ik} \qquad (1)$$

Similarly, let σ_{jk}; $k = 1, 2, ..., m$ be the standard deviation of the kth feature
values obtained from all the n documents of the cluster C_j.

i.e.,

$$\sigma_{jk} = \left[\frac{1}{n} \sum_{i=1}^{n} \left(f_{ik} - \mu_{jk} \right)^2 \right]^{1/2}$$ (2)

Now, we recommend capturing intraclass variations in each kth feature values of the jth cluster in the form of an interval valued features $\left[f_{jk}^-, f_{jk}^+ \right]$, where, $f_{jk}^- = \mu_{jk} - \sigma_{jk}$; $f_{jk}^+ = \mu_{jk} + \sigma_{jk}$. Thus, each interval $\left[f_{jk}^-, f_{jk}^+ \right]$ representation depends upon mean and its standard deviation of respective individual features of a cluster. The interval $\left[f_{jk}^-, f_{jk}^+ \right]$ represents the upper and lower limits of feature value of a document cluster in the knowledgebase. Now, the reference document for a cluster C_j, is formed by representing each feature $k = 1, 2, ..., m$ in the form of an interval and it is given by

$$RF_j = \left\{ \left[f_{j1}^-, f_{j1}^+ \right], \left[f_{j2}^-, f_{j2}^+ \right], ..., \left[f_{jm}^-, f_{jm}^+ \right] \right\}$$ (3)

Where, $j = 1, 2, ..., N$ represents the number of clusters of documents samples of a class. It shall be noted that unlike conventional feature vector, this is a interval valued features and this symbolic feature vector is stored in the knowledgebase as a representative of the jth cluster. Thus the knowledgebase has N number of symbolic vectors representing clusters corresponding to a class.

Here, to cluster symbolic feature vectors we used Fuzzy C Means clustering algorithm. The adaptive Fuzzy C Means clustering algorithm looks for a partition of a set of patterns in n documents of a document cluster say C_j; $j = 1, 2, 3, ..., N$ (N denotes the number of clusters), the corresponding N prototypes and the square of an adaptive squared Euclidean distance between vectors of intervals that is different between each class, such that a criterion W^2 measuring the fitting between the clusters and their representatives (prototypes) is locally minimized. This criterion W^2 is based on an adaptive squared Euclidean distance for each cluster and is defined as

$$W^2 = \sum_{i=1}^{N} \sum_{k=1}^{n} \left(u_{ik} \right)^m \psi \left(x_k, g_i \right)$$ (4)

Where x_k is vector of intervals, g_i is prototype (representative), u_{ik} is membership degree of patterns and ψ is now the square of an adaptive Euclidean distance defined for each class and parameterized by the vectors of weights, which change at each iteration. The algorithm starts from an initial membership degree for each document n in each cluster C_j and alternates a representation step and allocation step until the convergence, when the criterion W^2 reaches a stationary value representing a local minimum. Further, details of algorithm can be found in [26].

2.2   Document Classification

The document classification proposed in this work considers a test document, which is described by a set of m feature values of type crisp and compares it with the corresponding interval type feature values of the respective cluster stored in the knowledge base. Let, $F_i = [f_{i1}, f_{i2}, ..., f_{im}]$ be a m dimensional feature vector describing a test document.

Let RF_j be the interval valued symbolic feature vector of jth cluster of lth class. Now, each mth feature value of the test document is compared with the corresponding interval in RF_j to examine whether the feature value of the test document lies within the corresponding interval. The number of features of a test document, which fall inside the corresponding interval, is defined to be the degree of belongingness. We make use of Belongingness Count B_c as a measure of degree of belongingness for the test document to decide its class label.

$$B_c = \sum_{k=1}^{m} C\left(f_{ik}, \left[f_{jk}^-, f_{jk}^+\right]\right) \text{ and} \tag{5}$$

$$C\left(f_{ik}, \left[f_{jk}^-, f_{jk}^+\right]\right) = \begin{cases} 1; & if \left(f_{ik} \geq f_{jk}^- \text{ and } f_{ik} \leq f_{jk}^+\right) \\ \\ 0; & Otherwise \end{cases}$$

The crisp value of a test document falling into its respective feature interval of the reference class contributes a value 1 towards B_c and there will be no contribution from other features which fall outside the interval. Similarly, we compute the B_c value for all clusters of remaining classes and the class label of the cluster which has highest B_c value will be assigned to the test document as its label.

3   Experimental Setup

3.1   Experimentation

In this section, we present the results of the experiments conducted to demonstrate the effectiveness of the proposed method on all the five datasets viz., 20 Newsgroup Large, 20 Mini Newsgroup, Vehicles Wikipedia, Google Newsgroup and Research Article Abstracts. During experimentation, we used 60% of the documents of each class of a dataset to create training set and the remaining 40% of the documents for testing purpose. Experiments are repeated 5 times by choosing the training samples randomly. As measures of goodness of the proposed method, we

Table 1 Comparative Analysis of the Proposed Method with other State of the Art Techniques

Method		Datasets				
		20 Newsgroup Large	20 Mini-Newsgroup	Vehicles Wikipedia	Google Newsgroup	Research Article Abstracts
Probability Based Representation	Naïve Bayes Classifier	70.45	68.50	69.60	70.15	71.50
	KNN Classifier	72.55	69.05	70.25	70.40	72.40
	Centroid Based Classifier	73.90	70.10	71.50	71.95	73.55
	SVM Classifier	75.85	72.10	72.50	71.80	75.60
BOW Representation	Naïve Bayes Classifier	73.60	69.55	71.50	70.65	70.35
	KNN Classifier	75.45	70.10	71.85	72.50	72.90
	Centroid Based Classifier	78.10	72.50	72.80	72.95	74.60
	SVM Classifier	80.60	73.45	75.40	74.60	76.85
Proposed Method (Symbolic Representation + Symbolic Classifier)		**86.45 (5 Clusters)**	**79.30 (6 Clusters)**	**80.45 (2 Clusters)**	**84.55 (4 Clusters)**	**88.90 (3 Clusters)**

computed classification accuracy. The minimum, maximum and the average value of the classification accuracy of all the 5 trials are obtained. However, in Table 1 we present only the average results of five trials.

For experiments, we have randomly selected the training documents to create the symbolic feature vectors for each class. While conducting the experimentation we have varied the number of features m (empirically) selected through RLPI from 1 to 30 dimensions. For each obtained dimension we create cluster based interval representation as explained in section II A. At this juncture, we used adaptive Fuzzy C Means (FCM) clustering algorithms to create a cluster based symbolic representation. The Fuzzy C Means (FCM) clustering algorithms for symbolic data aims to furnish a fuzzy partitions of a set of pattern N clusters and a set of corresponding prototypes such that criterion measuring the fitting between the clusters and their representatives (prototypes) is locally minimized. This criterion is based on a non-adaptive squared Euclidean distance between vectors of intervals [25]. However, to overcome these limitations [26] proposed Fuzzy C Means clustering method for symbolic interval data based on adaptive squared Euclidean distances between vectors of intervals. The advantage of these adaptive distances is that the clustering algorithm is able to find cluster of different shapes and sizes. The more details of the algorithm can be found in [25].

It can be observed from the Table 1, Fuzzy C Means (FCM) clustering algorithm achieved a better results for varying number of clusters (Selection is adaptive). The reason behind this is FCM has its ability to discover cluster among the data, even when the boundaries among the data are overlapping. Also, FCM based techniques have the advantage over the conventional statistical techniques like NN classifier, maximum likelihood estimate etc, because its distribution is free and no knowledge about the distribution of data is required [25]. A comparative analysis of the proposed method with other state of the art techniques using standard classifiers and benchmark datasets is given in Table 1. From the Table 1 it is analyzed that the proposed method achieves better classification accuracy.

4   Conclusions

A simple and efficient symbolic text classification is presented. A text document is represented by the use of symbolic features. Term frequency vectors of each cluster are used to form a symbolic representation by the use of interval valued features. The main contribution of this paper is the introduction of adaptive fuzzy C-means clustering algorithm to cluster the term frequency vectors. To check the effectiveness and robustness of the proposed method, extensive experimentation is conducted on various datasets. The experimental results reveal that the proposed method outperforms the other existing methods. With the successful application of concepts of symbolic data to text classification, it is our firm belief that this work will open up new avenues to exploit the concepts of symbolic data for effective text representation.

References

[1]   Seabastiani, F.: Machine Learning in Automated Text Categorization. ACM Computing Surveys 34, 1–47 (2002)

[2]   Jiang, S., Pang, G., Wu, M., Kuang, L.: An improved K-nearest-neighbor algorithm for text categorization. Journal of Expert Systems with Applications 39, 1503–1509 (2012)

[3]   Guru, D.S., Harish, B.S., Manjunath, S.: Symbolic representation of text documents. In: Proceedings of Third Annual ACM Compute, Bangalore (2010)

[4]   Li, Y.H., Jain, A.K.: Classification of Text Documents. The Computer Journal 41, 537–546 (1998)

[5]   Hotho, A., Nürnberger, A., Paaß, G.: A Brief Survey of Text Mining. Journal for Computational Linguistics and Language Technology 20, 19–62 (2005)

[6]   Cavnar, W.B.: Using an N-Gram based document representation with a vector processing retrieval model. In: Third Text Retrieval Conference (TREC-3), pp. 269–278 (1994)

[7]   Milios, E., Zhang, Y., He, B., Dong, L.: Automatic term extraction and document similarity in special text corpora. In: Sixth Conference of the Pacific Association for Computational Linguistics (PACLing 2003), Canada, pp. 275–284 (2003)

[8]   Choudhary, B., Bhattacharyya, P.: Text clustering using Universal Networking Language representation. In: Proceedings of Eleventh International World Wide Web Conference (2002)

[9]   Craven, M., DiPasquo, D., Freitag, D., McCallum, A., Mitchell, T.M., Nigam, K., Slattery, S.: Learning to Extract Symbolic Knowledge from the World Wide Web. In: Proceedings of AAAI/IAAI, pp. 509–516 (1998)

[10]   Esteban, M., Rodrıguez, O.R.: A Symbolic Representation for Distributed Web Document Clustering. In: Proceedings of Fourth Latin American Web Congress, Cholula, Mexico (2006)

[11]   Isa, D., Lee, L.H., Kallimani, V.P., Rajkumar, R.: Text document preprocessing with the Bayes formula for classification using the support vector machine. IEEE Transactions on Knowledge and Data Engineering 20, 23–31 (2008)

[12]   Wan, C.H., Lee, L.H., Rajkumar, R., Isa, D.: A hybrid text classification approach with low dependency on parameter by integrating K-nearest neighbor and support vector machine. Journal of American Society of Information Science 41(16), 391–407 (1990)

[13]   Salton, G., Wang, A., Yang, C.S.: A Vector Space Model for Automatic Indexing. Communications of the ACM 18, 613–620 (1975)

[14]   Deerwester, S.C., Dumais, S.T., Landauer, T.K., Furnas, G.W., Harshman, R.A.: Indexing by Latent Semantic Analysis. Journal of the Expert Systems with Applications 39(15), 11880–11888 (2012)

[15]   He, X., Cai, D., Liu, H., Ma, W.Y.: Locality Preserving Indexing for document representation. In: Proceedings of International Conference on Research and Development I Information Retrieval (SIGIR 2004), UK, pp. 96–103 (2004)

[16]   Cai, D., He, X., Zhang, W.V., Han, J.: Regularized Locality Preserving Indexing via Spectral Regression. In: Proceedings of Conference on Information and Knowledge Management (CIKM 2007), pp. 741–750 (2007)

[17]   Kyriakopoulou, A., Kalamboukis, T.: Text classification using clustering. In: Proceedings of ECML-PKDD Discovery Challenge Workshop (2006)

[18]   Pereira, F., Tishby, N., Lee, L.: Distributional clustering of English words. In: Proceedings of the 31st Annual Meeting of the Association for Computational Linguistics, pp. 183–190 (1993)

[19]   Slonim, N., Tishby: The power of word clustering for text classification. In: Proceedings of the European Colloquium on IR Research, ECIR 2001 (2001)

[20]   Dhillon, I., Mallela, S., Kumar, R.: Enhanced word clustering for hierarchical text classification. In: Proceedings of the 8th ACM SIGKDD International Conference on Knowledge Discovery and Data Mining, Canada, pp. 191–200 (2002)

[21]   Takamura, H., Matsumoto, Y.: Two-dimensional clustering for text categorization. In: Proceedings of the Sixth Conference on Natural Language Learning (CoNLL-2002), Taiwan, pp. 29–35 (2002)

[22]   Raskutti, B., Ferr, H., Kowalczyk, A.: Using unlabeled data for text classification through addition of cluster parameters. In: Proceedings of the 19th International Conference on Machine Learning ICML, Australia, pp. 514–521 (2002)

[23]   Zeng, H.J., Wang, X.H., Chen, Z., Lu, H., Ma, W.Y.: CBC: Clustering based text classification requiring minimal labeled data. In: Proceedings of the 3rd IEEE International Conference on Data Mining, USA, pp. 443–450 (2003)

[24]   Bock, H.H., Diday, E.: Analysis of symbolic Data. Springer (1999)

[25]   Carvalho, F.D.A.T.: Fuzzy c-means clustering methods for symbolic interval data. Pattern Recognition Letters 28(4), 423–437 (2007)

[26]   Bezdek, J.C.: Pattern Recognition with Fuzzy Objective Algorithms. Kluwer Academic Publishers (1981)

Information and Relation Extraction for Semantic Annotation of eBook Texts

Ashraf Uddin, Rajesh Piryani, and Vivek Kumar Singh[*]

Abstract. This paper presents our algorithmic approach for information and relation extraction from unstructured texts (such as from eBook sections or webpages), performing other useful analytics on the text, and automatically generating a semantically meaningful structure (RDF schema). Our algorithmic formulation parses the unstructured text from eBooks and identifies key concepts described in the eBook along with relationship between the concepts. The extracted information is then used for four purposes: (a) for generating some computed metadata about the text source (such as readability of an eBook), (b) generate a concept profile for each distinct part of text, (c) identifying and plotting relationship between key concepts described in the text, and (d) to generate RDF representation for the text source. We have done our experiments on eBook texts from Computer Science domain; however, the approach can be applied to work on different forms of text in other domains as well. The results are not only useful for concept based tagging and navigation of unstructured text documents (such as eBook) but can also be used to design a comprehensive and sophisticated learning recommendation system.

1   Introduction

The rapid advancements in electronic devices, communication technologies and the World Wide Web have made available a huge volume of resources for students. Students nowadays have access to virtually unlimited amount of text from varied sources on any topic. The World Wide Web is now emerging as a major platform for knowledge dissemination. The proliferation of eBooks and online learning portals is another aspect of the technology transforming the learning environment. Availability of cheap readers (such as Kindle) and faster Internet connections are

Ashraf Uddin · Rajesh Piryani · Vivek Kumar Singh
Department of Computer Science, South Asian University, New Delhi, India
e-mail: vivek@cs.sau.ac.in
[*] Corresponding author.

S.M. Thampi et al. (eds.), *Recent Advances in Intelligent Informatics*,
Advances in Intelligent Systems and Computing 235,
DOI: 10.1007/978-3-319-01778-5_22, © Springer International Publishing Switzerland 2014

acting as a catalyst and there are now plenty of eBooks on every topic. This number is guaranteed to increase further. While on one hand, we now have more material, but at the same time it is becoming difficult to locate and identify appropriate and qualitative sources for finding knowledge about something. Here we present an algorithmic formulation that helps in this task. We have devised an algorithmic framework that takes eBooks (and possibly other texts) as inputs and generates a semantically tagged and easily navigable structure along with new interpretations of the text contained in the eBook.

We have used text processing techniques, such as information extraction and relation extraction for this purpose. The concepts extracted from an eBook are transformed through our program into an RDF structure. The RDF structure facilitates search and navigation tasks. The relationship between concepts described in an eBook are also extracted and visualized as a network. The most important aspect of the work is automated generation of RDF structure from unstructured texts. The rest of the paper is organized as follows. Section 2 gives a brief overview of the information extraction task and described our algorithmic approach for concept extraction. Section 3 briefly describes the relation extraction task along with the relation extraction algorithm used by us. Section 4 describes the semantic annotation and RDF generation. Section 5 illustrates the dataset used and the section 6 presents the experimental results. Section 7 is on conclusion and future work possibilities.

2   Information Extraction

Information extraction broadly refers to extracting knowledge (such as named entities) from unstructured text sources. The main motivation behind it is to allow semantic tagging of the text source as well allowing the possibility of machine reading of the text source. It is defined as an area of natural language processing that deals with finding factual information in free-form text (Piskorski & Yangarber, 2012; Banko & Etzioni, 2008). The process of extraction involves identifying objects or entities in text and consequently deriving the attributes around the entities. The information extraction goal may be categorized as targeted or open extraction. While in targeted extraction we look for information about some particular entities; in open extraction we aim to extract information about a general class of entities.

In targeted extraction, the domain is known and concepts or information to be extracted is predefined and hence irrelevant information can be ignored and specified information can be obtained in efficient manner. For example, for a given set of documents if we are interested to explore the information about the capitals and country name then the information extractor will look for capital and country name pair in each sentence of a document, and consequently it will run same process over the corpus. Bootstrapping and supervised learning are two main techniques used for this purpose (Brin 1998; Agichtein & Gravano, 2000; Kambhatla, 2004; GuoDong et al., 2005). Although these methods are effective but do not scale to large and heterogeneous corpora that contain large number of unknown relations.

Open information extraction uses a wider approach and is relation independent extraction model. An Open information extraction system extracts a diverse set of entities/ tuples from the unstructured text (Banko & Etzioni, 2008). It refers to techniques which discover all types of entities and tuples independent of corpus size and the domain. Open information extraction usually involves tasks like extracting named entities (such as persons, places, organizations, numerals etc.) and relation between entities. The typical approaches for this include machine learning algorithms, linguistics based methods and hybrid approaches. The information extraction algorithms have been applied and evaluated on English and other languages (Zhila & Gelbukh, 2013) and have also been used to measure informativeness of web documents (Horn et al., 2013).

2.1   Extracting Concepts from the eBook Text

The information of interest for us was concepts described in eBooks. We wanted to extract the learning concepts described in the eBook parts. Since most of the eBooks are in PDF format, we used a PDF to text API[1] for transforming the PDF to text. After converting the PDF to text, we extract various parts (preface, table of contents, chapters, sections etc.) of the eBook. From each such part of the eBook, we then extract identifiable learning concept. For learning concepts, we have used the idea proposed in (Justeson & Katz, 1995; Agrawal et al., 2011) for identifying terminological noun phrases. The assumption is that terminological noun phrases could be representative textual symbols for different learning concepts described in an eBook. We consider following three patterns (P1, P2, and P3) for determining terminological noun phrases:

$P1 = C*N$

$P2 = (C*NP)?(C*N)$

$P3 = A*N+$

where, N refers to a noun, P a preposition, A an adjective, and C = A|N. The pattern P1 corresponds to a sequence of zero or more adjectives or nouns which ends with a noun and P2 is a relaxation of P1 that also permits two such patterns separated by a preposition. Examples of the pattern P1 include "probability density function", "fiscal policy", and "thermal energy". Examples of the latter include "radiation of energy" and "Kingdom of Asoka". P3 corresponds to a sequence of zero or more adjectives, followed by one or more nouns. This pattern is a restricted version of P1, where an adjective occurring between two nouns is not allowed. The main target is that it is better to have more specific concepts than general concepts. A somewhat similar type of strategy was used in (Lent et al., 1997). It has been reported (Agrawal et al., 2010) that the pattern P1 performs better than P2. However, to identify these patterns, we were required to parse the

---

[1] iText Open Source PDF Library for JAVA, www.itextpdf.com

eBook text sentence-by-sentence. For each sentence, we do POS tagging of every word in it using the Stanford POS tagger[2] and then we look for phrases described by the above pattern forms. In addition to the concept extraction as stated above, we have also extracted some eBook metadata such as its price, authors, title, chapter titles etc. using a simple bookmark based text pattern extractor.

2.2   Concept Pruning

The concept extractor as described above returns all word patterns conforming to the desired terminological noun phrase pattern forms. However, not all of these extracted patterns are actually learning concepts in computer science domain. We have therefore pruned the list of concepts so extracted so as to use only useful and higher-level computer science learning concepts. For this purpose, we have used an augment version of ACM Computing Curricular framework[3] document. This document describes the entire computer science body of knowledge and groups it into 14 identifiable categories such as information management, programming etc. To make it more precise, we have taken all computer science terms from the IEEE Computer Society Taxonomy[4] and ACM Computing Classification System[5] documents and added them to the concept profile of the respective categories of ACM Computing Curricular Framework document. We extracted all terms occurring in these documents and merged them together to build a term profile for all 14 distinct areas of computer science listed in ACM Computing Curricular framework. The resultant document is hereafter referred to as augmented ACM-CCF.

After obtaining the augmented ACM-CCF, we started pruning undesirable terminological noun phrases so as to select only important computer science learning concepts described in the eBook text. An extracted concept is selected or discarded based on its similarity score obtained with the learning concepts described in the augmented ACM-CCF. For similarity computation we have used the following expression:

Similarity (TNP, LC) = |(TNP ∩ LC)| / |(TNP ∪ LC)|

where, TNP refers to a terminological noun phrase extracted and LC refers to a learning concept described in augmented ACM-CCF. The ∩ operator refers to the set of common words occurring in both TNP and LC; and the ∪ operator refers to the union of all words in both TNP and LC. The |S| operator as usual denotes the cardinality of a set S. We have used this similarity expression, to take care of words being used in different order in TNP and LC patterns. For example, consider the two patterns as 'complexity of computing' and 'computing complexity',

---

[2] http://nlp.stanford.edu/software/tagger.shtml

[3] http://ai.stanford.edu/users/sahami/CS2013/
ironman-draft/cs2013-ironman-v1.0.pdf

[4] http://www.computer.org/portal/web/publications/acmtaxonomy

[5] http://www.acm.org/about/class/2012

where first one is TNP and the second one an LC; the pruning system should allow the TNP to be recognized as a valid selected computer science learning concept despite the fact that they do not match exactly. We have set a similarity threshold of 0.6 to accept a TNP as a valid learning concept for further processing. Another example to illustrate it is when TNP is 'methods of numerical analysis' and LC is 'numerical analysis methods', this similarity expression allows the TNP to be selected (similarity value being 0.75), which is appropriate since they both refer to essentially the same thing. However, consider TNP as 'algorithm' and LC as 'clustering algorithm'; the similarity score obtained here is 0.5, less than the threshold of 0.6, and hence pruned. The similarity expression thus helps in identifying useful computer science learning concepts described in eBook text.

3   Relation Extraction

Relation Extraction (RE) is the task of identifying and extracting relationship between two or more entities occurring in a text source (Banko & Etzioni, 2011). Relations are usually in the form of tuple presented as {Arg1, Pred, Arg2}, where Arg1 and Arg2 are two entities, and Pred is the name of the relation between Arg1 and Arg2. For instance, given the sentence, "Mc-Cain fought hard against Barack Obama, but finally lost the election," an Open IE system should extract two tuples, (Mc-Cain, fought against, Barack Obama), and (McCain, lost, the election). The relation tuples are useful to understand the text in structured form without the requirement of reading it in detail. In the past several architectures have been proposed to find relation from texts. Some of the notable ones are TextRunner (Banko et al., 2007), WOEpos & WOEparse (Wu and Weld, 2010) and ReVerb (Etzioni et al., 2011).

The TextRunner system is based on Naive Bayes model with unlexicalized POS and NP-chunk features trained using examples heuristically generated from the Penn Treebank. The subsequent work showed that utilizing a linear-chain CRF (Banko & Etzioni, 2008) or Markov Logic Network (Zhu et al., 2009) can lead to further improvements. The WOEpos and WOEparse systems are two modifications proposed to the TextRunner system and use a model of relations learned from extractions heuristically generated from Wikipedia in two different ways. The ReVerb architecture is a general model of verb-based relation phrases, and has three important differences with previous methods. The first difference is that relation phrase is identified "holistically" rather than word-by-word. Secondly, potential phrases are filtered based on statistics over a large corpus that is actually the implementation of lexical constraint. The third one is that it is a "relation first" rather than "arguments first" system. ReVerb takes as input a POS-tagged and NP-chunked sentence and returns a set of (x; r; y) extraction triples.

A comparative study among TextRunner, WOEpos, WOEparse and ReVerb shows that ReVerb has the highest performance in terms of precision and recall

(Etzioni et al., 2011). We have therefore used the ReVerb extractor[6] to get relations from each sentence of some eBook part. The set of relations so identified are then pruned by seeing their arguments. Only those relations that have at least one valid learning concept as an argument are selected as valid relations.

4   Semantic Annotation and RDF

A semantic annotation approach tries to annotate unstructured text with appropriate tags so that it becomes more navigable and usable. Semantic annotation generates metadata that can be used for machine reading. RDF structures are essentially designed for reading by the machine. Therefore, though they look voluminous and difficult to understand for a normal user, computer systems are good at reading and interpreting them. If the data is represented in RDF format, it is easier to search, navigate and perform logical question answering on the data. It was this motivation that guided us in our attempt to process the unstructured text data from eBooks and generate equivalent and useful RDF schema, which allows better navigation and usage of the eBook content. This process in essence is equivalent to semantic tagging/ annotation of the unstructured eBook text.

RDF refers to Resource Description Framework and is created to support the semantic web in similar way that of HTML, which helped initiate the original web. The RDF specification includes syntax specification and RDF model. RDF syntax is set of triples which are known as RDF graph. RDF triples contains three components, subject, predicate and object. Subject refers to resource; predicate refers to features of the resource and thus represents the relationship between the subject and object. For example, consider the RDF example[7] given below:

```
<?xml version="1.0"?>
<RDF>
<Description about="http://www.w3schools.com/rdf">
<author>Jan EgilRefsnes</author>
<homepage>http://www.w3schools.com</homepage>
</Description>
</RDF>
```

Here, the statement represented in RDF is "The author of http://www.w3schools.com/rdf is Jan EgilRefenses." The subject in this case is "http://www.w3schools.com/rdf ", the predicate is "author" and the object is "Jan EgilRefsnes". We have devised an RDF schema to represent the extracted information from various parts of the eBooks. This will allow the more efficient search and navigation in the eBooks. The RDF schema we designed stores three kinds of information: (a) some metadata about the eBooks such as their title, author names, price etc.; (b) learning concepts and relations extracted from various parts of the eBook; and (c) some computed values from the extracted eBook data such as its

---

[6] http://reverb.cs.washington.edu/
[7] http://www.w3schools.com/rdf/rdf_rules.asp

readability, reviews and its sentiment rating. The entire data stored in our RDF schema is machine-readable and hence provides us with an efficient way of semantic search and navigation.

5   Dataset

We have performed our experimental work on a collection of about 30 eBooks, obtained mostly from the World Wide Web. We have not included the complete eBook identification information in the results shown in the next section, unless otherwise necessary. This has been done due to the fact that our work stores some qualitative evaluation information about eBooks in their RDF and also due to copyright issues. The dataset is comprised of eBooks from computer science domain, however this algorithmic formulation will work on eBooks in any domain. The only change required would be a reference document similar to the augmented ACM-CCF we used here.

6   Experimental Results

The following paragraphs present intermediate results obtained during different stages of execution of the algorithmic formulation. We have obtained results for all the 30 eBooks in our dataset. Here we present a part of the results for a particular eBook on Data Mining. The results shown here are for a particular part of the eBook and are representative sample of the results obtained. The results shown include TNPs identified, pruned concepts selected, valid relations, the RDF structure generated (along with its different nodes) and some graphical visualizations.

6.1   Extracted and Pruned Concepts

We have first extracted TNPs and then pruned them to select only valid computer science learning concepts. For example, for the chapter titled 'Introduction' for the eBook 'Data Mining- Concepts and techniques', we obtained a total of 1443 TNPs, out of which 96 are selected as valid computer science learning concepts. Some of these 96 learning concepts are:

> business intelligence, knowledge management, entity relationship models, information technology, database management system, semi supervised learning, cloud computing, web search, query processing.

6.2   Extracted and Selected Relations

The relations extracted using ReVerb are also pruned using the scheme described earlier. For the chapter 'Introduction' for the eBook 'Data Mining- Concepts and Techniques', we get a total of 596 relations, out of which we select 62 relations as involving valid computer science learning concept. A part of these relations (only 15 here), as visualized using Gephi[8], are shown in figure 1 below.

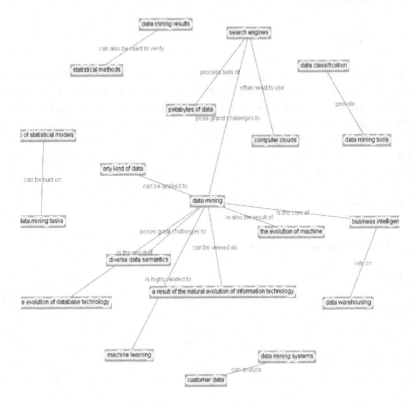

Fig. 1 Visualization of Relations Extracted from an eBook part

6.3   Generating the eBook RDF

The information about learning concepts, relations, externally obtained eBook information and computed values over extracted information from the eBook are written in the RDF schema programmatically. The RDF contains rdfs: resources for eBook metadata (title, author, number of chapters, number of pages, google eBook price & rating, its augmented ACM-CCF category, coverage score, readability score and sentiment polarity score); chapter information (title, top 20 concepts, and relations extracted); and the eBook review information (review text and computed sentiment). Thus, we are able to extract semantically useful information from eBook text and store it into a machine readable RDF structure. The eBook metadata representation is somewhat similar to a part representation as below:

---

[8] https://gephi.org/

```
<rdf:RDF
xmlns:rdf=http://www.w3.org/1999/02/22-rdf-syntax-ns#
xmlns:book="http://www.southasianuniversity.org/books/Data_Mining_Concepts_and_Techni
ques_Third_Edition#" >
<rdf:Description
rdf:about="http://www.southasianuniversity.org/books/Data_Mining_Concepts_and_Technique
s_Third_Edition#metadata">
<book:btitle>Data Mining Concepts and Techniques Third Edition</book:btitle>
<book:author>JiaweiHan,MichelineKamber,Jian Pei</book:author>
<book:no_of_chapters>13</book:no_of_chapters>
<book:no_of_pages>740</book:no_of_pages>
<book:bconcepts>rule based classification, resolution, support vector machines, machine learn-
ing,...</book:bconcepts>
<book:main_category>Intelligent Systems</book:main_category>
<book:main_cat_coverage_score>0.051107325</book:main_cat_coverage_score>
<book:related_category>Programming fundamentals</book:related_category>
<book:related_category>Information Management</book:related_category>
<book:googleRating>User Rating: **** (3 rating(s))</book:googleRating>
<book:readability_score>56 (Fairly Difficult)
</book:readability_score>
</rdf:Description>
```

The RDF representation contains the extracted concepts and relations, in addition to the eBook metadata. A representative sample of this information for a particular chapter node is given below:

```
<rdf:Description
rdf:about="http://www.southasianuniversity.org/books/Data_Mining_Concepts_and_Technique
s_Third_Edition#chapter6">
<book:cconcepts>transaction database, data mining, hash table, support count, decision analy-
sis, association rules,...............................................................................</ book:cconcepts>
<book:relation>the software display, may decide to purchase, a home security sys-
tem,0.3547584217041676</book:relation>
<book:relation>strong     association     rules,     satisfy,     both     minimum     sup-
port,0.44865456037003126</book:relation>
<book:relation>market basket analysis, may be performed on, the retail data of customer trans-
actions,0.9874720642457288</book:relation>
<book:relation>interesting     association     rules,     was     studied     by,     Omiecins-
ki,0.9628601549100813</book:relation>
.................................................................
</rdf:Description>
```

The eBook review information is stored in the RDF representation as follows. The information stored includes the review text as well as the sentiment polarity of the reviews computed using our earlier work (Singh et al., 2013a; 2013b)

```
<rdf:Description
rdf:about="http://www.southasianuniversity.org/books/Data_Mining_Concepts_and_Technique
s_Third_Edition#review-5">
<book:reviewLabel> positive </book:reviewLabel>
<book:reviewText> The content of the book seems pretty good...............</book:reviewText>
.............................................................
</rdf:Description>
```

The eBook metadata, which contains the data extracted from the eBook as well as data collected from the World Wide Web, is stored in the RDF. This metadata information can be visualized as an RDF graph as shown in figure 2 below.

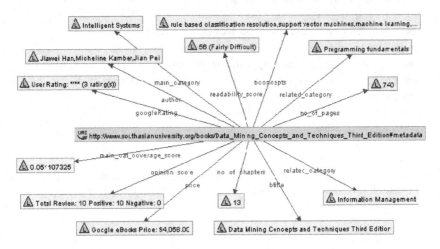

Fig. 2 RDF graph visualization of eBook metadata

7   Conclusion

We have demonstrated use of our algorithmic formulation for automated parsing of eBook text for concept and relation extraction, semantic annotation as RDF structure and visualizing relation network and eBook RDF. Our RDF schema contains information for eBooks, all of which are extracted through automated computational setup. We have identified the category of an eBook in the augmented ACM-CCF, mined a good amount of metadata for the eBook, extracted concepts and created a concept profile for every chapter, mined reviews from the Web and computed their sentiment polarity using our earlier work on sentiment analysis (Singh et. al, 2013a; 2013b). All the identified, extracted and mined information is then transformed into an appropriate machine readable RDF structure for the eBook. This conversion makes it very easy to search from the eBook, navigate through one or a large collection of eBooks, locate relevant chapter for obtaining knowledge about a learning concept, and organizing an eBook collection in an appropriate eBook management system. The entire analytics is efficient from the point of view of time complexity, since only one pass is needed to extract concepts from a sentence. The time complexity of our algorithmic approach is thus proportional to the number of sentences encountered in the eBook text.

In addition to the benefits and possibilities described above, we can use our system in many other fruitful ways. First of all, since we are doing an automatic information extraction and simultaneously classifying the book into one or more augmented ACM-CCF categories based on text analytics; the system can be used to automatically generate a set of eBooks for a particular set of students. Secondly, we can design a concept locator kind of user search system, which will guide the user in identifying the correct book and correct chapter of the eBooks for a detailed information about that concept. Thirdly, we have collected eBook reviews and mined their sentiment, we can utilize this information for recommending only those eBooks to users for which ratings are high and sentiment orientation labels are largely positive. Fourth, we can design a comprehensive learning recommendation system, that will not only recommend appropriate eBook (or a chapter from the eBook), but also suggests various other related eResources about it (Singh et al., 2013c). Some such resources could be top slides on the Web about a particular topic, top video-lectures on the topic, LinkedIn profiles of top persons working on that topic, top ranking web articles on the topic etc. The possibilities of future extension and application of this work are thus numerous.

References

Agichtein, E., Gravano, L.: Snowball: extracting relations from large plain-text collections. In: Proceedings of the ACM Conference on Digital Libraries, pp. 85–94. ACM (2000)

Agrawal, R., Gollapudi, S., Kannan, A., Kenthapadi, K.: Data Mining for Improving Textbooks. SIGKDD Explorations 13(2), 7–19 (2011)

Agrawal, R., Gollapudi, S., Kenthapadi, K., Srivastava, N., Velu, R.: Enriching textbooks through data mining. In: ACM DEV (2010)

Banko, M.: Open Information Extraction for the Web. Ph. D. dissertation, University of Washington (2009)

Banko, M., Etzioni, O.: The tradeoffs between open and traditional relation extraction. In: Proceedings ACL 2008, pp. 28–36 (2008)

Banko, M., Cafarella, M.J., Soderland, S., Broadhead, M., Etzioni, O.: Open information extraction from the web. In: Proceedings IJCAI (2007)

Bitton, D., Faerber, F., Haas, L., Shanmugasundaram, J.: One platform for mining structured and unstructured data: dream or reality? In: Proceedings 32nd VLDB, pp. 1261–1262 (2006)

Brin, S.: Extracting patterns and relations from the world wide web. In: Proceedings of International Workshop in the World Wide Web and Databases, pp. 172–183 (1998)

Etzioni, O., Fader, A., Christensen, J., Soderland, S., Mausam: Open Information Extraction: the Second Generation. In: Proceedings 22nd IJCAI, pp. 3–10 (2011)

GuoDong, Z., Jian, S., Jie, Z., Min, Z.: Exploring various knowledge in relation extraction. In: Proceedings ACL 2005, pp. 427–434 (2005)

Horn, C., Zhila, A., Gelbukh, A., Kern, R., Lex, E.: Using Factual Density to Measure Informativeness of Web Documents. In: Proceedings of the 19th Nordic Conference on Computational Linguistics (NODALIDA). Linkoping University Electronic Press, Oslo (2013)

Justeson, J.S., Katz, S.M.: Technical terminology: Some linguistic properties and an algorithm for identification in text. Natural Language Engineering 1(1) (1995)

Kambhatla, N.: Combining lexical, syntactic and semantic features with maximum entropy models. In: Proceedings 22nd ACL (2004)

Lent, B., Agarawal, R., Srikant, R.: Discovering trends in text databases. In: Proceedings KDD (1997)

Piskorski, J., Yangarber, R.: Multi-source, Multilingual Information Extraction and Summarization, Theory and Applications of Natural Language Processing. In: Poibeau, T., et al. (eds.) Information Extraction: Past, Present and Future. Introductory Survey, Springer, Heidelberg (2012)

Singh, V.K., Piryani, R., Uddin, A.: An eBook-based eResource Recommender System. In: Proceedings 5th International Conference on Pattern Recognition and Machine Intelligence, Kolkata, India. LNCS. Springer (in press, 2013c)

Singh, V.K., Piryani, R., Uddin, A., Waila, P.: Sentiment Analysis of Movie Reviews and Blog Posts: Evaluating SentiWordNet with different Linguistic Features and Scoring Schemes. In: Proceedings of 2013 IEEE International Advanced Computing Conference. IEEE Press, Ghaziabad-India (2013a)

Singh, V.K., Piryani, R., Uddin, A., Waila, P.: Sentiment Analysis of Movie Reviews: A new Feature-based Heuristic for Aspect-Level sentiment classification. In: Proceedings of the International Multi Conference on Automation, Computing, control, Communication and Compressed Sensing. IEEE Press (2013b)

Wu, F., Weld, D.S.: Open information extraction using Wikipedia. In: Proceedings 48th ACL, pp. 118–127 (2010)

Zhila, A., Gelbukh, A.: Comparison of Open Information Extraction for English and Spanish. In: 19th Annual International Conference Dialog 2013, Bekasovo, Russia, pp. 714–722 (2013)

Indexing Large Class Handwritten Character Database

D.S. Guru and V.N. Manjunath Aradhya

Abstract. This paper proposes a method of indexing handwritten characters of a large number of classes by the use of Kd-tree. The Ridgelets and Gabor features are used for the purpose of representation. A multi dimensional feature vectors are further projected to a lower dimensional feature space using PCA. The reduced dimensional feature vectors are used to index the character database by Kd-tree. In a large class OCR system, the aim is to identify a character from a large class of characters. Interest behind this work is to have a quick reference to only those potential characters which can have a best match for given unknown character to be recognized without requiring scanning of the entire database. The proposed method can be used as a supplementary tool to speed up the task of identification. The proposed method is tested on handwritten Kannada character database consisting of 2000 images of 200 classes. Experimental results show that the approach yields a good Correct Index Power (CIP) and also depicts the effectiveness of the indexing approach.

Keywords: Ridgelet Transform, Gabor Transform, Kd-tree, Handwritten Character Indexing.

1   Introduction

Optical Character Recognition (OCR) is one of the most fascinating and challenging areas of pattern recognition. Research on OCR is popular for its various potential applications in banks, postal departments, defense organizations etc. In such

D.S. Guru
Department of Studies in Computer Science, University of Mysore, Mysore
e-mail: dsg@compsci.uni-mysore.ac.in

V.N. Manjunath Aradhya
Department of Master of Computer Applications,
Sri Jayachamarajendra College of Engineering, Mysore
e-mail: aradhya.mysore@gmail.com

S.M. Thampi et al. (eds.), *Recent Advances in Intelligent Informatics*,   227
Advances in Intelligent Systems and Computing 235,
DOI: 10.1007/978-3-319-01778-5_23, © Springer International Publishing Switzerland 2014

applications, response time and search efficiency also have become important in addition to the measure of accuracy because of a large population. In conventional matching process, recognition of a character from a huge collection of characters requires matching of the character against all characters present in the database, which is essentially a more time consuming process. In order to speed up this process, a filtering process is usually brought up to select a minimum number of potential candidates for further matching operation. To select the minimum number of candidate hypotheses, one can think of classification or indexing approaches.

Indexing scheme is to ensure that the system gets a few potential candidates selected in a quick manner for a given query character, so that later a rigorous matching can be carried out in order to identify the correct character. Indexing can be either hash based or tree based. Compared to hash based indexing, a tree based indexing has attained more attention. Hence in this work, we present a mechanism of indexing characters using tree based indexing. Some of the interesting survey made on indexing can be seen in [5, 6, 7, 8].

Feature extraction is one of the most important stage in OCR system. Most of the feature extraction techniques fall into two or combination of three major categories: (i) Statistical (ii) Structural (iii) Global transformation [12]. Due to inherent advantages of using combination of transform techniques and subspace models, in this work we explore the concept of Ridgelet and Gabor transform as a feature extraction method. The proposed method is experimented on offline handwritten characters of Kannada. It is worth to note that the proposed indexing model is first of its kind in the literature on a character database. The main highlights of our work are (i) A speed up approach is used to select a minimum number of potential candidates for matching operation. (ii) An indexing approach based on Kd-tree is used. (iii) Explored the concept of Ridgelets and Gabor for better representation purpose.

The organization of the paper is as follows: Section II presents the proposed model. Experimental results are presented in section III. Section IV concludes the paper.

2 Proposed Model

In this section, we describe the proposed indexing model. The proposed model extracts features by ridgelet and Gabor transforms. The feature vectors obtained by these transforms are projected onto reduced dimensional space by PCA. The reduced feature vectors are then accommodated in Kd-tree for indexing purpose.

2.1 Ridgelet Features

Candes and Donoho [1] introduced a new system of representations named Ridgelets, which they showed to deal effectively with line singularities in 2-D.

In order to apply ridgelets to digital images, Do and Vetterli [2] proposed a Finite Ridgelet Transform (FRIT). FRIT is based on the Finite Radon Transform (FRAT), which is defined as summation of image pixels over a certain set of lines. Those lines are defined in a finite geometry in a similar way as the lines for the continuous Radon transform in the Euclidean geometry. Z_p is denoted as $Z_p = 0, 1, ..., p-1$. Where p is a prime number and Z_p is finite field with modulo p operations.

The FRAT of real discrete function f on the finite grid Z_p^2 is defined as:

$$FRAT_f(k,l) = \frac{1}{\sqrt{p}} \sum_{(i,j)\in L_{k,l}} f(i,j). \tag{1}$$

Here $L_{k,l}$ denotes the set of points that make up a line on the lattice Z_p^2, i.e.

$$L_{k,l} = \begin{cases} (i,j) : j = (ki+l)(mod\,p), i \in Z_p & \text{if } 0 \le k \le p \\ (l,j) : j \in Z_p & \text{if } k = p \end{cases}$$

Most of the energy information can be found in the low-pass of ridgelet image decomposition. Normally, feature vectors are typically several thousands elements wide. In order to reduce the dimension we used PCA [13] method. More details on ridgelets can be seen in [9].

2.2 Gabor Features

Gabor filter is a popular tool for extracting spatially localized features, useful for character recognition.

An even symmetric Gabor filter has the following general form in the spatial domain.

$$G(x,y,f,\theta) = exp\left[\frac{-1}{2}\left[\frac{x'}{\sigma_{x'}}\right]^2 + \left[\frac{y'}{\sigma_{y'}}\right]^2\right] cos(2*\pi*f*x') \tag{2}$$

Where $x'=x*sin\theta + y*cos\theta$ and $y'=x*cos\theta - y*sin\theta$. Design of the Gabor filter is accomplished by tunning the filter with a specific band of spatial frequency and orientation by appropriately selecting the filter parameter, such as the spread of the filter $\sigma_{x'}, \sigma_{y'}$, radial frequency f and the orientation of the field θ [3] [4]. Hence in this work, filtering is performed with frequency f value 4 and orientation parameters $\theta = 0, \frac{\pi}{4}, \frac{\pi}{2}, \frac{3*\pi}{4}$. The values of $\sigma_{x'}$ and $\sigma_{y'}$ were empirically determined to be 2 and 4 respectively.

The resultant Gabor filter orientation feature vectors is prohibitively high (2500 elements wide for each orientation). In order to compress the dimension of Gabor features, we used PCA [13] method. Detailed description on Gabor features can be seen in [10].

2.3 *Indexing Using Kd-Tree*

In the proposed indexing model, the obtained feature vectors are indexed using Kd-tree data structure [11]. Given a set of K dimensional data points, the Kd-tree organizes the points in a K-dimensional space, which is useful for searching / retrieving data similar to a query. The construction algorithm of Kd-tree is as follows, at the root, we split the set of points into two subsets of roughly the same size by a hyperplane perpendicular to the first coordinate of the points.

At the children of the root the partition is based on second coordinate and so on, until depth of K-1 at which partition occurs based on last coordinate, where K is the dimension of the feature space. After depth K, again, partitioning is based on first coordinate. The recursion stops only when one point is left, which is then stored at the leaf. Because a K-dimensional Kd-tree for a set of n points is a binary tree with n leaves, it uses O(n) storage with search time being O(nlogn). In addition to this, in Kd-tree there is no overlapping between nodes. When a query feature vector of dimension K is given, search is invoked using Kd-tree and top matches that lie within a specified distance from the query are retrieved.

3 Experimental Results

The proposed character indexing system is tested on offline handwritten Kannada characters. The database has been created with the assistance of 100 writers of different streams such as school children, degree students, university students of different age group. The dataset holds 200 classes (vowels, consonants and modifiers). In this work, we considered 2000 images (i.e., 10 images per class) for experimentation. We have carried out the experiment in four stages. In each stage, we have varied class size c (where c = 50, 100, 150, 200). In every stage of our experiment, the system was trained with 4 and 9 samples and 1 image is used for testing. All our experiments are carried out on a PC machine with P IV, 2.2GHz CPU and 1GB RAM memory under Matlab 7.0 platform.

After PCA process, feature vector of dimension K is indexed through Kd-tree. We varied the number of features K (where K = 5, 10, 15, 20, 25, 30, 40, 45 & 50). Since K has a considerable impact on Correct Index Power (CIP), we choose the value that corresponds to the best CIP on the image set. In this work, we use Correct Index Power (CIP) as a performance evaluation measure for indexing, where CIP is the ratio of the correctly answered queries to the total number of queries.

The graph of Top Matches v/s CIP under different class size are shown in Figure 1-4. The features obtained using Gabor performs quite better compared to Ridgelet features under 50 class size. The same is not observed for 100, 150 and 200 class size. The features of Ridgelet perform better compared to Gabor as class size increases. It is worth to note that, the results of Ridgelet gives better CIP when it is less trained (Refer Figure 1). In case of Gabor, we observed variations under different class size (Refer Figure 2, 3, and 4).

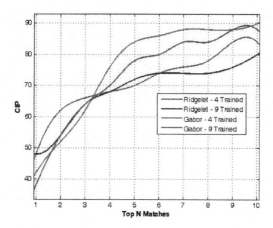

Fig. 1 Top Matches v/s CIP under 50 class size

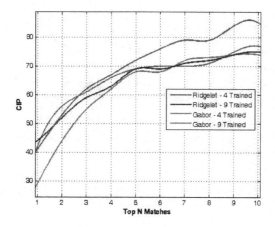

Fig. 2 Top Matches v/s CIP under 100 class size

The efficacy of the proposed indexing scheme lies in its efficiency from the point of search time. Table 1 shows the time analysis for indexing based and conventional identification method for 4 trained images. From this, it is noticed that, the proposed indexing method reduces the search time. It is also worth to note that, the percentage of time reduction is noticeable using Kd-tree based indexing model against the conventional identification method even when class size is increased.

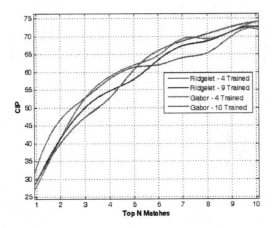

Fig. 3 Top Matches v/s CIP under 150 class size

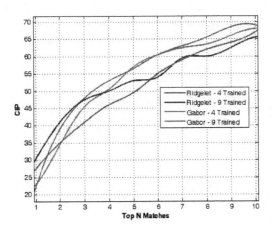

Fig. 4 Top Matches v/s CIP under 200 class size

Table 1 Time analysis for Indexing and Without Indexing

	Time in Secs.	
Class	With Indexing	Without Indexing
50	0.0019	0.016
100	0.0024	0.078
150	0.005	0.18
200	0.0057	0.29

4 Conclusion

We proposed an efficient indexing technique using Kd-tree to reduce the search space for a large class character database. As a part of feature extraction stage, Ridgelet and Gabor features are used to index the database by forming Kd-tree separately. Our first exploration on Kd-tree is proven to be suitable data structure for character identification system particularly in the analysis of execution of range search algorithm. The proposed method is recommended to be used as a supplementary tool to speed up the task of identification. From the experiment, it is clear that indexing prior to character identification is faster than conventional character identification. In near future the authors plan to work on different structural features and also on exploring different indexing schemes.

Acknowledgements. The authors would like to thank Dr. S. Manjunath and K.B. Nagasundara for their support rendered during the work.

References

1. Candes, E.J., Donoho, D.L.: Ridgelets: a key to higher-dimensional intermittency? Phil. Trans. R. Soc. Lond. A, 2495–2509 (1999)
2. Do, M.N., Vetterli, M.: Finite ridgelet transform for image representation. IEEE Transactions on Image Processing (2002)
3. Jain, A.K., Prabhakar, S., Hang, L.: A multichannel approach of fingerprint classification. IEEE Transactions on PAMI 21(4), 349–359 (1999)
4. Jain, A.K., Reisman, J.: A hybrid fingerprint matcher. Pattern Recognition 36(7), 1661–1673 (2003)
5. Jayaraman, U., Prakash, S., Gupta, P.: Indexing multimodal biometric databases using kd-tree with feature level fusion. In: ICISS 2008, pp. 221–234 (2008)
6. Lu, H., Ooi, B.C., Shen, H.T., Xue, X.: Hierarchical indexing structure for efficient similarity search in video retrieval. IEEE Trans. on Knowledge and Data Engineering 18, 1544–1559 (2006)
7. Mukherjee, R.: Indexing techniques for fingerprint and iris databases. Master Thesis, West Virginia University (2007)
8. Nagasundara, K.B., Guru, D.S., Manjunath, S.: Indexing of online signatures. International Journal of Machine Intelligence 3, 289–294 (2011)
9. Naveena, C., Manjunath Aradhya, V.N.: An impact of ridgelet transform in handwritten recognition: A study on very large dataset of kannada script. In: IEEE World Congress on Information and Communication Technologies (WCIT), pp. 622–625 (2011)
10. Naveena, C., Manjunath Aradhya, V.N., Niranjan, S.K.: The study of different similarity measure techniques in recognition of handwritten characters. In: ACM International Conference on Advances in Computing, Communications and Informatics (ICACCI), pp. 781–787 (2012)
11. Samet, H.: The Design and Analysis of Spatial Data Structures. Addison-Wesley (1990)
12. Tokas, R., Bhadu, A.: A comparative analysis of feature extraction techniques for handwritten character recognition. International Journal of Advanced Technology Engineering Research (IJATER) 2(4), 215–219 (2012)
13. Turk, M., Pentland, A.: Eigenfaces for recognition. Journal of Cognitive Neuroscience 3, 71–86 (1991)

Text Classification of Kannada Webpages Using Various Pre-processing Agents

N. Deepamala and P. Ramakanth Kumar

Abstract. Text classification of Webpages has wide applications and many techniques have been employed to achieve the same. In this paper, an attempt is made to classify Kannada webpages into pre-determined 6 classes or categories. Kannada is a morphologically rich Indian Language. Kannada Webpages are subjected to different pre-processing steps and machine learning techniques like Naïve Bayes and Maximum Entropy are applied to train models. All the pre-processing steps before classification are implemented as intelligent agents doing a particular task like Language Identification, Sentence Boundary detection and Term frequency calculation. It is observed that highest accuracy of 0.9 is achieved using both stemming and stopword removal.

1 Introduction

Classification of the given document into predefined categories of classes has many applications in current scenario. Text classification is used for content filtering, in digital libraries and other applications. Different methods are employed to achieve classification of text. The research work carried out till date to classify Kannada Language is limited. Kannada Language is morphologically rich and one of the 40 most spoken Languages in the world. In this paper, an attempt is made to classify the given text of a Webpage into one of the given 6 categories. The categories considered for classification are Astrology, Politics, Finance, Geography, Health and Sports.

The given Kannada Webpage content is subjected to different pre-processing agents before classification. Each pre-processing step is implemented using some intelligent method to achieve the requirements. The text classification involves

N. Deepamala · P. Ramakanth Kumar
R.V. College of Engineering, Bangalore
e-mail: {deepamalan,ramakanthkp}@rvce.edu.in

S.M. Thampi et al. (eds.), *Recent Advances in Intelligent Informatics*,
Advances in Intelligent Systems and Computing 235,
DOI: 10.1007/978-3-319-01778-5_24, © Springer International Publishing Switzerland 2014

two phases, training and testing. In training phase, models are trained using machine learning techniques like Naïve Bayes and Maximum Entropy. In testing phase, the category is identified using trained models. The training and testing phase are depicted in the Figure 1.

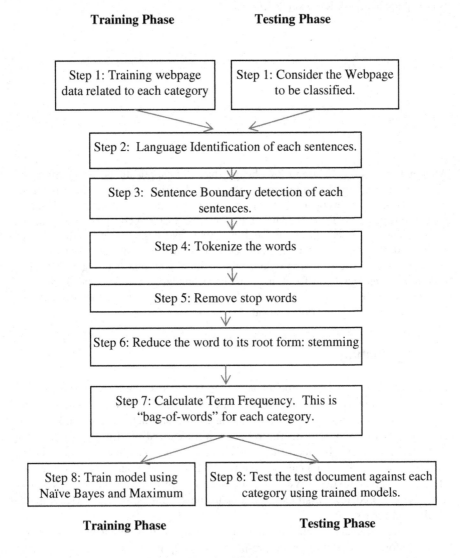

Fig. 1 Flow diagram for Training and Testing Webpages

2 Related Work

2.1 Text Classification Algorithms

Web page classification methods in machine learning are discussed in [1,2]. Different methods have been employed to achieve text classification or categorization. [3, 4, 5] uses Naïve Bayes classification to classify text documents. [6] Explains classification of twitter text to pre-defined classes based on features extracted from author's profile and text. [7] Discusses different approaches and the issues related to classifiers. Text categorization using different classifiers Bayesian and Decision tree are discussed in paper [8]. [9] Tests 5 different text categorization methods SVM, k-Nearest method, neural network, linear least squares fit and Naïve Bayes method. It shows that SVM, KNN And LLSF outperform Naïve Bayes. The classification of text based on Support Vector Machines (SVM) is discussed in [10, 11, 12, 13, 14, 15, 16] Feature selection for text classification using Gini coefficient of inequality is explained in [17]. Maximum Entropy is used for Arabic text classification in [18]. Graph based approach to text classification is discussed in [19]. The paper [20], improves performance by using both associative classifier and Naïve Bayes classifier for Chinese text.

2.2 Text Classification of Indian Languages

[21] Demonstrates the classification of 10 major Indian Languages using different classifiers like Naïve Bayes, k-Nearest Neighbor and SVM. It has been demonstrated that soft-margin SVMs outperform other techniques without use of any morphological analyzers or stemmers. [22] Proposes steps to classify Punjabi content into sports class. With stop words, stemming and term frequency, linguistic information is also added to achieve the results. [23] Shows performance improvement using Ontology based classification and Hybrid based classification on Punjabi text. [24] Discuss about lexical chain of related words for text classification using Word Net. [25] Achieve text classification using bag-of-words and Naïve Bayes at sentence level. The stop words are removed from the sentences, words that occur only once in the data are removed, term frequencies are measured. The models developed using bag-of-words and Naïve Bayes are applied on a corpus to classify Kannada sentences into 4 classes.

3 Proposed Work

The pre-processing steps are implemented as agents. Pre-processing agents and the method used to achieve the same are discussed below.

3.1   Training and Testing Corpus

There is no standard corpus like Reuters available for Kannada Language. The Kannada Webpages available in the internet are used as corpus for training and testing. The Webpages were formatted using a script to remove HTML tags, extra spaces etc. The total number of documents used for training and testing is shown in Table 1.

Table 1 Total Number of Train and Test documents

Category	Total documents	Total Number of Words
Astrology	100	82294
Finance	100	62193
Politics	100	163911
Geography	100	108525
Health	100	116929
Sports	100	132332

3.2   Language Identification

The webpages can contain sentences of different languages. It is common to see English sentences between Kannada sentences in a webpage. The Language Identification of each sentence is carried out using n-gram processing as discussed in [27]. Since majority of the sentences in Kannada are verb final sentences, n-gram processing of only the last word can identify Kannada sentences. [27] is a related work to this paper for text classification.

3.3   Sentence Boundary Detection

Next step is to identify the Kannada Sentence Boundary. The Majority of the Kannada sentences end with a verb. Verb takes different forms based on gender, tense and form (singular/plural). A suffix list of verbs (VERBS_SUFFIX file) and a list of abbreviations (ABBREVIATIONS file) were created. Algorithm discussed by Manning (et.al) [28] was modified to work for Kannada Language. The paper [29] discusses the method followed. The algorithm is as follows:

- Place putative sentence boundaries after all occurrences of punctuations. ? ! ; : -_ . Let this be **Sentence1**. If Sentence1 ends with ? ! ; : - _ , regard the putative sentence boundary as sentence boundary.
- Move the putative boundary after following quotation marks, if any, to next occurrence of .?!;:- Let this be **Sentence2**.
- Consider the last word of Sentence1 before period. Disqualify a period boundary of Sentence1 in the following circumstances

If period is preceded by a known abbreviation of a sort that does not normally occur word finally. Such abbreviations are listed in ABBREVIATIONS file.
- Regard the putative sentence boundary of Sentence1 as sentence boundary

  If it matches with any of the verb forms that can possibly end a sentence, such verb suffixes are listed in VERBS_SUFFIX file.
- Make Sentence2 as Sentence1 and Repeat from Step2.

3.4 Stemming

The Paice stemmer developed by Chris Paice [31] is used for stemming. The stemmer removes suffixes from the words based on the suffixes listed in the rules file. Kannada words take different forms based on suffixes. For Kannada 300 suffixes are listed in the rules file to stem the words. The rules file content is as shown below:

ಗಳ,?,stop   ;

ನನ್ನು,?,stop   ;

ಯೊಂದನ್ನು,?,stop   ;

ಯುವದದನ್ನು,?,stop   ;

ವದನ್ನು,?,stop   ;

E.g. ದೊರಕುತ್ತದೆ is stemmed to ದೊರಕು.

ಚಿತ್ರಗಳಿವೆ is stemmed to ಚಿತ್ರ.

With stemming, the number of words which act as features for text classification reduces.

3.5 Stop Word Removal

Nearly 202 stop words are listed which are common in any Kannada document and do not add any importance to the document category. These stop words are removed from the sentences.

E.g. ತಾನೆ ಎಂದು ಅವನ್ನು ನನ್ನನ್ನು ಅದರಿಂದ ಇವೂ ಅಂಥ ಅದಕ್ಕಾಗಿ ಈತನ ಏಕೆಂದರೆ ಅಷ್ಟೇ ಹಾಗೆ ಅಲ್ಲಿಗೆ ಎಂದೂ ಏನಾದರೂ ಅಂದು ಇರುತ್ತದೆ ಇದ್ದರೆ ಇವಳಿಗೆ ಮಾಡಿ

3.6 Term Frequency

Frequency of terms in each Webpage $TF_{t,d}$, are calculated before generating a model using the machine learning algorithm. Mallet is used for text classification. Before generating a model, the documents are converted into term-frequency feature set by the tool.[30]

3.7 *Text Classification*

Naïve Bayes and Maximum Entropy machine learning techniques are used to train models and test the test corpus. In Maximum Entropy, the word level features with corresponding weights are generated iteratively. The probability that the given document d belong to class c whose weights λ is

$$P(c|d, \lambda) = \frac{\exp \sum \lambda_i f_i (c, d)}{\sum_{c' c} \exp \sum_i \lambda_i f_i (\bar{c}, d)} \tag{1}$$

In Naïve Bayes, the probability that document d is classified into class c is based on the score calculated as below:

$$score (d, c) = P(c) * \prod_{i=1}^{n} P(w_i | c) \tag{2}$$

$$P(c|d) = \frac{score (d, c)}{\sum_{c' c} score (d, \bar{c})} \tag{3}$$

4 Experiment Results

The results of accuracy obtained by testing the trained models using Naïve Bayes and Maximum Entropy with different pre-processing techniques are shown in Table 2. The experiments were conducted on separate train and test sets and using 10-fold cross validation.

Table 2 Accuracies of Naive Bayes and Maximum Entropy with pre-processing agents

Sl. No	Stemming	Stop words	NB (Train and Test set)	ME (Train and Test set)	NB (10-fold cross validation)	ME (10-fold cross validation)
1	No	No	0.84	0.83	0.86	0.87
2	Yes	No	0.85	0.87	0.87	0.88
3	No	Yes	0.90	0.88	0.87	0.89
4	Yes	Yes	0.92	0.90	0.88	0.91

The results show that with stemming and removal of stopwords, the performance of Maximum Entropy and Naïve Bayes classifiers improves to a small extent. With stemming and stopword removal, the word list per document gets reduced. This improves the Maximum Entropy classifier since it uses word-level features. 10-fold cross validation by randomly splitting the corpus into train and test set gives steady increase in accuracy with stemming and stopword removal.

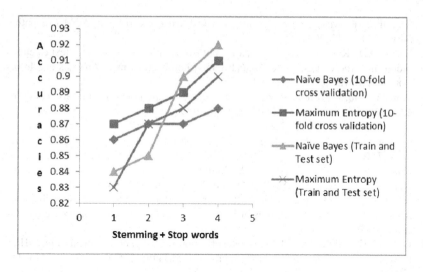

5   Conclusions

Text classification of Kannada Language webpages using Naïve Bayes and Maximum Entropy is discussed in this paper. Pre-processing steps like language identification, sentence boundary detection, stemming and stopword removal are applied on the webpage content before classification. In the results it is observed that classifier performance improves to a small extent with stemming and stopword removal in both Naïve Bayes and Maximum Entropy.

References

1. Qi, X., Davison, B.D.: Web page classification: features and algorithms. ACM Comput. Surv. 41(2), 1–31 (2009)
2. Tsukada, M., Washio, T., Motoda, H.: Automatic web-page classification by using machine learning methods. In: Zhong, N., Yao, Y., Ohsuga, S., Liu, J. (eds.) WI 2001. LNCS (LNAI), vol. 2198, pp. 303–313. Springer, Heidelberg (2001)
3. Dai, W., Xue, G.-R., Yang, Q., Yu, Y.: Transferring Naïve Bayes Classifiers for text classification. In: Proceedings of AAAI Conference on Artificial Intelligence, pp. 540–545 (2007)
4. McCallum, A., Nigam, K.: A Comparison of Event Models for Naive Bayes Text Classification. In: AAAI/ICML 1998 Workshop on Learning for Text Categorization, pp. 41–48 (1998)
5. Zhang, L., Zhu, J., Yao, T.: An evaluation of statistical spam filtering techniques. ACM Transactions on Asian Language Information Processing (TALIP) 3, 243–269 (2004)
6. Sriram, B., Fuhry, D., Demir, E., Ferhatosmanoglu, H., Demirbas, M.: Short Text Classification in Twitter to Improve Information Filtering. In: Proceeedings of 33rd International ACM SIGIR Conference, pp. 841–842 (2010)

7. Sebastiani, F.: Machine learning in automated text categorization. ACM Computing Surveys 34, 1–47 (2002)
8. Lewis, D.D., Knguette, M.: A comparison of two learning algorithms for text categorization. In: Proceedings of 3rd Annual Symposium on Document Analysis and Information Retrieval, pp. 81–93 (1994)
9. Yang, Y., Liu, X.: A re-examination of text categorization methods. In: Proceedings of 22nd Annual International SIGIR, pp. 42–49 (1999)
10. Joachims, T.: Text categorization with support vector machines: learning with many relevant features. In: Proceedings of 10th European Conference on Machine Learning, pp. 137–142 (1998)
11. Cristianini, N., Shawe-Taylor, J.: An Introduction to Support Vector Machines and Other Kernel-based Learning Methods. Cambridge University Press (2000)
12. Joachims, T.: Text categorization with support vector machines: learning with many relevant features. In: Proceedings of ECML 1998, 10th European Conference on Machine Learning (1998)
13. Leopold, E., Kindermann, J.: Text categorization with support vector machines. How to represent texts in input space? Journal of Machine Learning 46(1-3), 423–444 (2002)
14. Tong, S., Koller, D.: Support vector machine active learning with applications to text classification. Journal of Machine Learning Research, 999–1006 (2001)
15. Chang, C.C., Lin, C.J.: LIBSVM: A library for support vector machines. ACM Transactions on Intelligent Systems and Technology 2(3) (2011)
16. Sun, A., Lim, E.P., Ng, W.K.: Web classification using support vector machine. In: Proceedings of the Fourth International Workshop on Web Information and Data Management, McLean, Virginia, USA, pp. 96–99. ACM Press, New York (2002)
17. Ranabir Singh, S., Hema Murthy, A., Timothy Gonsalves, A.: Feature Selection for Text Classification based on Gini Coefficient of Inequality. In: Fourth International Workshop on Feature Selection in Data Mining, JMLR 2010, Hyderabad, pp. 76–85 (2010)
18. El-Halees, A.M.: Arabic Text Classification Using Maximum Entropy. The Islamic University Journal 15, 157–167 (2007)
19. Jiang, C., Coenen, F., Sanderson, R., Zito, M.: Text Classification using Graph Mining-Based Feature Extraction. The Journal of Knowledge Based Systems (2010)
20. Lu, S.H., Chiang, D.A., Keh, H.C., Huang, H.H.: Chinese text classification by the naïve Bayes classifier and the associative classifier with multiple confidence threshold values. Knowledge Based Syst. 23(6), 598–604 (2010)
21. Raghuveer, K., Murthy, K.N.: Text Categorization in Indian Languages using Machine Learning Approaches. In: Proceedings of 3rd International conference on Artificial Intelligence, pp. 1864–1883 (2007)
22. Nidhi, V.G.: Algorithm for Punjabi Text Classification. International Journal of Computer Applications 37, 30–35 (2012)
23. Nidhi, V.G.: Domain Based Classification of Punjabi Text Documents using Ontology and Hybrid Based Approach. In: Proceedings of the 3rd Workshop on South and Southeast Asian Natural Language Processing, COLING 2012, pp. 109–122 (2012)
24. Mohanty, S., Santi, P.K., Mishra, R., Mohapatra, R.N., Swain, S.: Semantic based text classification using wordnets: Indian languages perspective. In: Proceedings of the 3th International Global WordNejuh t Conf., South Jeju Island, Korea, pp. 321–324 (2006)

25. Jayashree, R., Srikanta Murthy, K.: An analysis of sentence level text classification for the Kannada language. In: Proceedings of International Conference of Soft Computing and Pattern Recognition (SoCPaR), pp. 147–151 (2011)
26. Deepamala, N., Ramakanth Kumar, P.: Language Identification of Kannada Language using N-Gram. International Journal of Computer Applications (0975-8887) 46(4), 24–28 (2012)
27. Manning, C.D., Schütze: Foundations of statistical natural language processing. The MIT Press, Cambridge
28. Deepamala, N., Ramakanth Kumar, P.: Sentence Boundary Detection in Kannada Language. International Journal of Computer Applications (0975-8887) 39(9), 38–41 (2012)
29. Mallet: A machine learning for language toolkit (2002), http://mallet.cs.umass.edu by A McCallum
30. Paice, C.: Paice Stemmer, http://www.comp.lancs.ac.uk/computing/research/stemming/general/paice.htm

A Novel Agglomerative Hierarchical Approach for Clustering in Medical Databases

Yogita Thakran and Durga Toshniwal

Abstract. Agglomerative hierarchical clustering plays vital roles in large number of application areas such as medical, bioinformatics, information retrieval etc. It generates clusters by iteratively merging sub clusters and hence merging criterion is very critical for its performance. But most of existing techniques do not consider global features of clusters to decide upon which clusters should be merged and are even unable to undo a wrong merge once it is done. These factors contribute to their poor performance. Also existing techniques face challenges on real world databases because these are proposed either for numeric or categorical data but real world data contains mixed attributes for example medical databases. To address the above mentioned drawbacks, in this paper, we propose a novel agglomerative hierarchical clustering method which avails the spread of data clusters as merging criterion. It is helpful in considering overall distribution of clusters in merging them. Variance and entropy are employed to measure the spread of a cluster in numeric and categorical attributes respectively. To counter the effect of a wrong merge, proposed method allows reallocation of data objects between clusters. We have experimented on real life medical databases and results show the efficacy of proposed approach.

1 Introduction

Clustering is a process of grouping data into classes or clusters such that objects within a cluster are highly similar to one other but are very dissimilar to objects of other clusters [4]. Hierarchical clustering techniques have vital role in large number of domains such as pattern recognition, medical diagnosis, bioinformatics etc [10]. These techniques do not require user to specify number of cluster and initialize cluster centres as oppose to partitioning clustering methods [8]. There are two basic categories of hierarchical clustering techniques [4, 10], first one is: Agglomerative

Yogita Thakran · Durga Toshniwal
IIT, Roorkee
e-mail: thakranyogita@gmail.com, durgafec@iitr.ernet.in

S.M. Thampi et al. (eds.), *Recent Advances in Intelligent Informatics*,
Advances in Intelligent Systems and Computing 235,
DOI: 10.1007/978-3-319-01778-5_25, © Springer International Publishing Switzerland 2014

which start with each object as individual cluster and successively merges clusters. Second category is: Divisive which start by taking whole data as a single cluster and successively divides a cluster into small cluster. Most of existing techniques [4, 10, 5] have their own limitation of using only selected data objects of clusters as representative of whole cluster to decide upon which clusters should be merged or split instead of considering global structure of clusters and even are unable to undo a wrong merge once it is done which leads to their poor performance.

The focus of clustering literature has been mainly on designing algorithms for either numeric datasets or categorical dataset [1, 7, 5, 2]. But real world applications generate mixed datasets which contains both numeric and categorical attributes. For an example a heart patient database from medical domain may contain following attributes: gender, chest pain type and defect type which are categorical attributes and age, serum cholesterol, maximum heart rate achieved, blood pressure which are numeric attributes. So existing clustering methods require either converting numeric attributes to categorical or vice- versa and face the challenge of information loss due attribute type conversion and result in poor quality clusters.

In this paper, we propose a novel agglomerative hierarchical clustering method which avails the spread of clusters as merging criterion because it represents the overall structure of a cluster. Variance and entropy [9] are employed to measure the spread of a cluster in numeric and categorical attributes respectively. Instead of focusing on two separate clusters, it temporally merges the clusters with one other and those two clusters of which spread is minimal are finally merged. To counter the effect of a wrong merge, proposed method allows reallocation of data objects between clusters.

The rest of this paper is organized as follows: Section 2 discusses related work. Preliminary concepts of agglomerative hierarchical clustering are given in Section 3. Proposed method is detailed in Section 4. Experimental results are presented in Section 5 and Section 6 concludes the paper.

2 Related Work

In this section, we will summarize the related literature on hierarchical clustering and mixed attribute data clustering.

A hierarchical clustering algorithm named as MPM is proposed in [7] for clustering of non-numeric data. It also presented a novel similarity measure that can capture the characterized properties from data. For better accuracy it applies matrix permutation and matrix participation partitioning to the similarity matrix. In [2] a heuristic method for converting categorical attributes to numerical values is presented which exploits the intra-attributes and inter-attributes information. It is based on the concepts of information theory and probabilistic model instead of just simply encoding categorical attributes or doing entropy based partitioning.

Hsu in [5] has extended the concept hierarchy to distance hierarchy by associating weight on each link for calculating distance between categorical data attributes. Also a distance hierarchy for numeric attributes is created to unify the distance calculation between data objects. Though it has shown effective results when integrated with hierarchical agglomerative clustering but it requires domain knowledge for distance hierarchy. CAVE an incremental partitioning clustering algorithm for mixed data has been proposed in [6]. It applies the concept of variance and entropy for handling mixed data along with distance hierarchy [5].

3 Preliminary Concepts - Agglomerative Hierarchical Clustering

An agglomerative hierarchical clustering method comprises following steps [10]:

- Step 1: Initialize each object as a cluster and compute the dissimilarity matrix
- Step 2: Repeat
- Step 3: Merge the two most similar cluster
- Step 4: Update the dissimilarity matrix
- Step 5: Until terminating condition or only single cluster remains

Merging criterion defines how to find the most similar clusters. Though most of agglomerative approaches follow the same basic approach but they differ in terms of merging criterion. Following are some popular variations [4, 10]:

- Single Link Hierarchical Agglomerative Clustering (Single Link HAC: It defines the similarity between two clusters as the similarity of closest objects that are in different clusters.
- Complete Link Hierarchical Agglomerative Clustering (Complete Link HAC): It defines the similarity between two clusters as the similarity of farthest objects that are in different clusters.
- Group Average Hierarchical Agglomerative Clustering (Group Average Link HAC): It defines the similarity between two clusters as the average of pair wise similarity of all pair of object from different clusters.

4 The Proposed Method

This section presents proposed agglomerative hierarchical clustering method:

In Algorithm 1, initially all clusters are temporally merged pair wise and their spread after merging is calculated and stored in a matrix. For later iterations only new cluster is temporally merged with other clusters to compute the spread, for other clusters spread from previous iterations is used. Proposed method uses minimum of temporally merged clusters spread as merging criterion because this criterion focus on a single cluster instead of two separate clusters. Let C is a given cluster which contains N data objects and each object is defined by P numeric and Q categorical attributes then spread of cluster is given by Equation (1):

$$Spread\ of\ Cluster(c) = \sum_{i=1}^{P} variance(x_i) + \sum_{j=1}^{Q} Entropy(x_j) \qquad (1)$$

Algorithm 1. Proposed Agglomerative Hierarchical Clustering Method

---

Input: Dataset, Threshold
Output: Clusters

1: Initialize each cluster to a single object $C_1, C_2, ..., C_n$
2: Temporally merge each cluster with one other and compute the spread of merged clusters as defined in Equation (1).
3: Let C_i and C_j are cluster that have minimum spread after merging and value of spread is SF.
4: IF SF < Threshold
5: Then Keep C_i and C_j merge as final and if number of clusters > 1 then Go to Step 2
6: Else Go to Step 7
7: Reallocation of data objects between clusters.

Where variance(x_i) represents the variance of ith numeric attribute and Entropy(x_j) is the entropy [9] of jth categorical attribute. x_i and x_j are the vector of data values of all N objects of cluster C corresponding i^{th} and j^{th} attribute. So spread of a cluster C of mixed attribute data objects is defined as the sum of variance of numeric attribute and entropy of categorical attributes for object of cluster C. Step 7 of Algorithm 1 check all the data objects for reallocation to other clusters. An object O is moved (reallocated) from cluster C_l to another cluster C_k if this movement of O decreases the spread of cluster C_l and increase in the spread of cluster C_k is also less than absolute decrement in the spread of cluster C_l. For setting the threshold value in experiments we have plotted the minimum spread of each iteration and set to the maximum value of spread in plotted graph.

Concept of variance and entropy has been used in literature to solve other problems also such as in [6], it has been used to address cluster centre initialization problem for mixed data. We have also used this concept for optimisation of merging criterion in agglomerative hierarchical clustering because variance for numeric data and entropy for categorical data represent the inconsistency between data objects of a cluster which we have named as spread of cluster.

5   Experiments

All experiments are performed on a system having Intel(R) Core(TM) 2 Duo T5750 processor and all implementations are done in matlab R2010a.

5.1   Real Life Medical Domain Datasets

To evaluate the performance of proposed method, we have worked on three benchmark real life datasets of medical domain. These are taken from UCI machine learning repository [3] and summarized in Table 1.

Originally there are 516 benign and 445 malignant cases in mammographic dataset but we have worked with 800 records out of 961 records other are discarded

Table 1 Characteristics of data sets used in experiments

Data set name	Types of Attributes	Number of instances	Number of attributes	Missing Values
BUPA liver disorders	Numeric, Categorical	345	7	No
Heart (Statlog)	Real , Binary, Nominal	270	13	No
Mammographic Mass	Mixed	961	6	Yes

during pre-processing step. Numeric attributes of all datasets has been normalized using z-score [4] method in our experiments and missing values are filled with most frequent value of that attribute.

5.2 Performance Measure

Two performance measures: accuracy rate and entropy have been examined to evaluate and compare the performance of proposed technique. Accuracy rate and entropy are defined as presented in [2].

Accuracy Rate:- It is the measure of extent to which a cluster contains object of a single class. Higher the accuracy rate more pure is the cluster. Let k is the number of clusters, M is the total number of data objects and M_i is the number of object of dominating class in cluster i then accuracy rate is defined by Equation (2):

$$Accuracy\,Rate = \frac{\sum_{i=1}^{k} M_i}{M} \tag{2}$$

Entropy:- It gives the randomness of clustering. Smaller the value of entropy better is the quality of clustering. Let l is the number of classes, k is the number of clusters and P(i,j) is the probability that object of cluster i belongs to class j then entropy of cluster i is defined by Equation (3). Entropy of clustering is the weighted sum of entropies of individual clusters as defined by Equation (4) where N_i is the size of cluster i and N is the size of dataset.

$$E(i) = -\sum_{j=1}^{l} P(i, j) \log_2 P(i, j) \tag{3}$$

$$E = \sum_{i=1}^{k} \frac{N_i}{N} * E(i) \tag{4}$$

5.3 Experimental Result

In this section, we present the experimental results of proposed method in terms of accuracy rate and entropy. Proposed method has been compared with all three

Fig. 1 Clustering Accuracy
Rate on Bupa Liver Disorder
Dataset

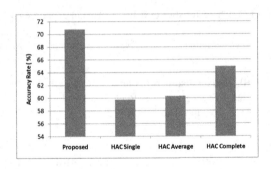

Fig. 2 Entropy of Clustering on Bupa Liver Disorder
Dataset

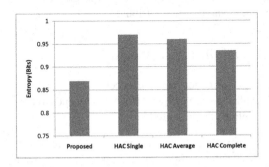

Fig. 3 Clustering Accuracy
Rate on Heart Dataset

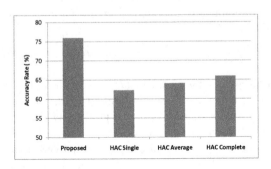

versions of hierarchical agglomerative clustering (HAC) that are: Single Link HAC,
Group Average HAC and complete link HAC. In these methods mixed Euclidean
distance has been used as a dissimilarity measure.

Accuracy rate and entropy of clustering on Bupa Liver Disorder Dataset is given
in Figure 1 and Figure 2 respectively. Results on Heart (Statlog) dataset are given in
Figure 3 and Figures 4. It can be analysed from Figure 5 and Figure 6 that performance of proposed method is better than other methods on Mammographic Mass
Dataset.

On all four medical datasets, accuracy rate of proposed method is higher than all
other methods and entropy of clustering is smaller than other methods. It shows the

Fig. 4 Entropy of Clustering on Heart Dataset

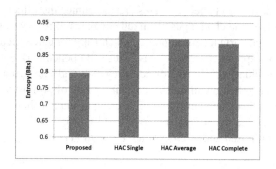

Fig. 5 Clustering Accuracy Rate on Mammographic Mass Dataset

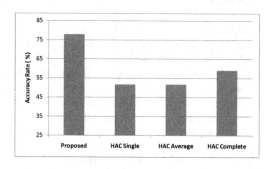

Fig. 6 Entropy of Clustering on Mammographic Mass Dataset

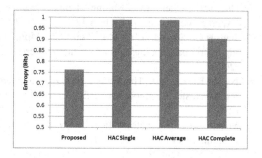

better quality of clusters generated by proposed method. This is due the fact that proposed method has considered the overall structure of clusters as merging criteria in terms of variance for numeric and entropy for categorical attributes. Instead of just considering a single object either nearest or farthest as representative of complete cluster structure for deciding which clusters should be merged.

6 Conclusion

In this paper, we have presented a novel agglomerative hierarchical approach for clustering which uses cluster spread as merging criterion and allows reallocation of data objects between clusters to counter the effect of a wrong merge. Keeping in view the challenges of mixed attributes of real world dataset, proposed method has used variance for numeric and entropy for categorical attributes as measure of cluster spread. This paper has focuses on benchmark medical datasets for performance evaluation in terms of accuracy rate and clustering entropy. Experimental results show that on all datasets, accuracy rate of proposed method is higher than all other methods and entropy of clustering is smaller than other methods. It shows the better quality of clusters generated by proposed method. In future we will apply the proposed approach in other application areas.

References

1. Barbará, D., Li, Y., Couto, J.: Coolcat: an entropy-based algorithm for categorical clustering. In: Proceedings of the Eleventh International Conference on Information and Knowledge Management, CIKM 2002, pp. 582–589 (2002)
2. Cai, D., Yau, S.S.T.: Categorical clustering by converting associated information. International Journal of Computer Systems Science and Engineering 3(4) (2006)
3. Frank, A., Asuncion, A.: UCI machine learning repository (2010),
   `http://archive.ics.uci.edu/ml`
4. Han, J., Kamber, M.: Data Mining: Concepts and Techniques, vol. 54. Morgan Kaufmann (2006)
5. Hsu, C.C., Chen, C.L., Su, Y.W.: Hierarchical clustering of mixed data based on distance hierarchy. Information Sciences 177(20), 4474–4492 (2007)
6. Hsu, C.C., Chen, Y.C.: Mining of mixed data with application to catalog marketing. Expert System with Application 32(1), 12–23 (2007)
7. Jiau, H.C., Su, Y.J., Lin, Y.M., Tsai, S.R.: Mpm: a hierarchical clustering algorithm using matrix partitioning method for non-numeric data. J. Intell. Inf. Syst. 26(2), 185–207 (2006)
8. Ng, M.K., Li, M.J., Huang, J.Z., He, Z.: On the impact of dissimilarity measure in k-modes clustering algorithm. IEEE Trans. Pattern Anal. Mach. Intell. 29(3), 503–507 (2007)
9. Shannon, C.E.: A mathematical theory of communication. SIGMOBILE Mob. Comput. Commun. Rev. 5(1), 3–55 (2001)
10. Tan, P.N., Steinbach, M., Kumar, V.: Introduction to Data Mining. Addison-Wesley (2005)

Accessibility Issues in Learning Management Systems for Learning Disabled: A Survey

Zainab Pirani and M. Sasikumar

Abstract. Learning Management System (LMS) is a major trend in distance education and educational technology and it will continue to grow in response to student needs and requirement. The elastic nature of LMS makes it suitable for almost any type of institutional academic structure. The major challenge faced by Learning Disabled (LD) users is to match their accessibility needs and preference in the existing LMS. The accessibility issues acts as a barrier in the growth of LMS. This paper provides comparative analysis of various LMS available in the market and the problems the LD faced to cope up with the educational technology.

Keywords: Learning Management Systems, Learning disabled, Accessibility issues.

1   Learning Management System

The basic description of a LMS is that it is a software application that automates the administration, tracking, and reporting of training events [1]. The purpose is to allow students to get complete training and share information in the convenience of their home or school. LMS provides a common platform for both teachers and students [2]. LMS give teachers the opportunity to deliver the entire course which includes the following features:

Zainab Pirani
MHSSCOE, Mumbai
e-mail: zainab.pirani@gmail.com

M. Sasikumar
CDAC, Mumbai

S.M. Thampi et al. (eds.), *Recent Advances in Intelligent Informatics*,
Advances in Intelligent Systems and Computing 235,
DOI: 10.1007/978-3-319-01778-5_26, © Springer International Publishing Switzerland 2014

- Providing different tools to run and manage an e-learning course
- Provide course content in a variety of media – including text and multimedia.
- Developing all learning activities and materials in a course
- Various options are provided for discussion forums, file sharing, management of assignments, lesson plans, syllabus, chat, etc.
- Monitor and manage communication between learners.
- Record assessment and provide feedback to the students
- LMS also provides following opportunity to the students community as well:
- Students can perform their self-evaluations
- Access chat rooms, discussion forum and whiteboard.
- Sharing of project ideas with peer groups.
- Interacting with teachers using synchronous or asynchronous mode.
- Completing and submitting of various assignments at any point of time.
- Can check their grade and progress status reports.

1.1   Categories of LMS

Basically, LMS can be classified into three categories based on to the relevant criteria as shown in the fig.1.

- **Source Code Availability.** It focuses on the programming code of the LMS which is further classified into two groups. First one is open source type in which all the files which make up the LMS are free for modifying, which allows customizing the system in the necessary way, for example – Moodle [3], Magic Tutor [4], LAMS [5], Sakai [6] etc. And the second is the commercial type in which the software vendors do not provide the source code and the learning system is distributed on payment basis. As a rule, the price for the LMS includes technical support which makes the software easy to implement and use for non-technicians and non-tech companies, e.g. Blackboard [7], WebCT [8].
- **Business orientation.** It provides the domain usage of LMS. It is further classified into three areas. First one is the education domain which is our domain of interest. It automates the administration of training events. These LMSs are designed for educational purposes only. The second area is the government domain which includes possibilities of educational and corporate LMSs, with a greater focus on security and some other features depending on the certain government structure. The last are focus on (evaluating) employee skills and competencies as well as providing tools for competency training.

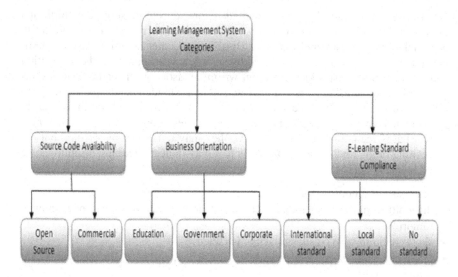

Fig. 1 Classification of LMS

- **E-learning standard compliance.** This category of LMS is also divided into three domains. The first one is with respect to various international standards like SCORM (Sharable Content Object Reference Model) - which is a set of specifications that, when applied to course content, produces small, reusable e-Learning objects [9]. AICC (Aviation Industry CBT [Computer-Based Training] Committee) standards apply to the development, delivery, and evaluation of training courses that are delivered via LMS [10]. IMS (Instructional Management System) specifications promote the adoption of learning and educational technology and allow selection of best of breed products that can be easily integrated with other such products[11]. If the LMS is those standards compliant, we can use the standard packages of those standards inside the LMS instead of (or together with) creating native LMS course content. It can be very useful if we don't want to tie ourselves using one particular LMS. The second division is local standard of a specific region or specific field of learning. The third division is not compliant with any content standards, and native content created in any LMSs is not reusable for other LMSs.

This paper is organized as follows: section 2 deals with detailed survey about the LD students and barrier they have to face in their academic life. Accessibility in existing LMS is discussed in 3. The actual evaluation of LMS is discussed in section 4. Finally section 5 concludes the paper.

2   Learning Disability

Learning disabilities (LD) are usually hidden disabilities that affect many individuals who usually have average or above average intelligence, but are unable to achieve their potential [12]. It is acquired before, during or soon after birth and affects an individual's ability to learn, all through his/her life. LD, according to National Adult Literacy and Learning Disabilities Center is defined as "a heterogeneous group of disorders manifested by significant difficulties in the acquisition and use of listening, speaking, reading, writing, reasoning, or mathematical abilities or of social skills"[13]. It is often referred to as "hidden handicaps" as they are difficult to identify. The kinds and severity of problems vary from individual to individual. They may do well in some areas, but very poorly in others. They may learn what is seen, but not what is heard; they may remember by writing, but not by reciting orally and so on. Some signs that may indicate learning disabilities are listed below [14] [15]:

- inconsistent school performance
- difficulty remembering today what was learned yesterday, but may know it tomorrow
- short attention span (restless, easily distracted)
- letter and number reversals (sees "b" for "d" or "p", "6" for "9", "pots" for "stop" or "post")
- poor reading (below age and grade level)
- frequent confusion about directions and time (right-left, up-down, yesterday-tomorrow)
- personal disorganization (difficulty in following simple directions/schedules; has trouble organizing, planning, and making best use of time; frequent loss or misplacement of homework, schoolbooks, or other items)
- deficits in study skills, such as test preparation, note taking, and listening comprehension
- impulsive and/or inappropriate behavior (poor judgment in social situations, talks and acts before thinking)
- failure on written tests but high scores on oral exams (or vice versa)
- speech problems (immature language development, trouble expressing ideas, poor word recall)
- difficulty understanding and following instructions unless they are broken down to one or two tasks at a time
- seems immature and has difficulty making friends
- low self-esteem
- difficulty interpreting body language, facial expression, or tone of voice
- higher dropout rates

Some of these problems though can be found in all children at certain stages of their learning development adding to the difficulties in relatives to LD.

2.1 Barrier Encountered by LD Children in Their Academic Life

Children with LD often feel that "I can't do anything right." "I'm no good." "I'm dumb." "Nobody likes me." "Everybody is picking on me" [16]. These feelings cause the individual to feel frustrated, discouraged, alone or angry and have a poor self-image. Due to such feelings LD student have lots of trouble to cope with their educational life. LD may also manifest itself in delayed conceptual development, difficulties in expressing ideas and feelings in words, a limited ability to abstract and generalize what they learn, limited attention-span and poor retention ability, slow speech and language development, and an underdeveloped sense of spatial aware-ness i.e. may have difficulty in taking notes, trouble with spelling, difficulty in orga-nizing thoughts on paper, understanding non-literal language like jokes or idioms [17]. Due to these inabilities these students are often ignored by their teachers and peers. Some LD students also experience major trouble with certain subjects causing the grades to drop drastically over time. Some LD students have difficulty in solving arithmetic problems and grasping math concepts. They may also face problems in understanding concepts related to time such as days, weeks, months, seasons, quar-ters, etc. Others may exhibits difficulties in recalling known words and may have large gap between written ideas and understanding demonstrated verbally which causes low self esteem and leads to failure in their academic life [18].

3   Accessibility in LMS

Accessibility describes materials that may be accessed by individuals with disabil-ities. In recent years, articles concerning disability accommodation and Americans with Disabilities Act (ADA) and Section 508 compliance have become more pop-ular[19]. However, the majority of those articles are focused on accommodating students with physical disabilities and do not specifically discuss learning disabili-ties in detail. LD students face the barriers (discussed in section 2.1) to achieve the same academic level as that of their peers. The kinds and severity of problems vary from student to student. They may do well in some areas, but very poorly in others. They may learn what is seen, but not what is heard; they may remember by writing, but not by reciting orally and so on [20]. Due to these inabilities they avoid going to school or seem to be disinterest in school which leads to failure in their academic life. Other problem that educators face in providing services to LD students in the postsecondary education is the vagueness in the federal regulations. Section 504 of the Rehabilitation Act (1973) requires that public school districts must provide proper services to disabled students in the K-12 environment. How-ever, postsecondary institutions are not covered directly by this requirement. Therefore it was necessary to make lots of modifications in the policies, practices and procedures of ADA and Section 504 to avoid discrimination between primary and post secondary LD students. Students with LD often struggle in a typical classroom setting and are too shy to speak during class time. Many adults with LD choose to avoid social situations because of inaccurate self-assessments of their social competence rather than their actual social abilities. There is no doubt that LMS is a major trend in distance education and educational technology. The use of LMS and its asynchronous nature have provided an independent outlet for these students. Studies have shown that students with LD have an increased willingness to self-disclose online. LMS can be perfect solution to improve the academic performance. But LMS consist of various features that could cause accessibility problems for LD students explained below;

- **Assignments.** One way of submitting work is to download the assignment and upload the completed assignment. Friendly and accessible Upload/Download feature needs to be provided.
- **Blogs.** Accessible sorting and filtering options are missing from most blogs.
- **Chat.** Composing entries and reading is difficult simultaneously. Also Scrolling through the messages and selecting or copying desired information is not possible.
- **Forums.** "Filling out Forms" issues are a potential problem along with being able to use sort, filter, and search features.
- **Quizzes.** Navigation between questions is needed. Also timed quizzes are problematic for assistive technology user's just takes more time to complete a quiz.
- **Emails.** Due to inaccessibility of web based e-mails, most LD students prefer to use their own e-mail clients. Most LMS e-mail features are not accessible.

There are many more features which are not yet designed with accessibility in mind. So if we generalized all the features of existing LMS, we have following core drawbacks:

- As we know every LD individual is different, teachers and instructional designers should build the course according to their requirements but existing LMS is designed in general without considering the students requirement.
- Ever LD student has its own learning style, so LMS should provide some sort of provision for different teaching styles. But there is no such accommodation as such in the existing LMS.
- Depending upon the type of LD student, screen content, layout and navigation has to vary. Again these accommodations are totally ignored in the existing LMS.

Therefore based on these core issues, the next section actually evaluates LD problems wrt to LMS.

4   Evaluation of LMS

The evaluation of LMS is done using following sections:

- **LMS Selection.** This section lists the name of the recommended LMS as well as how many times each one was recommended. We give one point for every recommendation; therefore if the study recommended two LMS then each one takes half point. For our evaluation purpose, we have taken into consideration only open source type of LMS which are as follows :

  - Magic Tutor
  - Moodle
  - LAMS
  - Sakai

Table 1 LMS Selection

LMS	Frequency
Magic Tutor	4
Moodle	8
LAMS	6
Sakai	7

From table 1, upon all the LMS – Moodle is the most recommended with 8 out of 10 points. See Figure 2.

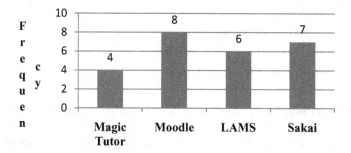

Fig. 2 LMS Selection

- **Criteria.** This section is based on our specific needs and requirements. Each criterion is given a weighing factor using scale of 1-5 where 5 is most important from our LD students accessibility point of view and 1 is the least important as shown in the table 2.

Table 2 Evaluation Criteria

Sr.No.	Evaluation Criteria	Criteria Weight
1.	Pedagogical Support	5
2.	Accessibility Compliance	4
3.	Content authoring	4
4.	Migration of existing courses	3
5.	Sections and groups	2
6.	E-portfolio	4
7.	Testing and assessment tools	4
8.	Training materials	4
9.	Gradebook and student tracking	4
10.	Score	34

The evaluation criteria are based on the factors which are of utmost important from the LD students point of view, see figure 3.

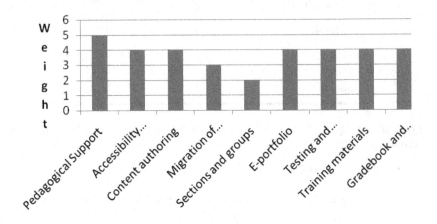

Fig. 3 Evaluation Criteria for LD students

- **Criteria Rating.** In this section, each criterion is compared with LMS and which is further weighted based on the scale factor of 1-5 where 5 is most satisfaction and 1 is no satisfaction. Fig 4. Shows the comparison of expected evaluation criteria present in Magic Tutor.

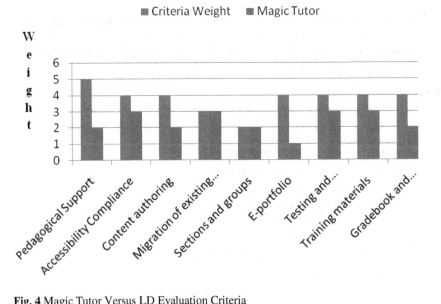

Fig. 4 Magic Tutor Versus LD Evaluation Criteria

In case of Magic Tutor, only one criteria matches exactly ie. Sections and groups. Fig. 5 shows the comparison of expected evaluation criteria present in Moodle.

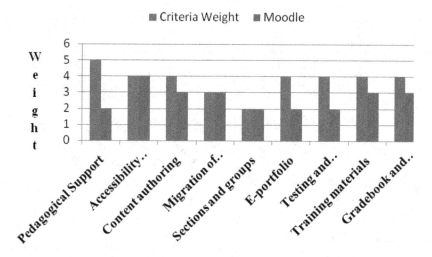

Fig. 5 Moodle Versus LD Evaluation Criteria

In case of Moodle, two criteria are matching exactly ie. Sections and groups and Migration of existing courses

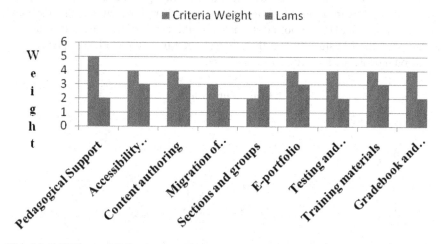

Fig. 6 LAMS Versus LD Evaluation Criteria

In case of LAMS, one criteria is above the evaluation criteria and ie. Sections and groups.

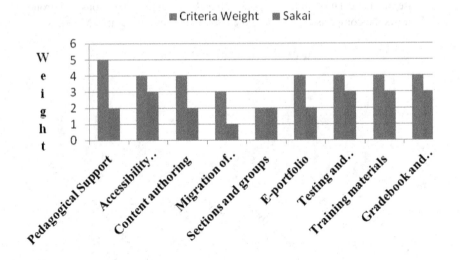

Fig. 7 Sakai Versus LD Evaluation Criteria

In case of Sakai also, only one criteria matches exactly the evaluation criteria and ie. Sections and groups

• **Evaluation Results.** In this section, we calculate a combined score for each LMS by combining the criteria weight and its rating respectively as shown in the fig 8. Thus from the above figure it is very clear that available open source LMS in the market are not at all matching the accessibility needs or requirements of the LD students. So there is real need for a unique LMS specially to accommodate the requirements of LD students.

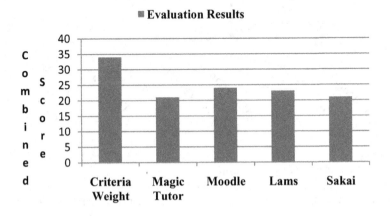

Fig. 8 LMS Evaluation Results

5   Conclusion and Future Work

The LD issues are attracting great attention since nowadays. Many solutions exist and many are evolving. Building an e-learning environment specifically for them is an active research area for academics and Industry. The analysis of accessibility issues in the area of LMS is done. In future we are trying to design an accessible LMS only for LD students.

References

1. Zhang, D., Zhao, J.L., Nunamaker, J.F.: Cane-learning replace classroom learning? Comm. of the ACM 47(5), 75–99 (2004)
2. Anido-Rifon, M.J., Fernandez-Iglesias, M., Llamas-Nistal, M., Caeiro-Rodriguez, J.: A Component Model for Standardized Web-Based Education. ACM Journal of Educational Resources in Computing 1(2), Article #1, 21 pages (summer 2001)
3. Moodle Online, http://www.moodle.org (retrieved on June 25, 2012)
4. Magic Tutor Online, http://www.magictutor.org (retrieved on June 28, 2012)
5. LAMS Online, http://www.lams.org (retrieved on June 26, 2011)
6. Sakai Online, http://www.sakaiproject.org (retrieved on June 26, 2011)
7. Blackboard Online, http://www.blackboard.com (retrieved on June 25, 2012)
8. WebCT Online, http://www.webct.org (retrieved on June 26, 2011)
9. SCORM (Sharable Content Object Reference Model) online, http://www.scorm.org (retrieved on June 15, 2002)
10. Woolf, B.P.: A Roadmap for Education Technology' funded by the National Science Foundation # 0637190, The Computing Community Consortium (CCC), managed by the Computing Research Association (CRA) with a sub-award to Global Resources for Online Education, Beverly Park Woolf
11. Kinshuk, S.G.: Considering Learning Styles in Instructional Management Systems: Investigating the Behavior of Children in an Online Course. In: First International Workshop on Semantic Media Adaptation and Personalization (SMAP 2006), pp. 46–55 (2006)
12. Health Care & Diagnostic Centre (2011), http://www.wrongdiagnosis.com/l/learning_disabilities/subtypes.htm (retrieved on May 2, 2011)
13. Skinner, Lindstrom: Bridging the gap between high school and college: Strategies for the successful transition of students with learning disabilities. Preventing School Failure 47, 132–138 (2003)
14. Learning Disabilities Online (2010), http://www.ldonline.org/ldbasics/signs (retrieved on May 21, 2011)
15. Health Care & Diagnostic Centre (2011), http://www.wrongdiagnosis.com/l/learning_disabilities/subtypes.htm (retrieved on May 2, 2011)
16. Resources for working with youth with special needs, extension of Illinios University, http://urbanext.illinois.edu/specialneeds/lrndisab.html (retrieved on October 6, 2011)

17. Anderson, J.: Primary Care Service Framework: Management of Health for People with Learning Disabilities in Primary Care (July 2007),
    `http://www.pcc.nhs.uk/uploads/primary_care_service_`
    `frameworks/primary_care_service_framework__ld_v3_final.pdf`
    (retrieved on September 16, 2011)
18. Jonassen, D.H., Grabowski, B.L.: Handbook of Individual Differences, Learning, and Instruction. Lawrence Erlbaum, Hillsdale (1993)
19. American disability act Online, `http://www.ada.org`
    (retrieved on June 26, 2011)
20. Horton, W.: Designing web-based training: how to teach anyone, anywhere, anytime. Wiley Computer Publications, New York (2000)

Mutagenicity Analysis Based on Rough Set Theory and Formal Concept Analysis

Mostafa A. Salama, Mohamed Mostafa M. Fouad, Nashwa El-Bendary, and Aboul Ella Otifey Hassanien

Abstract. Most of the current Machine Learning applications in cheminformatics are black box applications. Support vector machine and neural networks are the most used classification techniques in prediction of the mutagenic activity of compounds. The problem of these techniques is that the rules/reasons of prediction are unknown. The rules could show the most important features/descrpitors of the compounds and the relations among them. This article proposes a model for generating the rules that governs prediction through the rough set theory. These rules, which based on two levels of selection for the highly discriminating power features, are visualized by lattice generated using the formal concept analysis approach. That is, better understanding of the reasons that leads to the mutagenic activity can be obtained. The resulted lattice shows that lipophilicity, number of nitrogen atoms, and electronegativity are the most important parameters in mutagenicity detection. Moreover, experimental results are compared against previous researches for validating the proposed model.

Mostafa A. Salama
British University in Egypt (BUE), Cairo - Egypt
Scientific Research Group in Egypt (SRGE)
e-mail: mostafa.salama@gmail.com
http://www.egyptscience.net

Mohamed Mostafa M. Fouad · Nashwa El-Bendary
Arab Academy for Science, Technology, and Maritime Transport, Cairo - Egypt,
Scientific Research Group in Egypt (SRGE)
e-mail: mohamed_mostafa@aast.edu, nashwa.elbendary@ieee.org
http://www.egyptscience.net

Aboul Ella Otifey Hassanien
Information Technology Dept., Faculty of Computers and Information,
Cairo University, Cairo - Egypt,
Scientific Research Group in Egypt (SRGE)
e-mail: aboitcairo@gmail.com
http://www.egyptscience.net

S.M. Thampi et al. (eds.), *Recent Advances in Intelligent Informatics*,
Advances in Intelligent Systems and Computing 235,
DOI: 10.1007/978-3-319-01778-5_27, © Springer International Publishing Switzerland 2014

1   Introduction

Studying the role of chemical compounds in the biological systems has been further strengthened. The prediction of the effect these compounds on humans, and animals is one of the benefits that could be resulted from this study. An approach is proposed to find the common patterns, similarities, between compounds having the same re-actions, but this solution is an NP-complete problem [1]. Chemists provides a set of properties, descriptors, that describes the structure of the chemical compounds like the molecular type, atomic type, and bond type [2]. These descriptors are catego-rized as constitutional, topological, geometrical, charge related, semi-empirical, and thermodynamic [3]. Several approaches have been proposed to do a matching be-tween these descriptors and the activity of the corresponding chemical compounds. One of the problems that faced these approaches is the high number of molecular descriptors. The main concern of these approaches is the emerging and evolving of a huge number of molecular descriptors. Therefore, most of them are selecting a sub-set of these descriptors. The selection process reduces the computational complexity needed for too many descriptors, and also removes the redundancy of information. A main step in cheminformatics is to reduce the number of descriptors in order to apply the prediction of the activity of the compounds. The next step is to apply a classification technique that uses a set of compounds of known activity as a training set, then predict the activity of new compounds of un-known activity.

In order to apply an accurate feature selection of the descriptors and an accurate analysis of the mapping between the descriptors and the right decision, a new model should be proposed that differs from the classical mode.

This proposed model should handle all the features together when applying the feature selection technique, this to make use of the dependency among features. On the other hand, this model should clearly interprets the relationship among these fea-tures in a visual illustration for chemists. In this paper other machine learning tech-niques like descriptor/feature selection, rule generation and visualization techniques are applied. The descriptor/feature selection technique shows the most important de-scriptors that discriminate between the different activities of the drugs, for example, the difference between the mutagen and non-mutagen activity of drugs. Feature se-lection techniques select the features whose distribution correlates the distribution of the class labels. The rule generation techniques, shows the rules that governs the prediction of the aspects of the drugs. Finally the visualization technique shows the relation between the different descriptors and each other and between these descrip-tors and the predicted aspects.

The model proposed in this paper applies two phases of feature selection, the first phase uses a classical ChiMerge feature selection technique, while the second phase applies the rough set technique. This will compel a highly filtration of the unneeded descriptors and on the other hand, the rules will describe the relation among these important selected descriptors. The rest of this paper is organized as follows. Section 2 presents a machine learning based model for visualization. Section 3 presents experimental results, Section 4 presents analysis and discussion. Conclusion and future work are discussed in section 5.

2   Visualization Model

Machine learning and data visualization has a great contribution in knowledge discovery. The number of descriptors can be considered in cheminformatics is too high, it can reach 6122 descriptor like those extracted by PowerMV software [4] as shall be discussed later within the context. The corresponding expression of descriptors in machine learning is called features or attributes, and the effect (activity) of each chemical compound is called the target class. An example of this effect is whether this chemical compound is mutagenic or non-mutagenic. Set of compounds with a known target class are used as a training set, while other set of compounds will be test whether the prediction of the target class are correct or not. The prediction here will be refereed as the classification of the compounds in the training set.

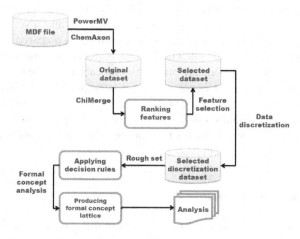

Fig. 1 Model of visualization of Mutagenicity in chemical compounds

Figure 1 shows the proposed model that illustrates the steps required for rule generation and visualization. First of all, the huge number of descriptors prevents an accurate analysis and prediction of the target class. This problem is named as the curse of dimensionality [5], where the data set contains high number of features (words dimensions). Therefore a forward feature selection technique is applied, where the features are ranked according to its importance in the discrimination between the different target classes. The ChiMerge technique is the most applicable ranking techniques, as it is capable of handling continuous data set [6]. Then these features are ordered, and tested in a series of tests, first test will be applied on data set of the highest feature, then applied on a data set of the highest two features, and so on until this data will include all the ranked features. The testing of the set of features selected is applied using any classification technique like Naieve Bayesian Tree (NBT) classifier. Only the set features that leads to the local and global maximum success in the tests will be selected [7]. This will shrink the original descriptors data set into a set of high level features those required to predict the target class. This shrinking process reduces the processing time required for knowledge learning process as well as enhancing the accuracy of the prediction of the target class of new compounds. The second step is to state the relation among these attributes through the extraction of rules governing this relation using the rough set theory. In order to apply rough set on the reduced data set, a discretization method is applied. The discretization method involves the creation of cuts dependent on the target class, in order to simplify the rule extraction and class prediction. Finally, the rules are extracted according to the rough set technique [8] to describe the range of the values of some/all features that indicates the corresponding predicted class. The third step is to visualize the relation between attributes. One of the visualization techniques applied is the formal concept analysis to generate a formal concept lattice [9]. The formal concept analysis is usually applied on the original highly dimensional data set. This leads to a high dense, unclear and non-informative formal concept lattice. Here, the formal concept analysis is applied on the generated rules from the rough set technique [12]. For each class, each extracted rule contains a set of attributes' range that leads this class. A list of all attributes' ranges is prepared, and for each rule, the attribute range that included in this rule is marked by one, otherwise it is marked by zero. This will generates a binary data for each class, the data is provided to the formal concept analysis technique for the generation of the formal concept lattice. The visualization of the rough set rules provides a simple and more descriptive formal concept lattice. Finally this formed lattice can be provided to the domain expert for analysis. This simplifies his task through the visualization of the relation among attributes for each target class.

3   Experimental Results

The Bursi Mutagenicity Dataset [10] were used in this paper. It consists of 2401 mutagenic compound and of 1936 non-mutagenic compound. In order to extract the descriptors of these compounds, two software are used the PowerMV software and the ChemAxon software [11]. PowerMV calculates a total of 6122 descriptors classified as 546 atom pair descriptors, 4662 Carhart descriptors, 735 fragment pair descriptors, 147 pharmacophore fingerprints, 24 Weighted Burden Number descriptor and 8 properties descriptors. ChemAxon also have a various kinds of descriptors, including structural and topological analyzer descriptors. Due to the huge number of descriptors extracted by PowerMV, a feature ranking and selection techniques are applied to select the most important features. This applied on only 200 compounds, divided equally between the two class. After applying ChiMerge technique, the WBN_EN_H_0.75 WEN, WBN_LP_H_0.75 WLP descriptors shows the highest ranks. When a test is applied on the 200 compounds data set, and based on these two descriptors only, the classification accuracy is 78.0% success. When the third ranked descriptor is applied the classification accuracy result is decreased. Therefore, the only two selected descriptors are the WEN and WLP, descriptors. These two descriptors are from the 24 Weighted Burden number-continuous descriptors. They refer to the electronegativity and the atomic lipophilicity property respectively.

Then, the rest of the descriptors are extracted from the ChemAxon software. The number of extracted descriptors from the ChemAxon software are 64 descriptor. The used descriptors are as shown in Table 1.

Table 1 The extracted descriptors for the mutagenic data set

Descr.	Description	Descr.	Description	Descr.	Description	Descr.	Description
MW	Molecular Mass	*MEM*	Exact Mass	*nBT*	#Bonds	*nAT*	#Atoms
nN	#N Atoms	*nS*	#S Atoms	*IMD*	Molecular Dim.	*nH*	#H Atoms
nP	#P Atoms	*HRC*	#Heterogenous	*nO*	#Oxygen Atoms	*nC*	#Carbon Atoms
nBR	#Br Atoms	*RC*	#Rings	*RBC*	#bonds in Rings	*nF*	#F Atoms
ALAC	#Aliphatic Atom	*TC*	Total Charge	*BL*	Avg. bond len.	*RAC*	#atoms in Rings
ALBC	#Aliphatic Bond	*nDB*	#double Bonds	*RBN*	#Rotatable Bond	*nCL*	#Cl Atoms
ALRC	#Aliphatic Ring	*ARAC*	#Aromatic Atom	*nAB*	#aromatic Bonds	*ARBC*	#Aromatic Bond
ARRC	#Aromatic Ring	*ASAC*	#Asymmetric	*wP*	wiener Polarity	*wI*	wiener Index
CRC	#Carbo Ring	*rI*	randic Index	*hwI*	hyperWiener Ind.	*nTB*	#trible Bonds
sI	szeged Index	*CBC*	#Chain Bond	*cN*	#cyclomatic	*CAR*	#Carbo-aromatic
CAC	#Chain Atom	*CCC*	#Chiral Center	*pI*	platt Index	*CAL*	#Carbo-aliphatic
hI	harary Index	*DB*	Double Precision	*fC*	#fragment	*bI*	balaban Index
FAL	#Fused Aliphatic	*FC*	#Fragment	*FAR*	#Fused Aromatic	*FRC*	#Fused

The rest of the descriptors describe the number of different types of bonds like the Number of AROMATIC Bonds Between "c and o" *nAbBco*, the Number Of Single Bonds Between "c and c" *n1bBcc*, and the Number Of double Bonds Between "c and o" *n2bBco*.

After the combination of these 64 descriptors to the selected descriptors from PowerMV descriptors, the data set resulted is now 66 features data set. Again the feature selection technique is applied on this data set. Figure 2 shows the classification accuracy of increasing set of features according to the ascending ranked features. The feature numbers are corresponding to the order of the maintained features. The selected descriptors from these set of features from both software are only 22 descriptors. When applying different classification techniques on the selected 22 features data set, the resulted classification accuracy varies according to the used technique. A 10-fold cross validation method is used to have an average results that

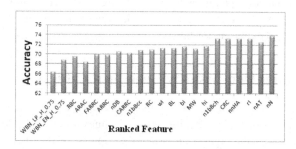

Fig. 2 Forward feature selection technique selectees 22 features out of 66 features

indicate the accuracy of the used classifier. Also, as discussed before, the input data set is nearly balanced among the two target classes. Naive Bayesian Tree classifier is used in the forward feature selection technique, the resulted classification accuracy of the selected data set is 75.78%. The resulted classification accuracy of other classifiers are as follows, Support Vector Machine shows 62.02%, Decision Tree shows 76.45% accuracy, the Lazy Classifier, Ib1, shows 77.15%, the Random Forest Classifier shows 77.28 % and finally Multi-Layer Perceptron Classifier shows 73.18% accuracy. The previous tests are performed using the WEKA software [13]. Then, rough set theory is applied on the 22 descriptors data set. The generated rules from the rough set theory is dependent only on 12 features. Those features are the most discriminating ones for compound classification whether it is mutagenic or not. These features are WLP, WEN, CAR, nN, nAT, hl, bI, wl, nDB, rl, $n1bBch$, CRC. The number of the generated rules for both mutagen and non-mutagen classes are more than 1000 rule. Each rule is applied to a number of compounds that varies according the covering of this rule. The following statement is an example of the generated rules:

(WLP=(3.205, 3.318) AND wl=(inf, 806.5) AND bI=(2.3, inf))
$\Rightarrow$ (class = mutagenicity) {covered by 71 compound}

Where this rule consists of three subrules, the first subrule stated that the value of the WLP descriptor lies between the two values 3.205 and 3.318. The second subrule stated that the value of the bI and wl descriptors lie between the values 2.3 to inf and inf to 806.5, respectively. If these three subrules are achieved, then the class of this compound is mutagenicity. This rule covered 71 compound and not leads to any compound whose class is non-mutagenicity. In order to visualize these rules a smaller subset of these rules is used, these selection of these rules is based on the highest number of compounds covered by each rule. Only 20 rules are selected for each class to simplify the visualization of relation and dependencies between descriptors. The visualization applied is based on the formal concept analysis which accepts only binary input. To convert this subset of rules into binary input to the formal concept analysis, the ranges of the 12 features will be tested either achieved or not for each rule. For example in the previously maintained rule, the range 3.205 and 3.318 of feature WLP is stated to 1 and the range (1.407 and 1.541 is stated to 1, while the rest of feature ranges will be assigned 0. The result of the formal concept analysis is a formal concept lattice. Two lattices are generated, Figure 3 shows the relationship between descriptors of mutagenicity class whereas Figure 4 illustrates the non-mutagenicity class' descriptors.

4   Observations from Visually Generated Lattices

Figures 4 and 3 show the relation among these descriptors, the degree of effectiveness of each one on the others, and the range of values for each one. WLP and nN descriptors appears as the highest important descriptors for the discrimination between the two mutagen and non-mutagen classes. The two range of values of these

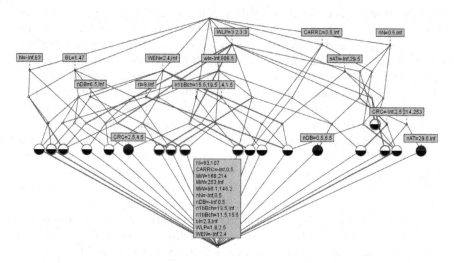

Fig. 3 Formal concept lattice for mutagenic data

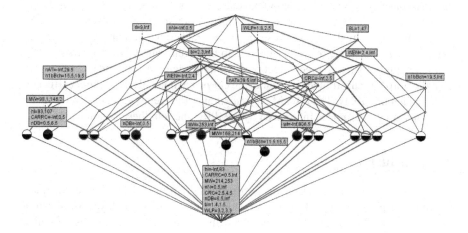

Fig. 4 Formal concept lattice for non-mutagenic data

two descriptors are different from each other in the two classes. This shows a connection between atomic lipophilicity and the number of nitrogen, and a connection between these two descriptors and the mutagenicity of the chemical compounds. It appears that research, since 1960s, shows an importance of lipophilicity for biological potency. It was thought that lipophilicity is an important descriptor in the QSAR research and required especially in the drug design [14]. On the other hand it appears that a good correlation exists between the resistance reversal activity and the intrinsic basicity of the nitrogen atom at the tricyclic ring system, frontier orbital energies, and lipophilicity [15]. This provides a proof that the connections between these two descriptors in the generated lattices is correct. Also, it gives an evidence that the influence of these two descriptors on the mutagenicity predication, as appeared in the generated formal concept lattice is correct. On the level of the values, Figures 4 and 3 show that as the values of the WLP and nN descriptors increases, as the compound becomes more mutagenic. This is because the ranges of WLP and nN descriptors in the case of mutagenic compounds are $[3.2, 3.3]$ and $[0.5, infinity]$ respectively. Another descriptors that show a high discrimination between the two classes is WEN. The range of values of the WEN descriptor in the case mutagenic class is $[2.4, infinity]$. While in the case of non-mutagenic compounds, the range of values of WEN descriptor is $[-infinity, 2.4]$. This means that as the value of the electronegativity of the compound WEN increases, as the possibility of the compound to be mutagenic increases. A study in 1995 shows that more electronegative compounds are more reactive to DNA and more capable of inducing mutation than other compounds [16]. Again, this is another proof that the generated lattices are correct. Also, there is a high correlation between the electronegativity and the number of atoms nAT which appears in the 12 selected features. The generated lattices show that in case of mutagenic class, this descriptor nAT range is $[-infinity, 29.5]$, while in case of non-mutagenic class objects, the range is $[29.5, infinity]$. This means that as the number of atoms in the compound decreases as the tendency of the compound to become non-mutagenic increases, and this is corresponding to its correlation to the electronegativity descriptor. Resistance reversal activity and intrinsic basicity of the nitrogen atom at the tricyclic ring system has a relation to the lipophilicity [15], while lipophilicity has a direct effect on the mutagenicity. A research in 2005 shows that the nitrogen-substituted position in the chrysene molecule, has a direct effect on the mutagenic activity. The ratio of participation of the metabolic activation enzyme isoforms of cytochrome is influenced by the differences in these positions [17].

5 Conclusions

The massive data generated from cheminformatics descriptors deteriorates the quality of the results' analysis. The proposed model applies different layers of filtration of the generated descriptors to come out with a small set of the highest effective descriptors. After applying feature selection and rule generations, the resulted set of descriptors includes the lipophilicity, the number of nitrogen atoms and electronegativity. These descriptors are the most important parameters required to detect mutagenicity. The generated lattice by the formal concept analysis shows a set of facts that were proved separately in previous medical researches. This proves the quality and preciseness of the proposed model.

References

1. Brown, N.: Chemoinformatics: an introduction for computer scientists. ACM Computing Surveys 41(2), Article 8 (2009)
2. Xu, J., Hagler, A.: Chemoinformatics and Drug Discovery. Molecules 7, 566–600 (2002)

3. Katritzky, A.R., Pacureanu, L., Dobchev, D., Karelson, M.: QSPR Study of Critical Micelle Concentration of Anionic Surfactants Using Computational Molecular Descriptors. Journal of Chemical Information and Modeling 47(3), 782–793 (2007)
4. Liu, K., Feng, J., Young, S.S.: PowerMV: a software environment for molecular viewing, descriptor generation, data analysis and hit evaluation. Journal of Chemical Information and Modeling 45(2), 515–522 (2005)
5. Salama, M.A., El-Bendary, N., Hassanien, A.E., Revett, K., Fahmy, A.A.: Interval based attribute evaluation algorithm. In: Proc. The Federated Conference on Computer Science and Information Systems, FedCSIS 2011, Szczecin, Poland, pp. 153–156 (2011)
6. Thabtah, F., Eljinini, M., Zamzeer, M., Hadi, W.: Naieve Bayesian based on Chi Square to Categorize Arabic Data. In: Proc. The 11th International Business Information Management Association Conference, IBIMA, on Innovation and Knowledge Management in Twin Track Economies, Cairo, Egypt, pp. 930–935 (2009)
7. Eid, H.F., Salama, M.A., Hassanien, A.E., Kim, T.-H.: Bi-Layer Behavioral-Based Feature Selection Approach for Network Intrusion Classification. In: Kim, T.-H., Adeli, H., Fang, W.-C., Garca-Villalba, L.J., Arnett, K.P., Khan, M.K. (eds.) Proc. Security Technology - International Conference, SecTech 2011, Part of the Future Generation Information Technology Conference, FGIT 2011, Jeju Island, Korea, pp. 195–203 (2011)
8. Al-Qaheri, H., Hassanien, A.E., Abraham, A.: A Generic Scheme for Generating Prediction Rules Using Rough Sets. In: Abraham, A., Falcan, R., Bello, R. (eds.) Rough Set Theory: A True Landmark in Data Analysis. SCI, vol. 174, pp. 163–186. Springer, Heidelberg (2009)
9. Motameny, S., Versmold, B., Schmutzler, R.: Formal Concept Analysis for the Identification of Combinatorial Biomarkers in Breast Cancer. In: Medina, R., Obiedkov, S. (eds.) ICFCA 2008. LNCS (LNAI), vol. 4933, pp. 229–240. Springer, Heidelberg (2008)
10. Kazius, J., McGuire, R., Bursi, R.: Derivation and Validation of Toxicophores for Mutagenicity Prediction. J. Med. Chem. 48(1), 312–320 (2005)
11. ChemAxon Software, http://www.chemaxon.com/ (last accessed: January 2013)
12. Kuznetsov, S.O.: Machine Learning and Formal Concept Analysis. In: Eklund, P. (ed.) ICFCA 2004. LNCS (LNAI), vol. 2961, pp. 287–312. Springer, Heidelberg (2004)
13. WEKA: Waikato Environment for Knowledge Analysis, version 3.5.9, http://www.cs.waikato.ac.nz/ml/weka/ (last accessed: January 2013)
14. Du, Q., Mezey, P.G., Chou, K.C.: Heuristic Molecular Lipophilicity Potential (HMLP): A 2D-QSAR Study to LADH of Molecular Family Pyrazole and Derivatives. Journal of Computational Chemistry 26(5), 461–470 (2005)
15. Bhattacharjee, A.K., Kyle, D.E., Vennerstrom, J.L., Milhous, W.K.: A 3D QSAR pharmacophore model and quantum chemical structure–activity analysis of chloroquine(CQ)-resistance reversal. Journal of Chemical Information and Computer Sciences 42(5), 1212–1220 (2002)
16. Rosenkranz, H.S., Klopman, G.: Relationships between electronegativity and genotoxicity. Mutation Research 328(2), 215–227 (1995)
17. Yamada, K., Hakura, A., Kato, T.A., Mizutani, T., Saeki, K.: Nitrogen-substitution effects on the mutagenicity and cytochrome P450 isoform-selectivity of chrysene analogs. Mutat Res 586(1), 87–95 (2005)

Particle Swarm Optimization with Lévy Flight and Adaptive Polynomial Mutation in *gbest* Particle

Nanda Dulal Jana and Jaya Sil

Abstract. In this paper, particle swarm optimization (PSO) with levy flight is proposed. PSO is a population based global optimization algorithm has faster convergence but often gets stuck in local optima due to lack of diversity in the population. In the proposed method, levy flight is applied on a percentage of particles excluding global best particle to create diversity in population. Adaptive polynomial mutation is applied on global best (*gbest*) particle to get it out from the trap in local optima. The method is applied on well-known benchmark unconstrained functions and results are compares with classical PSO. Form the experimental result, it has been observed that the proposed method performs better than classical PSO.

1 Introduction

Particle Swarm Optimization (PSO) is a population based global search technique having a stochastic nature and inspired by the social behavior of bird flocking and fish schooling [1][2]. It has good search ability with faster convergence speed for many optimization problems. Several variations [3][4][5][6] are proposed by researchers to improve performance and convergence behavior of the PSO algorithms. One class of variations includes the modification of velocity update equation in PSO. In [6], cognitive component and social component are replaced by two terms as the linear combination of global best of swarm and personal best of

Nanda Dulal Jana
Department of Information Technology, National Institute of Technology,
Durgapur, West Bengal, India
e-mail: nanda.jana@gmail.com

Jaya Sil
Department of Computer Science and Technology, BESU, West Bengal, India
e-mail: js@cs.becs.ac.in

S.M. Thampi et al. (eds.), *Recent Advances in Intelligent Informatics*,
Advances in Intelligent Systems and Computing 235,
DOI: 10.1007/978-3-319-01778-5_28, © Springer International Publishing Switzerland 2014

particle. However, PSO easily trapped into local optima while dealing with complex problems due to lacks in diversity in population. Different mutation strategies like Cauchy Mutation [7], Gaussian Mutation [8], Power Mutation [9] and Adaptive Mutation [10] are introduced into the varied version of PSO algorithms for solving local optima problem. Changhe Li et al. [7] introduced fast particle swarm optimization with Cauchy mutation and natural selection strategy. Xiaoling Wu et al.[9] introduced power mutation into PSO (PMPSO). Higashi [8] proposed a PSO algorithm with Gaussian mutation (PSO-GM). A new adaptive mutation by dynamically adjusting the mutation size in terms of current search space is proposed in [10]. Tapas Si et al. [11] introduced adaptive polynomial mutation in global best particles in PSO (PSO-APM). Nanda Dulal Jana et al. applied adaptive mutation and adaptive polynomial mutation in local best position in [12].

Lévy flight has been proposed in the paper to generate particles' positions based on random values. Adaptive polynomial mutation is applied on global best particle to get it out from the trap in local optima. Initial experimental results using benchmark set consisting of eight difficult high dimensional functions demonstrate that significance of the approach.

The rest of the paper is organized as follows: Section 2 describes the basic concept of the PSO. In section 3, the proposed PSO algorithm has been presented while in section 4, experimental results are analyzed. The paper concludes with a short discussion and some pointers for future work.

2   Particle Swarm Optimization

PSO is an optimization algorithm modeled by the behavior of fish schooling and flocking of birds [1]. Populations of individuals are called particles in PSO where each particle has its own positions and velocities for moving around the search space. Each particle has its fitness value in terms of objective function which is optimized to obtain best solution. Memory in each particle keeps its personal best position and corresponding fitness value. The best among personal best of all particles is called the global best of swarm. The basic concept of PSO technique lies in accelerating each particle towards its personal best and the locations at each time step t. At each generation, particles update their velocities and positions in the search space using equations (1) and (2) respectively, which govern the working principle of PSO.

$$v_{ij}(t+1) = w * v_{ij}(t) + c_1 r_1 \left(x_{ij}^{pbest}(t) - x_{ij}(t) \right)$$
$$+ c_2 r_2 \left(x_j^{gbest}(t) - x_{ij}(t) \right) \tag{1}$$

$$x_{ij}(t+1) = v_{ij}(t) + x_{ij}(t) \tag{2}$$

Where c_1 and c_2 are personal and social cognizance of a particle respectively while r_1 and r_2 are two uniformly distributed random numbers in the interval [0, 1].

3 Proposed PSO Algorithm

In this section, the strategies proposed to improve the performance of the PSO algorithm, applied in benchmark functions have been described.

3.1 Lévy Flight

In the proposed method, Lévy flight has been applied on particle's position and adaptive polynomial mutation in global best solution. Lévy flight and adaptive polynomial mutation are discussed below.

A Lévy flight is performed to generate a new solution from a particle's position x_i using equation (3).

$$x_j(t + 1) = x_j(t) + \alpha \times Lévy(\lambda) \tag{3}$$

Where $\alpha > 0$, 0, is the step size should be related to the scales of the problem of interest. Lévy flight provides a random walk drawn from a lévy distribution which has an infinite variance with an infinite mean.

$$Lévy \sim u = t^{-\lambda}, \; 1 < \lambda \le 3 \;\; 3 \tag{4}$$

In the proposed method, lévy flight is performed on particles except global best particle, on which adaptive polynomial mutation is applied.

3.2 Adaptive Polynomial Mutation (APM)

Polynomial mutation is based on polynomial probability distribution.

$$x_j(t + 1) = x_j(t) + \left(x_j^u - x_j^l\right) \times \delta \tag{5}$$

Where x_j^u is the upper bound and x_j^l is the lower bound of x_j. The parameter $\delta \in$ [-1, 1] is calculated from the polynomial probability distribution, given below.

$$P(\delta) = 0.5 \times (\eta_m + 1)(1 - |\delta|)^{\eta_m} \tag{6}$$

η_m is the polynomial distribution index.

$$\delta = \begin{cases} (2r)^{1/(\eta_m+1)} - 1, r < 0.5 \\ 1 - 2[(1 - r)]^{1/(\eta_m+1)}, otherwise \end{cases} \tag{7}$$

$$\eta_m = 100 + t \tag{8}$$

Where d is the dimension of the problem, t is the current iteration number and t_{max} is the maximum iteration number.

$$mx_j^{pbest} = x_j^{pbest} + \left(b_j(t) - a_j(t)\right) * \delta \tag{9}$$

Where δ is calculated using Eq.(7) and η_m is calculated using Eq.(8).

$$a_j(t) = \min (x_{ij}(t)) b(t) = \max (x_{ij}(t)) \qquad (10)$$

In adaptive polynomial mutation, mutation size is controlled dynamically depending on the current search space.

3.3 Algorithm: Lévy Flight PSO with APM

1. The next stage Initialize the particle's position and velocity
2. Calculate the fitness
3. While (stopping criteria)
4. For $i:=1$: *popsize*
5. $Ps = rand[0,1]$
6. If $rand[0,1] < Ps$
7. Perform levy flight
   Else
8. Update velocity and position
9. End if
10. Calculate the fitness
11. Update the local best
12. Update the global best
13. End for
14. Perform mutation in global best and calculate the fitness of muted solution
15. If muted solution is better than the global best, replace the global best
16. End while

4 Experimental Studies

4.1 Benchmark Problems

There are 8 different global optimization problems, including 4 unimodal functions $(f_1$-$f_4)$ and 4 multi-modals functions $(f_5$-$f_8)$, chosen in our experimental studies for minimization. These functions were used in an early study by X. Yao et al. [13]. The description of these benchmark functions and their global optima are given in Table 1.

4.2 Parameter Settings

1. Dimension=30
2. Population Size=20
3. $C_1=C_2=1.49445$
4. W=0.72894
5. Number of Function Evaluations(FEs)=1,00,000

Table 1 The 8 benchmark functions used in our experiments, where D is the dimension of the functions, *fmin* is the minimum value of the functions and $S \subseteq R^D$ in the search space

Fun. no.	Test Function	S	f_{min}
f_1	$\min f(x) = \sum\limits_{i=1}^{D} x_i^2$	$-100 \le x_i \le 100$	0
f_2	$\min f(x) = \sum\limits_{i=1}^{D} (\sum\limits_{j=1}^{i} x_j)^2$	$-100 \le x_i \le 100$	0
f_3	$\min f(x) = \sum\limits_{i=1}^{D} (10^6)^{\frac{i-1}{D-1}} x_i^2$	$-100 \le x_i \le 100$	0
f_4	$\min f(x) = \sum\limits_{i=1}^{D-1} [100(x_{i+1} - x_i^2)^2 + (x_i - 1)^2]$	$-100 \le x_i \le 100$	0
f_5	$\min f(x) = \sum\limits_{i=1}^{D} -x_i \sin(\sqrt{|x_i|})$	$-500 \le x_i \le 500$	-12569.5
f_6	$\min f(x) = \frac{1}{4000} \sum\limits_{i=1}^{D} x_i^2 - \prod\limits_{i=1}^{D} \cos\left(\frac{x_i}{\sqrt{i}}\right) + 1$	$-600 \le x_i \le 600$	0
f_7	$\min f(x) = -20\exp\left(-0.2\sqrt{\frac{1}{D}\sum\limits_{i=1}^{D} x_i^2}\right) - \exp\left(\frac{1}{D}\sum\limits_{i=1}^{D} \cos(2\pi x_i)\right)$ $+ 20 + e$	$-32 \le x_i \le 32$	0
f_8	$\min f(x) = \sum\limits_{i=1}^{D} [x_i^2 - 10\cos(2\pi x_i) + 10]$	$-5.12 \le x_i \le 5.12$	0

4.3   PC Configurations

1. System: Windows 7
2. CPU: AMD FX -8150 Eight-Core
3. RAM: 4GB
4. Software: Matlab 2010b

4.4   Results and Discussion

The proposed algorithm levy flight with PSO (LFPSO) and PSO are applied on unconstrained functions given in Table 1. Table 2 presents the mean and standard deviation of 50 runs of PSO algorithm and LFPSO algorithm considering 8 test functions with 30 dimensions. Same initial population is used for a single run of both algorithms to make a fair comparison in performances. Best-run-error values obtained from PSO and LFPSO are tabulated where best-run-error is the absolute difference of global optimum and best solution at the end of the run. The best results among the two approaches are shown in bold. From the results it has been observed that the proposed method performs better on both unimodal and multimodal problems and give better results except for functions f_2 and f_4. However, the standard deviations of some functions are better than the proposed algorithm. For functions f_6 and f_7, LFPSO algorithm provides significant results than PSO algorithms.

In Table 3, the mean and the standard deviation of number of function evaluations for PSO and proposed LFPSO are shown. From this table it has been observed that mean and standard deviation of function evaluations for the functions f_4, f_5, f_7 and f_8 are equal by both the algorithms. Function evaluations for the functions f_1, f_2 and f_3 by PSO algorithm are better than the proposed LFPSO algorithm with respect to the mean and standard deviation. Therefore, the proposed LFPSO algorithm has faster convergence speed in terms of number of function evaluations to obtain the solutions. The convergence characteristics of the functions f_5, f_6, f_7 and f_8 for 30D are shown in Fig. 1.

Table 2 Best-run-error values achieved by PSO and LFPSO

Test #	PSO		LFPSO	
	Mean	Std. Dev.	Mean	Std. Dev.
f_1	9.43e-004	6.89e-005	9.42e-04	6.16e-05
f_2	2.13e-04	2.46e-04	3.10e-04	2.86e-04
f_3	9.43e-04	5.24e-05	9.40e-04	6.05e-05
f_4	2.45e+01	3.14e+01	5.89e+01	8.99e+01
f_5	7.41e+03	6.05e+02	5.18e+03	7.58e+02
f_6	7.34e-02	1.46e-01	3.08e-02	4.84e-02
f_7	3.29e+00	2.04e+00	2.51e+00	9.06e-01
f_8	4.15e+01	1.25e+01	3.59e+01	1.55e+01

Table 3 Number of function evaluations for PSO and LFPSO

Test #	PSO		LFPSO	
	Mean	Std. Dev.	Mean	Std. Dev.
f_1	1.02e+04	1.74e+03	1.93e+04	2.69e+03
f_2	1.48e+03	9.64e+02	3.52e+03	2.93e+03
f_3	1.81e+04	3.02e+03	3.12e+04	4.17e+03
f_4	1.00e+05	0.00e+00	1.00e+05	0.00e+00
f_5	1.00e+05	0.00e+00	1.00e+05	0.00e+00
f_6	7.67e+04	3.97e+04	7.75e+04	3.65e+04
f_7	1.00e+05	0.00e+00	1.00e+05	0.00e+00
f_8	1.00e+05	0.00e+00	1.00e+05	0.00e+00

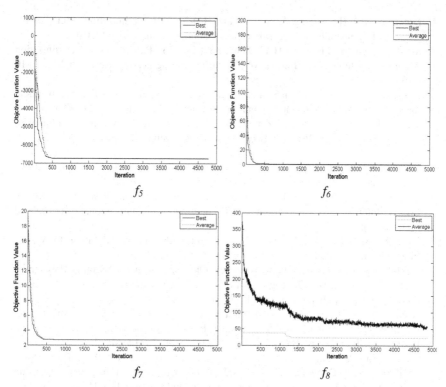

Fig. 1 Convergence characteristics of the functions f_5, f_6, f_7 and f_8

5   Conclusion and Future Work

In this work, lévy flight is done on particles position to update the position of the particles in PSO algorithms. Adaptive polynomial mutation is applied on global best of particle in the swarm to prevent local optima problem. The proposed algorithm LFPSO performed better than basic PSO. In future research, study of lévy flight to update the velocity equation of the PSO algorithm are to be considered and applied to more complex functions and real life problems.

References

1. Eberhart, R.C., Kennedy, J.: A New Optimizer Using Particle Swarm Theory. In: International Symposium on Micromachine and Human Science, pp. 39–434 (1995)
2. Kennedy, J., Eberhart, R.C.: Particle Swarm optimization. In: IEEE International Joint Conference on Neural Networks, pp. 1942–1948. IEEE Press (1995)
3. Chena, M.-R., Lia, X., Zhanga, X., Lu, Y.-Z.: A novel particle swarm optimizer hybridized with external optimization. Applied Soft Computing, 367–373 (2010)

4. Pedersen, M.E.H.: Tuning & Simplifying Heuristically Optimization, Ph.D. thesis, school of Engineering Science, University of Southampton, England (2010)
5. Singh, N., Singh, S.B.: One Half Global Best Position Particle Swarm Optimization Algorithm. International Journal of Scientific & Engineering Research 2(8), 1–10 (2012)
6. Deep, K., Bansal, J.C.: Mean Particle Swarm Optimization for function optimization. International Journal of Computational Intelligence studies 1(1), 72–92 (2009)
7. Wang, H., Liu, Y., Li, C.H., Zeng, S.Y.: A hybrid particle swarm algorithm with Cauchy mutation. In: Proc. of IEEE Swarm Intelligence Symposium, pp. 356–360 (2007)
8. Higashi, N., lba, H.: Particle Swarm Optimization with Gaussian Mutation. In: Proc. IEEE Swarm Intelligence Symposium, Indianapolis, pp. 72–79 (2003)
9. Wu, X., Zhong, M.: Particle Swarm Optimization Based on Power Mutation. In: ISECS International Colloquium on Computing, Communication, Control, and Management (2009)
10. Tang, J., Zhao, X.: Particle Swarm Optimization with Adaptive Mutation. In: WASE International Conference on Information Engineering (2009)
11. Si, T., Jana, N.D., Sil, J.: Particle Swarm Optimization with Adaptive Polynomial Mutation. In: Proc. World Congress on Information and Communication Technologies (WICT 2011), Mumbai, India, pp. 143–147 (2011)
12. Jana, N.D., Si, T., Sil, J.: Particle Swarm Optimization with Adaptive Mutation in Local Best of Particles. In: International Proceedings of Computer Science and Information Technology (IPCSIT 2012), Singapore, pp. 10–14 (March 2012)
13. Yao, X., Liu, Y., Lin, G.: Evolutionary programming made faster. IEEE Transactions on Evolutionary Computation 3, 82–102 (1999); Unger, R.: The genetic algorithm approach to protein structure prediction. Structure and Bonding 110, 153–175 (2004)

Enhanced MRAC Based Parallel Cascade Control Strategy for Unstable Process with Application to a Continuous Bioreactor

Rangaswamy Karthikeyan, Bhargav Chava, Karthik Koneru,
Syam Sundar Varma Godavarthi, Shikha Tripathi, and K.V.V. Murthy

Abstract. In this paper, Enhanced Model Reference Adaptive Control (E-MRAC) based Parallel Cascade Control strategy (PCC) is proposed for the control of unstable continuous bioreactor. This control system consists of secondary and primary loop. The secondary loop comprises of PID controller, which is designed based on the direct synthesis method. In order to achieve stable responses for unstable processes like continuous bioreactor, non linear control strategy in the primary loop would gain edge over linear control. Hence, the Enhanced MRAC (includes smith predictor) is introduced in the primary loop. The presence of Smith predictor has minimized the discrepancies due to dead times. This seems to be an added advantage over existing ones. From the simulation studies it is observed that Enhanced MRAC based PCC has shown better tracking performance when compared to the PID based PCC control strategy.

Keywords: Parallel Cascade Control (PCC), Bioreactor, Enhanced Model Reference Adaptive Control (E-MRAC), Model Reference Adaptive Control (MRAC), Smith predictor, Fuzzy Logic Control (FLC), Dead time compensator (DTC) and PID controller.

Rangaswamy Karthikeyan · Bhargav Chava · Karthik Koneru ·
Syam Sundar Varma Godavarthi · Shikha Tripathi
Department of Electronics and Communication Engineering,
Amrita Vishwa Vidyapeetham, Amrita School of Engineering,
Bangalore, Karnataka, India
e-mail: r_karthikeyan@blr.amrita.edu,
{chavab175,kkoneruk,syam.godavarthi}@gmail.com,
t_shikha@blr.amrita.edu

K.V.V. Murthy
Indian Institute of technology, Gandhi Nagar, Gujarat, India

S.M. Thampi et al. (eds.), *Recent Advances in Intelligent Informatics*,   283
Advances in Intelligent Systems and Computing 235,
DOI: 10.1007/978-3-319-01778-5_29, © Springer International Publishing Switzerland 2014

1   Introduction

The conventional Parallel Cascade Control (PCC), which was first introduced by Luyben [1], is one of the strategies that can be used to improve the system performance particularly in the presence of disturbances. The significant contributions on the tuning of PID controllers in cascade loops include Y. Lee [2]. The design of parallel cascade control for regulatory response and a method for selection of secondary measurement under different disturbances were addressed by C.C Yu [3]. The PCC model proposed by Seshagiri Rao et al [4] has a delay compensator in the primary loop which had a significant effect in reducing the instabilities caused by dead time. Another PCC strategy proposed by R. Karthikeyan et al [5] has fuzzy logic controller in the primary loop and this method has produced some significant results.

Mostly, all the PCC models have been proposed for linear control and delay compensations. An attempt to use non linear control in the primary loop of the PCC hasn't been made so far. Another factor that causes instability in the systems is dead time. Dead time occurs due to the distance velocity lags, recycle loops etc. To eliminate these instabilities, a smith predictor along with the non linear control strategy has been included in the primary loop so as to make the system robust. The performance of parallel cascade control depends mainly on the primary controller; more the efficacy of primary controller, higher would be the performance of the control system. Non linear control strategy along with delay compensation in the primary loop will have a significant effect in reducing the instability due to disturbances and dead time. MRAC belongs to a class of adaptive servo systems in which desired performance is expressed through a reference model. E- MRAC consists of a Fuzzy Logic Controller (FLC) and a Smith Predictor based Dead Time Compensator (DTC). The FLC provides adaptation gain to MRAC without human interference and the DTC is incorporated to solve the problem of instabilities caused by dead time. In this paper, an approach to compensate processes with large dead time with parallel cascade control strategy of using Enhanced MRAC in the primary loop is proposed and this model fetched better results.

The organization of the paper is as follows. Section 2 presents a short discussion on parallel cascade control and Smith predictor. Section 3 gives a description on Continuous bioreactor. Section 4 deals with the description of non linear control strategy i.e. the proposed model. Section 5 is a framework of simulation results of proposed model on application to the unstable continuous stirred tank bioreactor.

2   Parallel Cascade Control

PCC is one of the most successful methods known for its closed loop performance particularly when disturbances associated with the systems are more. It has been widely accepted in process industries for the control of temperature, flow and pressure loops while rejecting the disturbances swiftly. A PCC system is one in which both the manipulated variable and the disturbance affect the primary and the secondary output through the parallel actions while in a series cascade both actions on the primary output take place through secondary one. The PCC structure consists of two loops: primary (outer) loop and secondary (inner) loop as shown in Fig. 1, where G_{p1} and G_{p2} are the primary and secondary processes, G_{c1} and G_{c2} are the primary and secondary loop controllers, G_{d1} and G_{d2} are disturbance transfer functions respectively. y_1, y_2 are the primary and secondary process outputs, r_1 and d are the set point and disturbance inputs. In PCC, the secondary loop dynamics should be much faster than the primary loop because the disturbances entering into the secondary loop should be rejected immediately so that it reduces steady state error in the primary loop. This design has attracted relatively less attention despite the clear benefits of the parallel cascade control and its wide range applications in process industries. If a long time delay exists in the outer loop, the PCC may not give satisfactory closed-loop responses to set point changes. The most widely used dead time compensator is smith predictor based dead time compensator [6]. The use of non linear control strategy along with dead time compensation would definitely have more precise effect on disturbances and dead time effects.

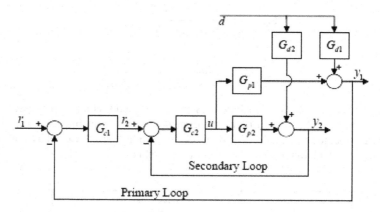

Fig. 1 Conventional PCC Structure

3   A Brief Description of Continuous Bioreactor

A continuous stirred tank bioreactor with a variable yield model is considered here whose model equations are given by Dibiasio et al [8].

$$\frac{dX}{dt} = \mu(s)X - DX \tag{1}$$

$$\frac{dS}{dt} = -\sigma(s) + D(S_f - S) \tag{2}$$

Where μ and σ are functions of S and are given by the expressions

$$\mu(s) = \frac{0.504s(1-0.204s)}{0.000849 + 0.406s^2}$$

$$\sigma(s) = \frac{s(1.32+3.86s-0.661s^2)}{0.000849+0.406s^2}$$

Here 'X' is the biomass concentration and 'S' is the substrate concentration. 'D' is the dilution rate and 'S$_f$' is the feed substrate concentration. 'μ' is the specific generation rate and 'σ' the specific composition rate. The reactor has multiple steady states, one at near wash out conditions, another one at high cell growth rate, and the third one as an unstable steady state at intermediate conditions. The desired operating point is unstable in the open loop conditions. The controlled variable is 'X', the manipulated variable is 'D', and the disturbance variable is 'S$_f$'. Here, the dilution rate effects both 'X' and 'S' in parallel manner and hence the parallel cascade control is expected to give improved performances. For the purpose of design of controllers, the primary process transfer functions G$_{p1}$ (= X(s)/D(s)) and secondary process transfer function G$_{p2}$ (= S(s)/D(s)) are derived around this steady state.

$$G_{p1}(s) = \frac{(-0.24s-0.29)e^{-2s}}{s^2+0.92s-0.1256} \; ;$$

$$G_{p2}(s) = \frac{(1.4s+0.57)e^{-0.5s}}{s^2+0.92s-0.1256}$$

G$_{p1}$ and G$_{p2}$ are the transfer functions of primary and secondary processes.

Simulation studies are carried out for unstable continuous bio reactor using the proposed method. Primary controller G$_{c1}$ of the proposed model is Enhanced MRAC The block diagram of the proposed model is shown in Fig. 2.

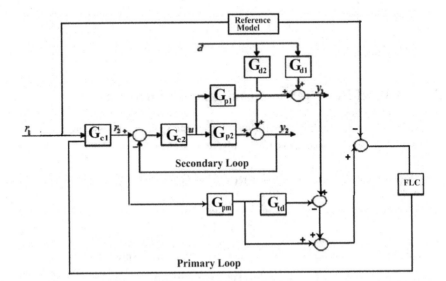

Fig. 2 E-MRAC based parallel cascade control strategy

4   Non Linear Parallel Cascade Control Strategy

The effective non linear control strategy that has less tuning parameters and higher efficacy is MRAC [7].   In general, MRAC system can be systematically represented by the block diagram as shown in the Fig. 3. It has four parts: plant, reference model, controller, and adaptation mechanism.

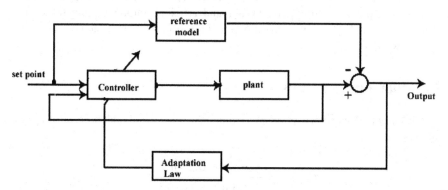

Fig. 3 Conventional MRAC

The plant is assumed to have a known structure although the parameters are unknown. A reference model is used to specify the ideal response of adaptive control system to external command. Intuitively, it provides the ideal plant response which the adaptation mechanism should follow in adjusting the parameters. The controller is usually parameterized by a number of adjustable parameters. The controller should have a perfect tracking capacity in order to allow the possibility of tracking convergence. The adaptation mechanism is used to adjust parameters in control law. In MRAC systems, the adaption law searches for parameters such that the response of the plant under adaptive control becomes the same as that of reference model, i.e. objective of adaptation is to make the tracking error converge to zero. The most commonly used method for designing MRAC is MIT rule.

4.1   Enhanced MRAC

In conventional MRAC a particular gain value γ can yield effective set point tracking only if the average change in the set point or reference signal is within a band that a given γ can handle. Violation of the above value requires another γ to allow the needed tracking. This task of redesign can get sufficiently tedious and time consuming. To overcome this problem, exploitation of fuzzy inference has provided potential solution [10]. It shall be worthwhile to note that, while MRAC drives the plant's response in a manner that makes it mimic the output of the model, the incorporated fuzzy logic control claims the responsibility of providing the needed gain, allowing the plant to follow the specified model. Also, no necessity of tuning the MRAC in the advent of large step input is the merit of this fuzzy-MRAC hybrid genre.

Adaptive control plays a significant role in stabilizing the nonlinearities caused due to dead times. Plants with long time delays can't often be controlled effectively using above models. The main reason is that, additional phase lag is contributed by the time delay, which tends to destabilize the closed loop system. To overcome this problem a smith predictor based dead time compensator is incorporated with the Fuzzy MRAC. This insertion has trimmed the instability caused by the dead times for various processes. The efficient tracking of MRAC, performance of dead time compensator and efficient automatic gain tuning by FLC all together have given a jaw-dropping performance when tested on unstable process with dead time.

5 Simulation Results

The primary and secondary processes of the unstable stirred tank bioreactor are as follows.

$$G_{p1}(s) = \frac{(-0.24s - 0.29)e^{-2s}}{s^2 + 0.92s - 0.1256} \ ;$$

$$G_{p2}(s) = \frac{(1.4s + 0.57)e^{-0.5s}}{s^2 + 0.92s - 0.1256}$$

The secondary and primary processes are unstable hence it is necessary to stabilize the inner loop first by a suitable controller. PID controller is designed for the secondary loop. The parameters of PID are obtained using direct synthesis method [9]. The parameters of PID are identified as $k_c = 2.2106$ $\tau_i = 9.4932$ $\tau_d = 1.538$. With the designed PID controller, the overall primary process model is obtained using Least square technique. Hence the overall process model of the primary loop is identified as,

$$G_m(s) = \frac{(-3.37s - 0.52)e^{-2.28s}}{20.85s^2 + 6.72s + 1}$$

The identified process model of the primary loop is used to design smith predictor based DTC. E-MRAC is designed for primary controller as discussed in section 4.1. The FLC has two inputs, error (e) and change in error range (Δe). The FLC rule base is given in Table 1 and the parameters used to design the FLC are given below:

FLC Input1: error range (e): [-1, 1]
FLC Input2: change in error range (Δe): [-1, 1]
FLC Output range: [0, 0.2]

Table 1 Rule Base for FLC

e ▲ ė	VL	L	ME	H	VH
VL	VL	VL	ME	ME	H
L	L	L	ME	H	H
ME	ME	ME	ME	H	H
H	H	H	H	VH	VH
VH	ME	H	VH	VH	VH

where, VL -Very Low, L - Low, ME-Medium, H-High, VH- Very High.

With above designed controller, simulation results are obtained for different test inputs. Fig. 4 shows the response of the primary process for unit step input with a disturbance of 0.5 given at 300 sec. From the response it can be inferred that the settling time is less when compared to PID based PCC strategy. For further analysis the proposed model has been tested with varying step, sinusoidal and pulsating inputs as shown in Fig.5, Fig.6 and Fig.7 respectively, and the proposed model has given better tracking ability when compared to PID controller.

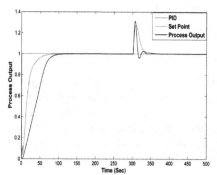

Fig. 4 Responses of the proposed model for the unit step input

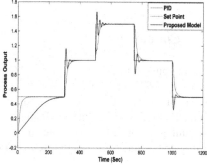

Fig. 5 Responses of the proposed model for the varying step input

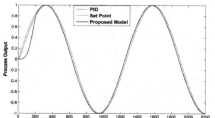

Fig. 6 Responses of the proposed model for the sinusoidal input

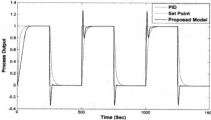

Fig. 7 Responses of the proposed model for the pulse input

5.1   Robustness Analysis

As the design of controllers is based on the process models, when there are modeling errors (uncertainties), one should carry the sensitivity analysis. It is necessary to analyze the robustness of the system in the presence of model uncertainties. The type of uncertainty considered here is the parametric uncertainty in time delay. Robustness test has been conducted for step and sinusoidal inputs. It is evident from the responses shown in Fig.8 and Fig.9 that settling time is less and also the proposed model has shown precise tracking ability.

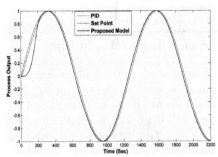
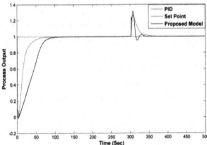

Fig. 8 Responses of the proposed model for the sinusoidal input

Fig. 9 Responses of the proposed model for the step input

6   Conclusion

An integrated framework of E- MRAC and parallel cascade control is proposed for improved control of unstable processes with time delay with application to bioreactor. There are two main features that make the proposed model more stable. Firstly, the adaptability features of the MRAC. Secondly, the automatic gain tuning using FLC. From the simulation results it is evident that, tracking and settling ability has been ameliorated with the use of proposed model.

References

1. Luyben, W.: Parallel cascade control. Ind. Eng. Chem. Fund. 12, 463–467 (1973)
2. Lee, Y., Park, S., Lee, M.: PID Controller Tuning to Obtain Desired Loop Responses for Cascade Control Systems. Ind. Eng.Chem. Res. 37, 1859–1865 (1998)
3. Yu, C.C.: Design of parallel cascade control for disturbance rejection. AIChE J. 34(11), 123–128 (1988)
4. Rao, A.S., Seethaladevi, S., Uma, S., Chidambaram, M.: Enhancing the performance of parallel cascade control using Smith Predictor. ISA Transactions (2009)
5. Karthikeyan, R., Pahikkala, T., Virtanen, S., Isoaho, J., Manickavasagam, K., Murthy, K.V.V.: Fuzzy Logic Based Control for Parallel Cascade Control. ICGST-ACSE Journal 10(1), 39–48 (2010)

6. Saravanakumar, G., Wahidabanu, R.S.D., Nayak, C.G., Thirunavukkarasu: Design and analysis of modified smith predictors for self-regulating and non-self-regulating processes with dead time. Indian Institute of Chemical Engineers 50, 1–16 (2008)
7. Slotine, J.-J.E., Li, W.: Applied Linear Control. Prentice Hall, Eaglewood Cliffs (1991)
8. Dibiasio, D., Lim, H.C., Weigand, W.A., Tsao, G.T.: Phase plane analysis of feedback control of unsteady states in biological reactor. AIChE 24(4), 686–693 (1978)
9. Vanavil, B., Uma, S., Seshagiri Rao, A.: Smith Predictor Based Parallel Cascade Control Strategy for Unstable Processes with Application to a Continuous Bioreactor. Chemical Product and Process Modelling 7(1) Article 10
10. Karthikeyan, R., Yadav, R.K., Tripathi, S., Hemanth Kumar, G.: Analyzing Large Dynamic Set-point change Tracking of MRAC by exploiting Fuzzy Logic based Automatic Gain Tuning. In: IEEE Control and System Graduate Research Colloquium, ICSGRC 2012 (2012)

Fuzzy Fractional Order PID Based Parallel Cascade Control System

Rangaswamy Karthikeyan, Sreekanth Pasam, Sandu Sudheer,
Vallabhaneni Teja, and Shikha Tripathi

Abstract. Parallel cascade controllers are used in process and control industries to improve the dynamic performance of a control system in the presence of disturbances. In the present work, fuzzy set point weighted Fractional Order Proportional Integral Derivative (FOPID) controller is designed for the primary loop of the parallel cascade control system. The secondary controller is designed using the internal model control (IMC) method. Also, a smith predictor based dead time compensator is designed to compensate large time delay in the process. Several case studies are considered to show the advantage of the proposed method when compared to other recently reported methods. The proposed method provides robust control performance which significantly improves the closed loop response with less settling time when compared to conventional PID controller based parallel cascade control system.

Keywords: Fractional Order Proportional Integral Derivative (FOPID) control, Fuzzy Set-point Weighting(FSW), Parallel Cascade Control, Smith Predictor, Internal Model Control (IMC), PID.

1   Introduction

Although the fractional calculus theory was established more than 300 years ago, it has been mainly used for theory research. With the development of computer science, the applications of Fractional Calculus have been gradually penetrated

Rangaswamy Karthikeyan · Sreekanth Pasam · Sandu Sudheer · Vallabhaneni Teja ·
Shikha Tripathi
Amrita Vishwa Vidyapeetham, Amrita School of Engineering,
Department of Electronics and Communication Engineering,
Bangalore, Karnataka, India
e-mail: {r_karthikeyan,t_shikha}@blr.amrita.edu, {psrikanth.988,
       sudheer.sandu,vallabhaneniteja242}@gmail.com

S.M. Thampi et al. (eds.), *Recent Advances in Intelligent Informatics*,   293
Advances in Intelligent Systems and Computing 235,
DOI: 10.1007/978-3-319-01778-5_30, © Springer International Publishing Switzerland 2014

into many fields of engineering such as various kinds of material's memory, the description of mechanics and electrical characteristics, damping and fractional theory and automatic control. Application of Fractional Calculus in control is still an innovative idea. Cascade control is one of the most successful methods for enhancing single loop control performance particularly when the disturbances are associated with the manipulated variable [1]. In series cascade control, both the manipulated variable and the disturbance affects directly on the intermediate (secondary) variable and this in-turn affects the primary controlled variable, whereas, in parallel cascade control, the manipulated variable and the disturbances affect both the primary and secondary outputs simultaneously.

FOPID controller is the promotion of concept of traditional integer-order PID controller, and the integer-order PID controller is a special case of FOPID controllers. when compared to PID, FOPID is a robust controller which does not vary its response for disturbance. Fuzzy FOPID is the improved version of the FOPID for which the settling time for the conventional PID and FOPID is more and this has been reduced by fuzzy FOPID. Introduction of fuzzy FOPID has given an output response which has less settling time and is overshoot free. There are no overshoots in fuzzy FOPID when compared to PID and FOPID. Time delays occur frequently in industrial processes due to the distance velocity lags, recycle loops and composition analysis loops. The main difficulty with the time delays is in increased phase lag, thereby decreasing the gain and phase margin of the transfer function, which imposes a limit on the controller gain, leads to instability of the control system and thus limits the achievable closed loop performance. A cascade control strategy alone is not enough if a long time delay exists in the outer loop, since it may result in poor responses as explained earlier. Hence time delay compensation strategy in the outer loop of cascade control should be incorporated in order to achieve satisfactory performance for both set-point changes and disturbance rejection.

Hence in the present work, fuzzy FOPID is incorporated in the primary loop of parallel cascade control system. To obtain improvised control, Internal Model Control (IMC) is designed for the secondary controller in the inner loop A Smith predictor [2] is used to compensate the time delay based in the process. For clear illustration, Fuzzy set-point weighting for fuzzy FOPID controller (FSW) is given in Section 2. The design of controllers is explained in Section 3. Theoretical developments of parallel cascade control are explained in section 4. The simulation results are provided and the final conclusion is given in Section 5.

2   Fuzzy Set-Point Weighting for FOPID Controller

The last two decades, fractional calculus has been rediscovered by scientists and engineers and applied in an increasing number of fields, namely in the area of control theory. The success of fractional-order controllers is unquestionable with a lot of success due to emerging of effective methods in differentiation and integration of non-integer order equations. PID controllers belong to dominating industrial controllers and therefore are objects of steady effort for improvements of their quality and robustness. One of the possibilities to improve PID controllers is to use fractional-order controllers with non-integer derivation and integration parts. The Fuzzy based FOPID can be used for some of the complex processes for which the FOPID cannot be used. Fractional order PID controllers reduce the overshoot and rise and settling times when compared to integral PID controllers. They add more flexibility in designing systems and expressing the observations in a more easy to follow linguistic notation.

FOPID controller is the promotion of concept of traditional integer-order PID controller, and the integer-order PID controller is a special case of FOPID controllers. when compared to PID, FOPID is a robust controller which does not vary its response for disturbance. A fractional order continuous-time linear time-invariant dynamical system can be described by a fractional order Differential equation [3]

Riemann-Liouville version has the form of differential equation,

$$_aD_t^\alpha f(t) = \frac{1}{\Gamma(n-\alpha)}\frac{d^n}{dt^n}\int_a^t \frac{f(\tau)}{(t-\tau)^{\alpha-n+1}}d\tau \tag{1}$$

$$\begin{aligned}
&a_n D^{(\alpha_n)} y(t) + a_{(n-1)} D^{(\alpha_n-1)} y(t) + \ldots + a_0 D^{(\alpha_0)} y(t)\\
&= b_m D^{(\beta_m)} u(t) + b_{(m-1)} D^{(\beta_{m-1})} u(t) + \ldots + b_0 D^{(\beta_0)} u(t)
\end{aligned} \tag{2}$$

where $u(t)$ is the input signal and $y(t)$ is the output signal. $D^\gamma = {}_0D_t^\gamma$ represents fractional derivative, a_k with $(k = 0,....,n)$ and b_k with $(k = 0,....,m)$ denote constants, and α_k with $(k = 0,....,n)$ and β_k with $(k = 0,.... m)$ are arbitrary real numbers. According to [3] one can assume inequalities, $\alpha_n > \alpha_{n-1} > .... > \alpha_0$ and $\beta_n > \beta_{n-1} > .... > \beta_0$ without loss of generality. Another option for fractional order system description is in the form of incommensurate real orders transfer function [3]

$$G(s) = \frac{B\left(s^{\beta_k}\right)}{A\left(s^{\alpha_k}\right)} = \frac{b_m s^{\beta_m} + b_{m-1} s^{\beta_{m-1}} + \ldots + b_0 s^{\beta_0}}{a_n s^{\alpha_n} + a_{n-1} s^{\alpha_{n-1}} + \ldots + a_0 s^{\alpha_0}} \tag{3}$$

The symbols in (3) have the same meaning as in (2). Every incommensurate order system (3) can be expressed as a commensurate one by means of a multivalued transfer function. If the transfer function (3) is supposed for $\alpha_k = \alpha k$, $\beta_k = \beta k$, $0 < \alpha < 1$, $k \in \mathbb{Z}$, it represents the specialized case of commensurate order system. Then the corresponding transfer function is [3]

$$G(s) = K\frac{B\left(s^\alpha\right)}{A\left(s^\alpha\right)} = K\frac{\sum_{k=0}^m b_k (s^\alpha)^k}{\sum_{k=0}^n a_k (s^\alpha)^k} \tag{4}$$

The analytical solution of fractional order differential equations in time domain (e.g. for the purpose of step or impulse responses computation) can be expressed for example with the assistance of functions of Mittag-Leffler type [3]

The FOPID concept is obtained from[3]. Fuzzy set point weighting for PID is proposed by Visioli [4] and to improve the response the proposed method is been implemented as Fuzzy setpoint weighting for FOPID control.which has two additional parameters involved which helps in attenuating good load disturbance and the two parameters involved are λ and μ which are tuned manually. In this way, the control law can be written as

$$u(t)=k_P (b(t)y_{sp}(t) - y(t)) + K_I D^{-\lambda} e(t) + K_D D^{\mu} e(t) \tag{5}$$

$$b(t)=w + f(t) \tag{6}$$

where w is a positive constant parameter less than or equal to 1 [5], and f(t) is the output of the fuzzy inference system, which consists of five triangular membership functions for the two inputs $e(t)$ and $\Delta e(t)$ and five triangular membership functions for the output. Table 1 shows the overall control scheme. Fuzzy mechanism parameters undertake the most significant role and the parameters are somewhat intuitive, and it is very similar to the one in the typical fuzzy FOPID controller, for which tuning procedure have been established. Hence, the task of tuning is simplified. In any case, a simple empirical procedure for the manual tuning of the fuzzy module has been proposed. Fuzzy logic can be used in the above context to vary the FOPID parameter values during the transient response, in order to improve the step response in more efficient manner. (e.g. the fine-tuned Ziegler-Nichols parameters) [6]. In this way, it is possible to implement a nonlinear controller so that the rise time and the overshoot are reduced and the settling time is significantly reduced. This system shows the methodology which is robust to varying parameters of the Ziegler-Nichols, and therefore its practical implementation seems to be quite simple.

Table 1 Linguistic rule base of the inference engine

Δe \ e	VL	L	ME	H	VH
VL	VL	VL	ME	ME	H
L	L	L	ME	H	H
ME	ME	ME	ME	H	H
H	H	H	H	VH	VH
VH	ME	H	VH	VH	VH

3   Proposed Method – Fuzzy Set Point Weighting for FOPID Based Parallel Cascade Control

A parallel cascade system is the one in which both the manipulated variable and the disturbances affect the primary and secondary outputs through parallel actions.

It is commonly used in the control of product quality in chemical processes to strictly control a process variable which is closely related to the property of interest which is readily available from online measurements. In general, parallel cascade control is appropriate when the secondary loop has a faster dynamic response and the rejection of the disturbance in the secondary output reduces the steady state output error in the primary loop. The parallel cascade control structure is shown in Fig. 1 where P_1 and P_2 are the transfer functions of the primary and secondary processes respectively. C_1 and C_2 are the primary and secondary controllers. P_{d1} and P_{d2} are the transfer functions of the disturbances for primary and secondary loops respectively. In parallel cascade control, the secondary loop dynamics should be much faster than the primary loop because the disturbances entering in to the secondary loop should be rejected immediately so that it reduces steady state error in the primary loop. Parallel cascade control is also beneficial when measurements of the primary output are sampled infrequently and with long time delays.

The closed loop transfer function between the secondary output (y_2)and secondary set point (u) and disturbance (d) is obtained by assuming perfect model conditions ($P_2 = P_{2m}$) as

$$y_2 = P_{2m}C_2q_{f2}u + (1 - P_{2m}C_2)P_{d2}d \tag{7}$$

The relation between primary output .(y_1) and secondary set point (y_{sp}) is obtained as

$$y_1 = P_1C_2q_{f2}\,u \tag{8}$$

The closed loop transfer function between primary output (y_1)   and primary set point y_{sp} and disturbance (d) is obtained as

$$y_1 = \frac{q_{f2}P_1C_1C_2}{1+P_pC_1 - P_pC_1\tau_Fe^{-\theta_ms} + P_1C_1C_2\tau_Fq_{f2}}y_{sp} + \frac{(1+P_pC_1 - \tau_FP_pC_1e^{\theta_ms})(P_{d1} - C_2P_1P_{d2})}{1+P_pC_1 - P_pC_1\tau_Fe^{-\theta_ms} + P_1C_1C_2\tau_Fq_{f2}}d \tag{9}$$

From Eq. (8), by assuming perfect model conditions, the primary process model is obtained as

$$P_p = P_1C_2q_{f2} \tag{10}$$

Substituting Eq. (10) in Eq. (9) gives

$$y_1 = \frac{P_pC_1}{1+P_pC_1}y_{sp} + \frac{(1+P_pC_1 - \tau_FP_pC_1e^{\theta_ms})(P_{d1} - C_2P_1P_{d2})}{1+P_pC_1}d \tag{11}$$

From Eq. (11), it is clear that the primary controller can be designed based on the delay free primary model (Pp) as the characteristic equation does not contain any time delay. Usually, the primary model (Pp) will be of high order and should be reduced to first order form for the purpose of design of controller. In the present work, Skogestad's Half-rule [7] is adopted for model reduction. As discussed in section 2, FSW based FOPID gives improved performance over conventional PID controller. Hence FSW based FOPID is designed for the primary controllers. Usually in parallel cascade control, the secondary process has no or a negligible time delay while the primary process has a large time delay compared to secondary loop. To compensate for this effect, smith predictor based delay compensation technique is incorporated. As shown in Fig. 1, C_1 is the primary controller i.e fuzzy FOPID controller and C_2 is the IMC controller for secondary loop. P_{2m} is the secondary process model and P_p is the primary process model without time delay and m is the overall time delay of the primary process model. $q_{f\,2}$ is the set-point filter for inner loop. The robustness of the delay compensator can be increased by using a filter for the predicted disturbance [8]. Here, a first order filter (F) is used for the predicted disturbance.

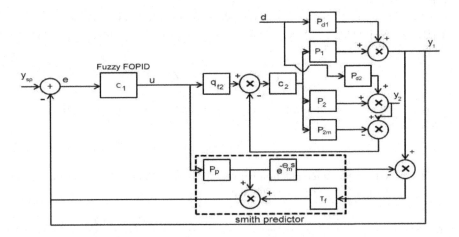

Fig. 1 Implementation of parallel cascade control with fuzzy FOPID

4   Design of Controllers

Here, controller design is addressed both for the secondary loop and primary loop separately.

4.1   Design of Secondary Controller

The controller in the secondary loop should be designed in such a way that it should reject the disturbances entering the secondary process immediately. To achieve this condition, the secondary variable should follow its set-point as good as possible. The secondary controller is designed based on IMC principles and is designed based on slow and fast dynamics of the inner loop [9]. According to IMC design, the secondary loop controllercs is designed as,

$$C_2 = \frac{(\tau_2 s + 1)}{k_2} \frac{(\alpha_1 s + 1)}{(\lambda_2 s + 1)^m} \tag{12}$$

and the corresponding set-point filter for slow dynamics is,

$$q_{f2} = \frac{1}{(\alpha_1 s + 1)} \tag{13}$$

where $\alpha 1$ is calculated,

$$\alpha_1 = \tau_2 \left[1 - \left(1 - \frac{\lambda_2}{\tau_2} \right)^2 e^{-\theta_2/\tau_2} \right] \tag{14}$$

set-point filter for fast dynamics is,

$$q_{f2} = 1 . \tag{15}$$

4.2 Design of Primary Controller

With the designed secondary controller, a primary process model Pp is found out from the closed loop transfer function between primary output y1 and set-point u. (assuming the secondary loop process mode exactly describes the behavior of the secondary loop process).

The primary process model $\tilde{P}_p$ can be written as $\tilde{P}_p = P_p e^{-\theta_1 s}$ where P_p is the primary process model without time delay and θ_1 is the time delay of the primary loop. With the obtained process model, the dead time compensator is designed.

And the primary controller as fuzzy FOPID controller of the form,

$$C_1 = k_P (w + f(t)) y_{sp}(t) - y(t)) + K_I D^{-\lambda} e(t) + K_D D^{\mu} e(t) \tag{16}$$

f(t) is the output of the fuzzy inference system, which consists of five triangular membership functions for the two inputs $\Delta e(t)$ and $e(t)$ and nine triangular membership functions for the output. The fuzzy rules are based on theMacvicar-Whelan matrix (the meaning of the linguistic variables is described in Table 1). It is worth stressing that in this method the role of the fuzzy mechanism parameters is somewhat intuitive, and it is very similar to the one in the typical fuzzy PD-like controller, for which tuning procedure have been established.

5 Simulation Results

For the purpose of simulation, a first order process with dead time (FOPDT) is considered from A.Seshagiri Rao et.al [10]

$$P_1 = P_{d1} = \frac{e^{-80s}}{20s+1} \tag{17}$$

$$P_2 = P_{d2} = \frac{1}{10s+1} \tag{18}$$

IMC controller designed for the above process model is,

$$C_2 = \frac{(10s+1)(1.9s+1)}{(s+1)^2} \tag{19}$$

The setpoint filter,

$$q_{f2} = \frac{1}{1.9s+1} \tag{20}$$

With the designed IMC controller, the FOPID parameters used to design the primary controller is given below.

FOPID parameters:
$K_c = 0.666$, $\lambda = -0.4$
$\tau_i = 0.01$, $\mu = -0.1$
$\tau_d = 0$

Fuzzy parameters:
FLC Input1: error range= [-20, 20]
FLC Input2: change in error range= [-20, 20]
FLC Output range= [0, 16.5]

Fuzzy FOPID (Proposed) controller is designed with above parameters and simulation results are observed for the step, perturbation in step, varying step and pulse inputs. Fig. 2, Fig. 3, Fig. 4 and Fig. 5 shows the response of the process for a step input with a disturbance of magnitude -10 is given at 500 sec. For comparison purpose, the results obtained is compared with that of A.Seshagiri Rao et. al. [10]. As an inference from Fig. 2, the settling time for the conventional PID is more and this has been reduced by fuzzy FOPID. Introduction of fuzzy FOPID has given an output response which has less settling time and is overshoot free. Robustness of the proposed model is tested by a secondary influence on a system that causes its time constant to deviate slightly and dead time by 20%, this simulation has given us satisfactory results shown in Fig. 3. There are no overshoots in fuzzy FOPID when compared to PID. With reference to Fig. 4, Fig. 5 variable step input and pulsed input has tracked efficiently by proposed model when compared to PID Controller.

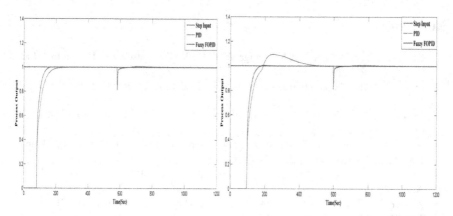

Fig. 2 Responses for the step input model with disturbance

Fig. 3 Control action responses for step input with perturbation of +20%

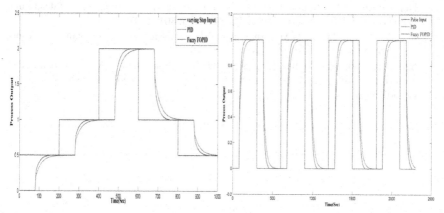

Fig. 4 Response for varying step input

Fig. 5 Tracking response for the pulsed input

6 Conclusion

Fuzzy setpoint weighted FOPID is designed for the primary controller in Parallel cascade control system. Time delay compensation is incorporated in the primary loop of the parallel cascade control system. The proposed method is simple and analytical relations are provided for tuning of both secondary and primary controllers. Simulation study was done on first order process with time delay. This implementation of fuzzy logic to the controller has demonstrated significant potential in handling both set point tracking and disturbance rejection. This concept has reduced the settling time and has significantly improved the tracking ability of the controller when compared to conventional PID based Parallel cascade control system.

References

1. Krishnaswamy, P.R., Jha, R.K., Rangaiah, G.P.: When to use cascaded control. Ind. Eng. Chem. Res. 29, 2163 (1990)
2. Smith, O.J.M.: Closer control of loops with dead time. Chem. Eng. Prog. 53, 217 (1957)
3. Matusu, R.: Application of fractional order calculus to control theory. IJMMMAS 5(7) (2011)
4. Visoli, A.: Fuzzy logic based set-point weighting for PID controllers. IEEE Trans. Syst. Man, Cybern. - Pt. A 29, 587–592 (1999)
5. Visoli, A.: Tuning of PID controllers with fuzzy logic. IEE Proc.-Control Theory Appl. 148(I) (January 2001)
6. Ziegler, J.G., Nichols, N.B.: Optimum setting for automatic Controllers. ASME Trans., 759–768 (1942)
7. Skogestad, S.: Simple analytic rules for model reduction and PID controller tuning. J. Process Control 13, 291–309 (2003)
8. Normey-Rico, J.E., Bordon, C., Camacho, E.F.: Improving the robustness of dead time compensating PI controllers. Control Eng. Pract. 5, 801 (1997)
9. Lee, Y., Skliar, M., Lee, M.: Analytical PID controller design for parallel cascade control systems. J. Process. Control 16, 809–818 (2006)
10. Seshagiri Rao, A., Seethaladevi, S., Uma, S., Chidambaram, M.: Enhancing the performance of parallel cascade control using Smith predictor. ISA Trans. Pt, 00190578 (2008)

A Novel Fault Detection and Replacement Scheme in WSN

Indrajit Banerjee, Anirban Datta, Sonalisa Pal, Soujanya Chatterjee, and Tuhina Samanta

Abstract. Wireless Sensor Network (WSN) is a collection of nodes, which are limited in energy content and interact with each other to complete a job assigned to them. Performance of WSN may degrade because of faulty nodes present in the network. The strategy in the paper discusses about a fault detection technique, which depends upon three components of the node. The strategy exploits a node to the fullest extent possible before declaring it as faulty. Moreover the faulty-node replacement technique replaces a faulty node with a healthy node, based on a regression plain model. Optimal use of sensor node is done by reallocation of jobs of the faulty nodes to the active nodes.

Keywords: Fault detection, Fault replacement, Regression plain model, 3 dimensional WSN.

1   Introduction

Wireless Sensor Network (WSN) has emerged as a computer platforms for environmentmonitoring,health, power, inventorylocation, factory and process automation,and seismic and structural features[1]. The network iscomposed of a large number of tiny sensor nodesequipped withlimited computing and communication capabilities. Since low-costsensor nodes are often deployed in an uncontrolled or even harshenvironment, they are prone to several faults. It isthus desirable todetect, and locate the faulty sensor nodes, and exclude them from thenetwork during normal operation unless they can be used as communicationnodes [2].So the network requires an efficient fault detection and replacement technique.The problem of fault tolerant multimodal sensor fusion for digital binary sensors has been addressed in

Indrajit Banerjee · AnirbanDatta · Sonalisa Pal · SoujanyaChatterjee · TuhinaSamanta
Bengal Engineering and Science University, Shibpur, Howrah, India
e-mail: {ibanerjee,t_samanta}@it.becs.ac.in,
        {anirban.datta.24,sonalisapal,
        soujanya.chatterjee}@gmail.com

S.M. Thampi et al. (eds.), *Recent Advances in Intelligent Informatics*,
Advances in Intelligent Systems and Computing 235,
DOI: 10.1007/978-3-319-01778-5_31, © Springer International Publishing Switzerland 2014

[3]. In [4], a localized fault detection strategy is developed, which ensures thenetwork quality of service by detecting the faults and taking the actions to avoid furtherdegradation of the service. In [5] the problem for distributed fault detection and recovery technique in WSNs is addressed.

We have developed a strategy to properly classify the faults in a network and relocate the task of a faulty node to a healthy node with optimal parameters if possible. In our strategy we divide a sensor node in three sections: the sensor, the battery, transmitter circuit and the receiver circuit. A state-graph model is generated to identify the different states of the deployed nodes, and variation in the states of the nodes with time. The classification of the faulty nodes is dependent on the proposed state graph. We then propose a strategy to deal with the faults independently and according to the requirements of the network. In our strategy we also propose a regression model based faulty node replacement policy. Nodes are deployed in a three dimensional model. Depending on the space coordinates of the faulty nodes, regression planes are generated recursively, and faulty nodes are replaced with the non-faulty nodes by allocation the jobs of the faulty nodes to the selected active nodes to increase network life time.

The remainder of the paper is organized as follows. Section 2 describes the necessary problem formulation for fault-detection and replacement scheme. Section 3 deals with the fault detection strategy in detail. Section 4 explains the regression model and the fault replacement strategy using that model. Experimental results are shown in Section 5. Finally, Section 6 concludes the paper.

2   Problem Statement

Wireless sensor network is a collection of nodes, generally represented as a connected graph G (V, E). Our proposed scheme detects faulty nodes, and replaces the task of those faulty nodes with the other active nodes, located in the near neighbor. Let Vs be the node set that has sensor circuit fault, Vr be the node set that has receiver fault, Vp be the node set that has peripheral fault where Vs , Vr, Vp are the subsets of V; $Vs \cap Vr \cap Vp \neq \emptyset$. So the remaining non-faulty nodes are represented by $|S|-|Vs \cup Vp \cup Vr|= |S| - (|Vs| + |Vp|+|Vr|-|Vs \cap Vp| - |Vs \cap Vr| - |Vr \cap Vp|+|Vs \cap Vp \cap Vr|)$.

There are certain unique properties of our proposed scheme, enumerated as follows,

i) Our proposed network model is a three dimensional model. ii) All the networks in which our proposed fault detection techniques have been used have shown low functional time, with much improvement compared to the other recent works. iii) Our scheme is capable of finding uniform coverage based on user specified parameters.

3   Proposed Fault Detection Model

3.1   Basic Control Units of WSN

In the proposed strategy, we analyze a sensor node as a three component device comprising of: i) Peripheral unit (PU): The first unit comprising of the three essential components, without which a sensor node is considered faulty, (a) Battery or limited power source, (b) Processor and memory unit, (c) Transmitter circuit used to transmit data. ii) A sensor circuit (S). iii) A receiver circuit (R).

The strategy detects a sensor node as faulty based on the conditions of these three components. For this purpose, we maintain a three parameter table for each node, each column depicting a bit value for each of the components of the sensor node.

3.2 Fault Detection Procedure

Primary focus of the fault detection procedure is maximum utilization of the resources available in the sensor network. The strategy declares a node faulty based upon the combination of the parameter-bits rather than on the basis of any one of the parameter-bits. This means that even if one of the parameter-bits gets its value zero, the strategy looks for other possible options, to keep a node working based on the values of other parameter-bits. Following are the criteria based upon which the strategy takes its decision about a node:

(a) If the first parameter-bit (PU-bit) is zero, the node is detected as faulty. It is because without the power a sensor node ceases to function normally.
(b) With the PU-bit set to one, it is necessary for any one or both the next bits to be set to one. A node with second parameter-bit (sensor-bit) and third parameter-bit (receiver-bit) set to zero is detected as faulty. A node with only sensor-bit set to one can continue to sense the surroundings. A node with only transceiver bit set to one can serve as an intermediate-node in the network by transmitting and receiving the data sensed by other nodes from one node to other.

Initially when a wireless sensor network is yet to be deployed, the parameter-bits of all the sensor nodes are set to one, assuming that all the components of all the sensor nodes are alright prior to its deployment. Figure 1(a) shows a state of nodes with initial deployment.

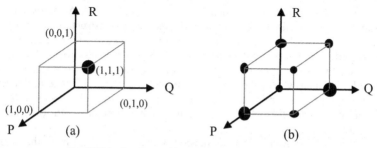

P-> PU-bit, Q-> Sensor-bit, R-> Receiver-bit

Fig. 1 (a) Initial State-graph of a sensor network before deployment. (b) State-graph of the network after certain time of deployment

Now as the proposed strategy continuously checks for each of the components of the nodes for their proper functioning. The state for a node changes from (1,q,r) to (0,q,r), as PU dies down below threshold. Hence, with time, the nodes deviate from the ideal non-faulty state to the other faulty states, occupying other vertices of the cube, as is depicted in Figure 1(b).

The energy loss due to transmission and receiving of data is calculated on the basis of the technique described in detail DFDNM [2]. For the sensor component, the algorithm continuously validates the reading of a node with the readings of the nodes of its surroundings. There is a difference-threshold (D_N) for the network that is set to decide upon the authenticity of a reading. If the difference between the reading of a node 'n' (R_n) and average of the readings of the surrounding nodes (A_n) exceeds D_N, the reading of the node is detected as faulty and the state of the node changes from (x,1,z) to(x,0,z). The threshold D_N is determined according to the specifications mentioned by the manufacturers of the wireless sensor motes.

$$R_{n-}A_n > D_N \tag{1}$$

If in-equation (1) holds, the sensor-bit is set to zero.

Transmitter fault detection has been done by implementing sampled signal. When a node receives the signal, it sends back an acknowledgement signal. If after certain time no acknowledgment signal is received by the sending node, a fault is detected in the transmitter circuit. The nodes that are detected as faulty as per the criteria mentioned are handled by the fault-tolerance strategy based upon a three dimensional regression model.

4   Fault Replacement Strategy

The strategy for fault replacement based on regression model selects the nearest optimal functional node to the faulty node and assigns the task of the faulty node to it. Next, detail analysis of regression plain construction and fault replacement is performed.

4.1   Finding Regression Plane

Space coordinates of the faulty nodes are retrieved and stored. A regression plane based upon these coordinates is constructed. The plane constructed passes through the zone of the network that has the highest number of faulty nodes. After the plane is constructed, the nearest faulty node to the plane is selected for reallocation of task to a non-faulty node. As the faults are corrected, the regression plane changes accordingly. So as a whole, the regression model is a dynamic system, where the equation of the plane changes according to the position of faulty nodes in the network. This guarantees the selection of faulty nodes for reallocation of task in an optimal way.

For a three-dimensional linear regression model, we will estimate the value of the coordinate z based on the values of independent variables x and y by expressing z as a mathematical function f(x,y).Let the linear regression equation of z on x,y be

$$Z = a + bX + cY \tag{2}$$

For a given data set (x_1,y_1,z_1), (x_2,y_2,z_2) . . . , (x_m,y_m,z_m), where $m \geq 3$, the best fitting curve has the least square error, i.e.,

$$\pi = \sum_{i=0}^{m}[z_i - f(x_i, y_i)]^2 = \sum_{i=0}^{m}[z_i - (a + bx_i + cy_i)]^2 = \min \qquad (3)$$

Please note that a, b, and c are unknown coefficients, while all xi, yi, and zi are given. To obtain the least square error, the unknown coefficients a, b, and c must yield zero first derivatives.

We solve the regression plain equations with best curve fitting model. With each iteration of the algorithm, the equation of the plane changes according to the region with the maximum dense region of the faulty nodes. The regression equation is used to fit a predictive model to an observed data set of z, y and x values.

4.2 Fault Replacement Procedure

For fault tolerance, we collect the coordinates of faulty nodes and use the data set for obtaining the regression equation. The regression plane mainly represents the average plane which consists of the faulty nodes. Then it tries to find a healthy node according the procedure mentioned next and replace the faulty node.The node replacement technique deals with the replacement of a faulty node's activities with a healthy node's activity, so that the network continuesworking in an optimal state. The healthy nodes are classified as optimal nodes based on an optimal parameter, and are called optimal nodes.For selecting an optimal healthy node, we take into account the following four parameters.

i)State Parameter: The State-parameter (S) allows the strategy to select a healthy node on the basis of the healthier state of the node. As we see from Table I, the decimal values corresponding to the three valid states of the nodes are five, six and seven respectively.The nodes with higher decimal values of states are healthier than nodes with lower decimal value of states.

Table 1 Table showing the decimal values corresponding to the different valid states of the nodes

Decimal value of the state (D)	Peripheral Unit (PU)	Sensor (S)	Receiver (R)
7	1	1	1
6	1	1	0
5	1	0	1

ii) Power Parameter: *The Power-parameter (P) defines the power level available in a node. A node with higher level of power is given higher priority for selection for node replacement.*

iii) Communication Parameter:*The Communication-parameter (C) gives the communication cost of thecandidate healthy node for communication with the five nearest working nodes. The communication cost is calculated as a function of the distance of the node from the candidate healthy node.*

iv) Covrage Parameter: *The Coverage-parameter (Co) defines the distance of the node to be replaced from the candidate node by which it is to be replaced..*

4.3   Three Dimensional Coverage Model

For the coverage purpose in a three dimensional space coordinate, the nodes should be deployed at the vertices of a platonic solid. This is because placing the nodes at the vertices of platonic solids ensures that all the nodes are at approximately equal distances to each other. Of all the platonic solids available, tetrahedron is the most suitable, as node placed at its vertex covers a greater volume of other nodes than any other platonic solids like cubeoroctahedron.

The nodes arranged at the vertices of tetrahedron cover maximum space than other platonic solids like cube or octahedron.For a given radius of coverage 'r' for a node, nodes arranged at the vertices of a tetrahedron cover a larger number of neighboring nodes than any other platonic solids. Thus, we consider the tetrahedron for the coverage purpose in our strategy. Therefore, we have assumed that the nodes are always at the vertices of an approximate regular tetrahedron.

5   Performance Analysis

The proposed scheme is implemented in C language. Sensor nodes are distributed throughout a specified region in the three dimensional space. The simulation was carried out for different sizes of a three dimensional grid structure, where the sensor nodes are placed at the grid points. The simulation results shown below are considered for an edge size of twenty-one nodes. As a result our sample cubical network consists of $21 \times 21 \times 21 = 91$ nodes. The initial energy of each of the nodes of the network is 0.9 Joules (J). Each node's transmitter circuit energy loss is 40 nJ/bit and the amplifier circuit energy loss is 10 pJ/bit.

The first set of results (Figure 2(a)) show the rate of consumption of energy in the network. The proposed strategy shows a network life-time increase of 74.8% compared to thatof DFDNM [2] and an increase of 81.2% from the cellular algorithm [5]. The graph corresponding to the proposed strategy does not reach zero, as each of the nodes is declared as faulty when the energy for that node falls below a certain pre-determined threshold. This is mainly done so that the network remains functional with full capacity, i.e. the node replacement strategy finds a suitable candidate node by the time to-be-faulty node gets expired. So there is no need for the network to function with fewer nodes till the time the faulty nodes get replaced. So there is a possibility that a fraction of the total initial energy remains in the network unused, when the network expires.

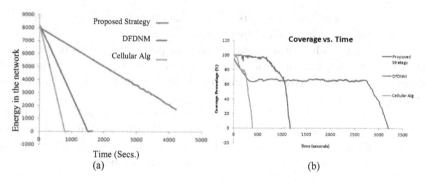

Fig. 2 (a)Energy in the network with respect to time, (b)Variation of coverage percentage of the network with respect to time

Figure 2(b) shows the variation of coverage percentage with the passage of time. The coverage percentage is the ratio of number of nodes covered by the deployed nodes to the total number of nodes in the network multiplied by hundred. We see that for the proposed strategy, for the given test case, the coverage percentage declines gradually from initial coverage percentage to a value around 62.5% within a small duration of time, a value of coverage percentage which the network can maintain, while increasing the lifetime of the network to an optimum value. The DFDNM and the cellular algorithm, unlike the proposed algorithm, tries to maintain the maximum possible coverage percentage at all times, and thus gets exhausted quite early. The proposed strategy shows a coverage improvement of 61.6% over DFDNM and an improvement of 87.7% over cellular algorithm.

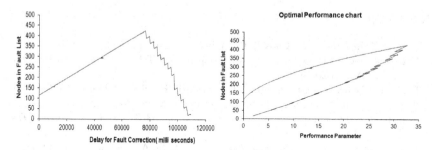

Fig. 3 (a) Time required for correcting a fault with respect to number of faults in the fault list, (b) Performance variation of the Proposed Srategy with Node Fault

Figure 3(a) shows the time required in milli-seconds to correct a fault i.e. replace a faulty node with a suitable healthy node with respect to the number of nodes in queue in the faulty node list that are to be replaced. Initially when every node of the network is healthy, the strategy takes relatively greater amount of time to find an optimal node for replacement of the faulty node than later. The reason is, initially the algorithm has to scan through a greater number of nodes, checking for their parameters. But as the nodes of the network begin to die, the algorithm has to scan through fewer numbers of nodes and thus the fault correction delay decreases with the aging of the network.

Here in Figure 3(b), we show the variation of performance with respect to the number of nodes queued up in the faulty node list. We calculate the performance parameter 'P' by:P = (Nodes in fault list * fault correction delay)/Number of nodes in the network.

Therefore we see that when the number of nodes in the fault list is the maximum possible number of fault nodes that can occur simultaneously, the performance of the strategy is optimal.Thus the above results show that the proposed strategy performs better than all the existing strategies discussed before.

6   Conclusion

In this paper, the strategy proposes an innovative model for fault tolerance in a three dimensional sensor network, which is mainly aimed at utilizing the nodes to maximum possible extent. Thus the state of each of the nodes is determined on the basis of its three components: Peripheral Unit, sensor and transmitter circuit. The optimal position of the faulty node that is to be replaced is found out with the help of three dimensionalregression model. After that the optimal node that is to be used for replacement of a faulty node is selected on the basis of four parameters: state, power, communication and coverage. For the coverage purpose we deploy the nodes in the three dimensional network on the vertices of a regular tetrahedron. The results show that our strategy out-performs all the existing algorithms in the field. Moreover the performance of the network is also enhanced due to the self-adjusting nature of the algorithm which ensures the network longevity.

References

1. Yick, J., Mukherjee, B., Ghosal, D.: Wireless sensor network survey" Computer Networks. The International Journal of Computer and Telecommunications Networking 52(12) (August 2008)
2. Banerjee, I., Chanak, P., Sikdar, B.K., Rahaman, H.: DFDNM: A Distributed Fault Detection and Node Management Scheme for Wireless Sensor Network. In: Advances in Computing and Communications, vol. 192, Part 2, pp. 68–81 (2011), doi:10.1007/978-3-642-22720-2_7
3. Koushanfar, F., Potkonjak, M., Sangiovanni-Vincentelli, A.: Fault tolerance in wireless sensor networks. In: Mahgoub, I., Ilyas, M. (eds.) Handbook of Sensor Networks
4. Chen, J., Kher, S., Somani, A.: Distributed Fault Detection of Wireless Sensor Networks. In: Proceeding DIWANS 2006 Proceedings of the 2006 Workshop on Dependability issues in Wireless ad Hoc Networks and Sensor Networks (2006) ISBN:1-59593-471-5
5. Asim, M., Mokhtar, H., Merabti, M.: A cellular approach to fault detection and recovery in wireless sensor networks. In: IEEE Third International Conference on Sensor Technologies and Applications (2009)

Mobile Sink Management for Nonuniformly Distributed Sensor Node Coverage Using a Game Theoretic Approach

Santanu Datta, Indrajit Banerjee, and Tuhina Samanta

Abstract. The paper proposes a novel approach to explore the controlled and coordinated sink movement for Wireless Sensor Network (WSN). Past research showed that using mobile agents for single hop data gathering has several boons over the classical cluster based multi hop data forwarding and collection scheme. In this paper, we explore the mobility of multiple mobile sinks in a network to achieve optimum network coverage. We have incorporated the cooperative behavior among the mobile sinks using the basic idea of game theory. The entire data collection process is portrayed as a game, where the mobile sinks are the players. Players try to increase their payoffs by collecting data from maximum number of sensor nodes. The game is played repetitively. The cooperative nature of the game motivates a mobile sink to move in such a way that it minimizes the common strategies played by the other players, and maximizes the cumulative overall payoff. Our proposed algorithm gives substantial performance enhancement in terms better network coverage, and lower network energy consumption in a randomly deployed sensor environment.

Keywords: Wireless Sensor Network, Mobile Sink, Grid Structure, Game Theory, Cooperative Game, Nash Equilibrium.

1 Introduction

Wireless sensor network (WSN) is a collection of a large number of small, energy constrained sensor nodes that harvest data from their surroundings and collectively send those gathered data to a central depository often called base station for

Santanu Datta · Indrajit Banerjee · Tuhina Samanta
Bengal Engineering and Science University, Shibpur, Howrah, India
e-mail: santanu.datta007@gmail.com,
       {ibanerjee,t_samanta}@it.becs.ac.in

S.M. Thampi et al. (eds.), *Recent Advances in Intelligent Informatics*,
Advances in Intelligent Systems and Computing 235,
DOI: 10.1007/978-3-319-01778-5_32, © Springer International Publishing Switzerland 2014

analysis of those data. Data collection is the basic objective of WSN. Multi hop data forwarding is a common scheme but often suffers from a severe drawback named Hot Spot problem. Recently, with the introduction of mobile sink, satisfying improvement in network life maximization has been achieved. Sinks are those nodes, which collects data from different static sensor nodes. By making sinks movable, single hop data transmission has been made extremely simple. According to this approach, sensor nodes directly send their harvested data to one of the mobile sinks when it comes to the node's vicinity.

Movement pattern of mobile sinks are of two types, random and defined. In [2], C. Konstantopoulos et al. proposed Mobicluster, a five phase Rendezvous based algorithm for energy efficient data collection from WSN with mobile sink of uncontrolled mobility. However, it is seen that controlled sink movement or constrained random walk gives better data collection ability due to the introduction of constrains [3]. Since WSNs can be spread across a huge geographical area, employing only one mobile sink per WSN may not be enough to cover the entire network. Using multiple mobile sinks gives better network coverage in those cases. To employ multiple mobile sinks in a network, proper coordination and controlled mobility are desired. In [1], Kinalis et al. came up with a set of new protocols for scalable data collection. In [4], Saad et al. proposed two novel algorithms, named Path-points identification algorithm, and Bees algorithm to detect mobile sink movement path. In [6], J. Luo and J. P. Hubaux explored the joint effect of controlled sink movement and multi-hop data routing within a WSN. The optimization process in terms of network lifetime is carried out by moving sink in an optimized trajectory.

Using game theory to solve different problems in WSN has been a popular choice among the researchers. Game theory has been used extensively to solve optimization problems in distributed cross layer framework, finding reliable routing path, designing an intrusion detection scheme using game theory, enforcing security, network lifetime maximization etc [7]. However, to the best of our knowledge, not much work is done in the field of mobile sink movement and its path optimization based on game theory. In this paper we propose an algorithm to implement game theory, while moving multiple mobile sinks in a controlled and coordinated manner. Initially, random and uneven distribution of sensor nodes over a ground is considered. The mobile sinks move in such a way that total number of sensor nodes visited by all the sinks gets maximized. The algorithm aims to balance work load among all the mobile sinks, as well as cover all the deployed sensors in a time and energy efficient manner.

Rest of the paper is organized as follows. The current problem is formulated in section 2, followed by a mathematical game formulation in section 3. Section 4 presents our proposed algorithm. Simulation results are illustrated in section 5. The paper finally concludes the work in Section 6.

2   Problem Formulations

In this paper, we assume existence of two mobile sinks ms_1 and ms_2 to complete the task of network coverage. Each of the mobile sink can go towards one of these possible eight directions; North, North-East, East, South-East, South, South-West, West and North-West. At each step, the proposed algorithm decides in which directions ms_1 and ms_2 will go. Assuming the position of ms_1 be (x_1,y_1) and ms_2 be (x_2,y_2) at time t_1, the objective is to find the location coordinates (x_1',y_1') of ms_1 and (x_2',y_2') of ms_2 at time t_2. The change of position for each move is given by a shift through two consecutive grids towards any one among the eight specified directions. The Coordination between two mobile sinks is the most important challenge here. We use fundamentals of Game Theory to study the sink coordination problem. Presuming sensor data collection a common goal for both the sinks, the idea of Cooperative Game is incorporated here.

3   Mathematical Formulations

In this section, we illustrate a mathematical model for our proposed problem. Since we have to explore the cooperative behavior of the mobile sinks, we formulate a cooperative game in current context.

Let, ms_1 and ms_2 are at grid locations (x_1,y_1) and (x_2,y_2) respectively at time t_1. To change their respective locations after time Δt, each of ms_1 and ms_2 can choose one of the strategies from their respective strategy profiles $strat_1$ and $strat_2$ where,

$$strat_1 \rightarrow \{N, \ NE, \ E, \ SE, \ S, \ SW, \ W, \ NW\} \tag{1}$$

$$strat_2 \rightarrow \{N, \ NE, \ E, \ SE, \ S, \ SW, \ W, \ NW\} \tag{2}$$

The strategies belong to set D which is set of eight possible directions for movement i.e.

$$D = \{N, \ NE, \ E, \ SE, \ S, \ SW, \ W, \ NW\} \tag{3}$$

Suppose, ms_1 and ms_2 choose directions d_i and d_j to reach grid locations (x_1',y_1') and (x_2',y_2') respectively at time instance t_2. Now, the number of sensor nodes covered by ms_1 at time t_2 depends on the strategy d_i and node density for the grids that will be covered by ms_1. However, the number of grids covered by ms_1 is proportional to its transmission range r_1. Since the location of ms_1 at time t_2 is (x_1',y_1'), we characterize the node density of the network region covered by ms_1 by grid $G_{(x1',y1')}$ and denote it with $\rho_{(x1',y1')}$. Since, nodes which have already been covered by either of the two mobile sinks are marked as "covered"; they are not supposed to send any data to ms_1 until the timer values associated to them are decremented down to zero.

Hence, our concern will be to find cov_1, the number of "uncovered" nodes that can be covered by ms_1 located at (x_1',y_1') at time t_2. Now, we can construct equation (4),

$$\text{cov}_1 = \text{fun}_1\left(d_i, r_1, \rho_{(x1', y1')}\right) \tag{4}$$

Similarly for ms_2 we can form equation (5),

$$\text{cov}_2 = \text{fun}_2\left(d_j, r_2, \rho_{(x2', y2')}\right) \tag{5}$$

Now, if there is a situation where parts of the coverage area of ms_1 and ms_2 overlaps, then at time t_2, there can be one or more than one common grids under the transmission range of both the mobile sinks. Let, set_1 and set_2 be the set of nodes that fall under the transmission range of ms_1 and ms_2 respectively and $com_{1,2}$ be the set of nodes that fall under the overlapping zone. Mathematically,

$$com_{1,2} = set_1 \cap set_2 \tag{6}$$

Nodes that belong to $\{com_{1,2}\}$ can be covered by any of the two mobile sinks. In that case, ms_2 has to sacrifice for ms_1 and nodes of $com_{1,2}$ are visited only by ms_1. Hence, equation (5) can be modified to,

$$\text{cov}_2 = \text{fun}_2\left(d_j, r_2, \rho_{(x2', y2')}\right) - \left|com_{1,2}\right| \tag{7}$$

Player B (ms$_2$)

Player A (ms$_1$)	N	NE	E	SE	S	SW	W	NW
N	$P_{N,1},$ $P_{N,2}$	$P_{N,1},$ $P_{NE,2}$	$P_{N,1},$ $P_{E,2}$	$P_{N,1},$ $P_{SE,2}$	$P_{N,1},$ $P_{S,2}$	$P_{N,1},$ $P_{SW,2}$	$P_{N,1},$ $P_{W,2}$	$P_{N,1},$ $P_{NW,2}$
NE	$P_{NE,1},$ $P_{N,2}$	$P_{NE,1},$ $P_{NE,2}$	$P_{NE,1},$ $P_{E,2}$	$P_{NE,1},$ $P_{SE,2}$	$P_{NE,1},$ $P_{S,2}$	$P_{NE,1},$ $P_{SW,2}$	$P_{NE,1},$ $P_{W,2}$	$P_{NE,1},$ $P_{NW,2}$
E	$P_{E,1},$ $P_{N,2}$	$P_{E,1},$ $P_{NE,2}$	$P_{E,1},$ $P_{E,2}$	$P_{E,1},$ $P_{SE,2}$	$P_{E,1},$ $P_{S,2}$	$P_{E,1},$ $P_{SW,2}$	$P_{E,1},$ $P_{W,2}$	$P_{E,1},$ $P_{NW,2}$
SE	$P_{SE,1},$ $P_{N,2}$	$P_{SE,1},$ $P_{NE,2}$	$P_{SE,1},$ $P_{E,2}$	$P_{SE,1},$ $P_{SE,2}$	$P_{SE,1},$ $P_{S,2}$	$P_{SE,1},$ $P_{SW,2}$	$P_{SE,1},$ $P_{W,2}$	$P_{SE,1},$ $P_{NW,2}$
S	$P_{S,1},$ $P_{N,2}$	$P_{S,1},$ $P_{NE,2}$	$P_{S,1},$ $P_{E,2}$	$P_{S,1},$ $P_{SE,2}$	$P_{S,1},$ $P_{S,2}$	$P_{S,1},$ $P_{SW,2}$	$P_{S,1},$ $P_{W,2}$	$P_{S,1},$ $P_{NW,2}$
SW	$P_{SW,1},$ $P_{N,2}$	$P_{SW,1},$ $P_{NE,2}$	$P_{SW,1},$ $P_{E,2}$	$P_{SW,1},$ $P_{SE,2}$	$P_{SW,1},$ $P_{S,2}$	$P_{SW,1},$ $P_{SW,2}$	$P_{SW,1},$ $P_{W,2}$	$P_{SW,1},$ $P_{NW,2}$
W	$P_{W,1},$ $P_{N,2}$	$P_{W,1},$ $P_{NE,2}$	$P_{W,1},$ $P_{E,2}$	$P_{W,1},$ $P_{SE,2}$	$P_{W,1},$ $P_{S,2}$	$P_{W,1},$ $P_{SW,2}$	$P_{W,1},$ $P_{W,2}$	$P_{W,1},$ $P_{NW,2}$
NW	$P_{NW,1},$ $P_{N,2}$	$P_{NW,1},$ $P_{NE,2}$	$P_{NW,1},$ $P_{E,2}$	$P_{NW,1},$ $P_{SE,2}$	$P_{NW,1},$ $P_{S,2}$	$P_{NW,1},$ $P_{SW,2}$	$P_{NW,1},$ $P_{W,2}$	$P_{NW,1},$ $P_{NW,2}$

Fig. 1 A sample payoff matrix containing two players, their full strategy profiles and payoff against different strategies.

Now, cov_1 and cov_2 are eventually payoffs for the round game played at time t_{int} where,

$$t_1 < t_{int} < t_2 \tag{8}$$

We denote cov_1 and cov_2 with $P_{i,1}$ and $P_{j,2}$ as they are the payoffs for ms_1 and ms_2 for choosing strategies d_i and d_j respectively. Since, the nature of the game is a repetitive cooperative game, ultimate goal is to maximize net utility at each round of game. The net utility $U_{i,j}$, for ms_1 playing strategy d_i and ms_2 playing strategy d_j is given by the following equation.

$$U_{i,j} = P_{i,1} + P_{j,2} \qquad (9)$$

Since each of the player can play on from the set of eight possible strategies, there will be $|K|$ number of strategy pairs, such that,

$$|K| = |strat_1 \times strat_2| \qquad (10)$$

Also, the set K has all possible pair of simultaneous movement direction.

$$K = \{(N, N)...., (SW, NE)...., (W, E)...., (NW, NW)\} \qquad (11)$$

Figure 1 represents a sample payoff matrix for the game.

Now, if at any time instance t_{equi}, for a pair of strategy (a,b) from K, the net utility $U_{i,j}$ for that round of game, cannot be increased further, we can conclude that the game has achieved an equilibrium state. Here, at time t_{equi}, the individual payoffs $P_{i,1}$ and $P_{j,2}$ and the net utility $U_{i,j}$ are maximum. Any attempt to further increase $U_{i,j}$ at this point of time by opting any other strategy pair (c,d) from K will eventually degrade $U_{i,j}$. The game is said to be at Nash Equilibrium at t_{equi}. The detail mathematical proof is omitted here.

4   Proposed Algorithm

In this section we state and discuss the working algorithm for our paper. Since, our problem is a repetitive game, the procedure runs iteratively.

At the beginning, states of all the nodes are uncovered. Each mobile sink starts from an initial grid. Sinks calculate their expected individual payoffs for all possible directions for movement. Expected net utilities are calculated for each pair of strategies (pairs formed by taking one strategy from each of the strategy profiles of two mobile sinks). While calculating an expected net utility, summation of individual payoffs of the players are considered. Since, existence of overlapping zone increases the number of common sensor nodes falling under transmission ranges of ms_1 and ms_2, the sinks cooperatively try to choose their next move, such that the overlapping zone decreases. After making decisions, mobile sinks move towards chosen directions. Their locations are updated and they collect data from new sensor nodes. Newly visited nodes are marked as covered. The procedure is executed repeatedly for further movement of mobile sinks. We can find out the movement pattern of each mobile sink by joining its consecutive grid positions one by one. A formal description of our proposed strategy is presented in Algorithm 1. $G_1(i)$ and $G_2(j)$ are set of grids covered by ms_1 on going towards direction d_i and ms_2 on going towards direction d_j respectively, as is mentioned in the algorithm. G_{comm} is the set of grids common between $G_1(i)$ and $G_2(j)$.

Algorithm 1:

1. *construct set K= strat$_1$×strat$_2$*
2. *Start Procedure*
3. *Initialize PAYOFF of order (8×8)*
4.   *For each element (d$_i$,d$_j$) in K*
    4.1. *construct set G$_1$(i) for ms$_1$*
    4.2. *construct set G$_2$(j) for ms$_2$*
    4.3. *find out G$_{comm}$:= {G$_1$(i)∩G$_2$(j)}*
    4.4. *Net utility (U$_{i,j}$):= {G$_1$(i) ∪ G$_2$(j) − G$_{comm}$}*
    4.5. *PAYOFF := store | U$_{i,j}$ | in cell (i,j)*
5.   *End For*
6. *choose the cell (a,b) in PAYOFF having the highest numerical value present*
7. *set movement direction of ms$_1$ as dir$_1$:=a*
8. *set movement direction of ms$_2$ as dir$_2$:=b*
9. *move ms$_1$ and ms$_2$ accordingly*
10. *update current grid location of both ms$_1$ and ms$_2$*
11. *collect data from the nodes reachable from new location of ms$_1$ and ms$_2$*
12. *mark all the newly visited nodes as COVERED*
13. *set timers of all the COVERED nodes to a maximum value T*
14. *start decrementing timers of all the COVERED nodes*
15. *make all other nodes of network UNCOVERED if their timers reach zero*
16. *Repeat Procedure*

5   Simulation Results

We have simulated our proposed algorithm with two mobile sinks and have obtained satisfactory results in terms of percentage of nodes covered, remaining average energy of nodes, load balancing between two mobile sinks.

Energy Model: Since, sinks are placed on mobile agents driven by powerful motors; there is not much to worry about the energy consumption due to sink movement. Also, mobile sinks are assumed to have infinite buffers and they are attached to high energy batteries. We focus on the energy dissipation of static sensor nodes. Sensor nodes dissipate energy according to the following equation.

$$\phi_{total} = \beta\left(\phi_{trans} + \phi_{amp}.\ell^{n}\right) + \phi_{recv} \tag{12}$$

Here, Φ_{total}, Φ_{trans}, Φ_{amp}, Φ_{recv} denote total energy consumption, energy dissipation for data transmission, energy loss due to amplification for a static sensor and energy expense for receiving a *Hello* message from a mobile sink, respectively. β is the size of data packet sent. ℓ signifies the distance of a source node to a mobile sink. Lastly, n is a constant whose value, depending on the nature of transmission medium, ranges from 2 to 4. We have considered n=2, here. Parameters used for energy calculation in our work are taken from [5], and are listed in the table 1.

Network Coverage: Simulation results show that our algorithm performs well, in terms of network coverage. By playing the game iteratively, almost 99.75% can be covered by the mobile sinks cumulatively. The graph in Figure 2(a) describes how more numbers of nodes are getting covered with the increase in number of iterations.

Table 1 Parameters used for energy calculation

Parameter	Value
Initial energy of node (Φ_{total})	.5J
Size of data packet sent (β)	800bit
Energy consumed in transmission (Φ_{trans})	50nJ/bit
Energy consumed in amplification (Φ_{amp})	10pJ/bit/m^2
Energy consumed in receiving (Φ_{recv})	10nJ/bit

Energy Consumption:

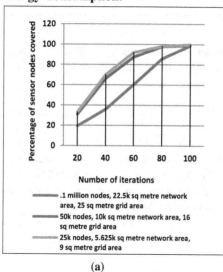

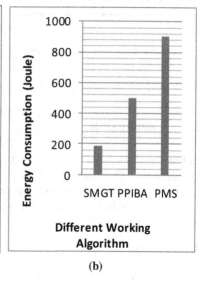

(a)   (b)

Fig. 2 (a) Percentage of nodes covered with increasing number of iterations considering different network arrangement, (b) Comparisons with different existing algorithms in terms of energy consumption

Since, our algorithm does not require any clustering or inter node communication for multi hop data forwarding, the energy consumption is pretty less here. Sensors only deplete energy while receiving *Hello* messages from mobile sinks or transmitting data packet direct to a sink or using amplification circuit. We have compared our obtained results with two existing algorithms proposed in [4] and [6] using same simulation environment and we have seen that our algorithm outperforms them. Figure 2(b) gives the comparative study among our algorithm;

Sink Movement using Game Theory (SMGT) and two existing algorithms: Path-point Identification with Bees Algorithm (PPIBA) [4] and Periphery Moving Strategy (PMS) [6]. In our work, there are no such concepts of intra cluster communication like [4] or multi hop communication like [6]. Hence, the energy consumption for our work is much lesser compared to [4] and [6].

Load Balancing: The fairness of the current game can be evaluated by the distribution of data collection work load among the mobile sinks. Simulation results, briefed in Table 2 clearly indicate that the percentage of nodes visited by ms_1 and ms_2 are close enough.

Table 2 Load balancing between mobile sinks

Percentage of static nodes covered by the sinks individually	ms_1	ms_2
	53.5%	46.5%

It can be seen that ms_1 obtains slightly better payoffs than ms_2. This is because; ms_1 gets priority over ms_2 when there is any overlapping area of coverage at a particular time.

6   Conclusions

In this paper we have explored the mutual and cooperative behavior between two mobile sinks. The working algorithm is simple and easy to implement. The work has been made realistic by making the sensor distribution random. In future, this work can be further extended for more than two mobile sinks.

References

[1] Kinalis, A., Nikoletseas, S.: Scalable Data Collection Protocols for Wireless Sensor Networks with Multiple Mobile Sinks. In: Proceedings of the 40th Annual Simulation Symposium, ANSS 2007, pp. 60–72 (2007)
[2] Konstantopoulos, C., Pantziou, G., Gavalas, D., Mpitziopoulos, A., Mamalis, B.: A Rendezvous-Based Approach Enabling Energy-Efficient Sensory Data Collection with Mobile Sinks. IEEE Transactions on Parallel and Distributed Systems 23(5), 809–817 (2012)
[3] Goldenberg, D.K., Lin, J., Morse, A.S., Rosen, B.E., Yang, Y.R.: Towards Mobility as a Network Control Primitive. In: MobiHoc 2004, Proceedings of the 5th ACM International Symposium on Mobile Ad Hoc Networking and Computing, pp. 163–174 (2004)

[4] Saad, E.M., Awadalla, M.H., Saleh, M.A., Keshk, H., Darwish, R.R.: A Data Gathering Algorithm for a Mobile Sink in Large-Scale Sensor Networks. In: MMACTEE 2008, Proceedings of the 10th WSEAS International Conference on Mathematical Methods and Computational Techniques in Electrical Engineering, pp. 288–294 (2008)

[5] Banerjee, I., Chanak, P., Sikdar, B.K., Rahaman, H.: EERIH: Energy Efficient Routing via Information Highway in Sensor Network. In: IEEE International Conference on Emerging Trends in Electrical and Computer Technology, Kanyakumari, India, March 23-24, pp. 1057–1062 (2011)

[6] Luo, J., Hubaux, J.P.: Joint Mobility and Routing for Lifetime Elongation in Wireless Sensor Networks. In: Proceedings of IEEE INFOCOM 2010, pp. 1735–1746 (2010)

[7] Machado, R., Tekinay, S.: A Survey of Game-theoretic Approaches in Wireless Sensor Networks. Published in The International Journal of Computer and Telecommunications Networking 52(16), 3047–3061 (2008)

Pricing of Cloud IaaS Based on Feature Prioritization – A Value Based Approach

Arpan Kumar Kar and Atanu Rakshit

Abstract. Today, cloud computing is transforming the IT industry. An increasing number of providers are offering various services where computing, storage and application resources, can be dynamically provisioned on a pay per use basis. However the demands and requirements of different users are diversifying and growing sharply. In order to maximize the revenue, a proper pricing model is required. Thus systemic approaches need to be explored to estimate the potential value of such services to specific users for a specific context. The current study proposes a value based pricing approach for Infrastructure as a Service (IaaS), one of the important delivery models. The decision support approach prioritizes and aggregates the dimensions of IaaS for the migration to cloud, from the users' perspective by integrating Fuzzy Set Theory and Analytic Hierarchy Process for group decision making.

1 Introduction

Cloud computing is defined as *"A computing Cloud is a set of network enabled services, providing scalable , Quality of Service guaranteed, normally personalized, inexpensive computing platforms on demand, which could be accessed in a simple and pervasive way". Cloud computing is a natural evolution for data and computation centers with automated systems management, workload balancing, and virtualization technologies* [34]. This technological trend in Information Technology (IT) has enabled the realization of new computing models in which resources are provided as utilities that can be leased and released by users on-demand. Cloud computing delivers infrastructure, platform, and software that are made available as subscription-based services to consumers. Clouds [36] aim to power the next-generation data centers as the enabling platform for dynamic and flexible application provisioning. No wonder cloud computing is heralded with so

Arpan K. Kar · Atanu Rakshit
Indian Institute of Management - Rohtak, MDU Campus, Rohtak 124001, India

S.M. Thampi et al. (eds.), *Recent Advances in Intelligent Informatics*,
Advances in Intelligent Systems and Computing 235,
DOI: 10.1007/978-3-319-01778-5_33, © Springer International Publishing Switzerland 2014

much promise for revolutionizing the way industries use IT [7, 23, 24]. However, like any emerging technology, pricing of cloud offerings remains a challenge.

This study examines the pricing strategies of cloud computing service providers while focusing on the IaaS market. Based on a literature review and interviews conducted with experts, six user service dimensions were identified. Fuzzy set theory and Analytic Hierarchy Process (AHP) have been integrated for the prioritization and aggregation of the users' preferences for these six dimensions. Subsequently a sigmoid curve has been used to estimate the actual trade-off among the priorities for estimating the overall perceived value, based on these dimensions.

2 Review of Literature

2.1 Cloud Computing and Its Pricing Strategies

Even in the 1960s, it was envisioned that computing facilities can be provided to the general public like a utility [27]. The increasing adoption of cloud computing is currently driving a significant increase in both the supply and the demand side of this new market for IT utilities. Cloud computing can be delivered through different delivery models such as Infrastructure as a Service (IaaS), Platform as a Service, Software as a Service, Business Process as a Service, Data as a Service. In this study, the focus is on providing a decision support approach for pricing IaaS.

In recent times, the number of IaaS providers has increased drastically. Many cloud service providers have a pay-as-you-go pricing scheme (e.g. Amazon Web Services, Windows Azure, Google AppEngine) and a need is there to analyze these pricing schemes. Sometimes a provider sets a static or infrequently updated per-unit price, and users pay for only what they use. Along with the pay-as-you-go offer, there are two additional pricing schemes widely adopted in cloud markets: *subscription* [35] and *spot market* [36]. In the former scheme, a user pays a one-time subscription fee to reserve one unit of resource for a certain period of time. The spot market, on the other hand, is an auction-like mechanism where users periodically submit bids to the provider, who in turn posts a series of spot prices. In fact, some providers use multiple pricing schemes simultaneously for IaaS.

2.2 Pricing of IT and IT Based Services

IT vendors have mostly focused on *cost based pricing* approaches using metrics like function point count, code size, development effort, development time and development complexity [28]. However although cost based pricing is often used for information goods, it is limited in addressing the customer's price sensitivity and potential competitor action [5]. Pricing literature [17] argue that all effective pricing strategies should be mapped to the *customer's value perception* about the offering. However linking perceived value to pricing strategies is a challenge, for intangible goods (*like IT*). Also, the marginal cost of information goods is often so negligible that cost based pricing is often viewed as being a viable strategy [16]. Literature [19, 38, 18] has explored different popular IT pricing strategies and the major issues faced by technology developers, like cost based approaches, flat fee pricing, nonlinear usage-based pricing, two-part tariff pricing and many other strategies. However traditional non-linear pricing theory [25, 37, 3] argues that the optimal pricing strategy for a seller's market is most rewarding when it is based on usage. For emergent technologies like cloud computing, the seller is more likely to be operating in a market with low direct competition but wider and costlier range of substitutes. When such an offering is being introduced in the market, the provider may adopt a single pricing scheme to simplify management [12, 37]. However it was established after comparing cost based pricing against usage-based pricing and established that when users are characterized by heterogeneous consumption levels, incorporating usage in pricing generates greater revenue for the service provider. This is why a pricing strategy for an emergent technology like cloud computing needs to be mapped with the perceived value from usage, to generate greater revenue [39]. Studies [20, 21, 22] have proposed systematic approaches for pricing based on perceived value for the end users, depending on usage, whereby prioritizations have been used to estimate tradeoffs. However, these studies failed to address the subjectivity in prioritization of different perspectives.

2.3   The Fuzzy Extension of the Analytic Hierarchy Process

The AHP was developed for use in multi-hierarchical, multi-criteria decision making problems [31, 32]. It decomposes the problem into a hierarchy of more easily comprehended sub-problems, each of which can be analyzed independently through comparative independent judgments. AHP is extremely suitable for usage in multi-criteria decision making problems where prioritization is required from multiple decision makers on different dimensions. AHP has appropriate measures for estimating consistency of priorities of expert decision makers [32, 1, 2, 13]. There are also approaches to improve the consistency of priorities systematically [15, 8]. Further AHP provides extensive theories for the aggregation of group preferences when the priorities of multiple users need to be combined [4, 10, 14]. While the earlier studies using AHP were conducted using crisp AHP theory, the recent studies have started integrating AHP with fuzzy set theory to accommodate the subjectivity in the human decision making process in complex problems. Fuzzy set theory [41, 42, 43, 44] has increasingly found its application in complex problems where the subjectivity in the decision making process impacts the possibility of choices and thus affects the outcome. Fuzzy set theory has been widely integrated to AHP [30, 40, 26, 33] and has been applied in different business domains in applications involving prioritization of criteria. In this study, the integrated approach has been adopted using fuzzy set theory and AHP for the prioritization of the dimensions for the migration to Cloud, by extending Kar & Pani [22] by using the Geometric Mean Method for the prioritization and aggregation.

3 Contribution

While many studies have argued that perceived value should be considered while pricing any product, few studies have attempted to operationalize this concept while pricing emergent IT offerings with multiple dimensions of value. In fact, there has been no systematic approach to provide decision support to the pricing problem for cloud computing. Within cloud computing, the current study focuses on the pricing of IaaS from a value based approach, from the perspective of multiple users but for the same context. The current study follows the line of existing literature [22] and extends the same by incorporating AHP theory where the prioritization has been done with Geometric Mean Method. The earlier work focused on prioritization based on the Eigen Vector Method. However, for group decision making, the Geometric Mean Method is more suitable for prioritization and aggregation of preferences since it maintains the reciprocal properties and extreme estimates in aggregated hierarchies better. The proposed method also uses fuzzy set theory while estimating subjective preferences thereby accommodating the subjectivity of multiple decision makers, while estimating trade-offs. Finally, a sigmoid curve has been used to manage the trade-offs between different dimensions based on its priority to estimate the total perceived value of the IaaS.

4 Computational Method

In this study, the users' priorities have been captured using an integrated decision support approach. The criteria or dimensions which were prioritized for the same context (i.e. evaluation of IaaS for migration) were derived from literature [9, 29, 24], after checking their contextual suitability through focused interviews with 5 Cloud computing consultants with over 10 years of expertise.

- **(C1) Flexibility**: The ability to respond quickly to changing requirements
- **(C2) Costs**: The capital expenditure and working expenditure, like acquisition and maintenance costs for servers, licenses and other hardware and software.
- **(C3) Scope & performance**: Factors include the degree of fulfillment of specific requirements, knowledge about the service and performance quality.
- **(C4) IT security & compliance**: Factors like government, industry and firm specific needs in the areas of security, compliance and privacy are covered.
- **(C5) Reliability & trustworthiness**: Factors like service availability and fulfillment of the Service Level Agreements
- **(C6) Service & cloud management**: Factors like offered support and functions for controlling, monitoring and individualization of the web interface.

The prioritization of these dimensions is addressed through an integrated approach. In this approach, first the linguistic judgments of users are captured and mapped to quantifiable fuzzy judgments. Subsequently, these fuzzy linguistic judgments are converted to crisp priorities using AHP theory. These crisp priorities have been further combined using geometric mean method for the aggregation of priorities for estimating the trade-offs for the different dimensions.

Let $U=(u_1,...u_n)$ be the set of n *users* having a relative importance of ψ_i such that $\psi=(\psi_1,...\psi_n)$ is the weight vector of the individual users and $\sum \psi_i = 1$. These are the users who are prioritizing the important dimensions for migrating to Cloud. Comparative fuzzy judgments $A=(a_{ij})_{k \times k}$ would be coded as illustrated in Figure 1.

Definition		Fuzzy sets for the fuzzy AHP
Equal importance	$\tilde{1}$	{(1,0.25), (1,0.50), (3,0.25)}
Moderate importance	$\tilde{3}$	{(1,0.25), (3,0.50), (5,0.25)}
Strong importance	$\tilde{5}$	{(3,0.25) (5,0.50) (7,0.25)}
Very strong importance	$\tilde{7}$	{(5,0.25), (7,0.50), (9,0.25)}
Extreme high importance	$\tilde{9}$	{(7,0.25), (9,0.50), (9,0.25)}

Fig. 1 Scale for the conversion of linguistic preferences

A triangular fuzzy function has been used for coding the judgments since there is equal probability of the response of the next level as is to the response of the previous level, when a comparative judgment is made by Cloud user.

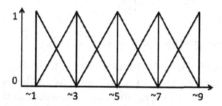

Fig. 2 Triangular fuzzy function for coding judgments

The simple pairwise comparison approach [6] for fuzzy set operations has been used for the fuzzy sets $\tilde{a}_i = (a_{i,1}, a_{i,2}, a_{i,3})$ and $\tilde{a}_j = (a_{j,1}, a_{j,2}, a_{j,3})$ as illustrated:

$$\tilde{a}_i + \tilde{a}_j = ((a_{i,1} + a_{j,1}), (a_{i,2} + a_{j,2}), (a_{i,3} + a_{j,3})) \tag{1}$$

$$\tilde{a}_i - \tilde{a}_j = ((a_{i,1} - a_{j,1}), (a_{i,2} - a_{j,2}), (a_{i,3} - a_{j,3})) \tag{2}$$

$$\tilde{a}_i \times \tilde{a}_j = ((a_{i,1} \times a_{j,1}), (a_{i,2} \times a_{j,2}), (a_{i,3} \times a_{j,3})) \tag{3}$$

$$\tilde{a}_i \div \tilde{a}_j = ((a_{i,1} \div a_{j,1}), (a_{i,2} \div a_{j,2}), (a_{i,3} \div a_{j,3})) \tag{4}$$

$$\tilde{a}_i^k = (a_{i,1}{}^k, a_{i,2}{}^k, a_{i,3}{}^k) \text{ for } (k \in N) \tag{5}$$

$$\tilde{a}_i^{1/k} = (a_{i,3}{}^{1/k}, a_{i,2}{}^{1/k}, a_{i,1}{}^{1/k}) \text{ for } (k \in W) \tag{6}$$

The individual priorities are obtained by solving the following system:

$$\min \sum_{i=1}^{k} \sum_{j>i}^{k} (\ln \tilde{a}_{i,j} - (\ln \tilde{p}_i - \ln \tilde{p}_j)^2) \text{ s.t. } \tilde{a}_{ij} \geq 0; \ \tilde{a}_{ij} \times \tilde{a}_{ji} = 1; \ \tilde{p}_i \geq 0, \sum \tilde{p}_i = 1. \quad (7)$$

The individual priority vector [11] is obtained by $\displaystyle \tilde{p}_i = \frac{\sqrt[1/k]{\prod_{j=1}^{k} \tilde{a}_{i,j}}}{\sum_{i=1}^{n} \sqrt[1/k]{\prod_{j=1}^{k} \tilde{a}_{i,j}}}$ \quad (8)

Where $\tilde{p}_i$ is the priority of the decision criteria i such that $\tilde{P}_i = \{\tilde{p}_1, \tilde{p}_2, ..., \tilde{p}_7\}$ for *user i*. In the subsequent step, before the crisp aggregation rules are computed, consistency of these priorities needs to be evaluated. The Geometric Consistency Index (GCI) is used to estimate consistency of the individual priorities [2].

$$\text{GCI} (A^{d_i}) = \frac{2}{(k-1)(k-2)} \times \sum_{j>i}^{k} (log \ |\tilde{a}_{i,j}| - (log \ |\tilde{p}_i| - log \ |\tilde{p}_j|)^2) \quad (9)$$

GCI $(A^{d_i}) \leq \overline{GCI}$ is the criteria for consistency. For $k=6$, $\overline{GCI}$ is 0.37. Collective preferences of the group for deriving the decision vector can be estimated subsequently by the aggregation of individual priorities such that the aggregate priorities (i.e. the collective priority vector) are defined as $\tilde{P}^{(c)} = (\tilde{p}_1^{(c)}, \tilde{p}_2^{(c)}, ... \tilde{p}_r^{(c)})$ where $\tilde{p}_i^{(c)}$ is obtained by the aggregation of priorities.

$$\tilde{p}_i^{(c)} = \frac{\prod_1^n (p^{(k)}{}_i)^{\psi_i}}{\sum_1^r \prod_1^n (p^{(k)}{}_i)^{\psi_i}} \quad (10)$$

For crisp conversion of priority $|\tilde{p}_i| = p_{i,2} \cdot 0.25 + p_{i,2} \cdot 0.5 + p_{i,3} \cdot 0.25$ \quad (11)

Now the dimension with the highest importance is fixed as p_{max}. Then to evaluate the collective value after trade-off of the dimensions for the users, arising from the 6 dimensions, each dimension's priority may be discounted using a preference function. A sigmoid curve has been used since such curves represent the deterministic behavior of users better than linear approximations. Further it has an increasing rate of utility in the initial stages and a decreasing rate of utility in the final stages and thus corresponds to the utility theory of expectations from marginal returns. An exponential sigmoid function has been used as it is real-valued, positive and differentiable. Stretching p_i over the range [±6] ensures that the range of the priorities captures 99.6% of the curve. Also, for normalizing the weights for 0 to 1, p_{min} is assumed to be 0. Now, for each quantified benefit denoted as B_i, the perceived benefit PB_i from that dimension is estimated as $PB_i = B_i \times t_i$ \quad (12)

Where $t_i = f(x_i) = \frac{1}{1+e^{x_i}}$ where $x_i = -6 + 12 \times \frac{p_i - p_{min}}{p_{max} - p_{min}}$ \quad (13)

Here x_i is calculated by stretching the range of p_i over the range [±6] such that 99.6% of the curve is achieved by the normalization with a 0.2% cut-off at each tail. Now, total perceived value (TPV) for the technology as estimated by the client would be calculated as $\text{TPV} = \sum_k PB_i = \sum_k B_i \times t_i$ \quad (14)

This TPV should be further discounted by the operating profit margin (percentage) of the firm, to get the upper limit of what the IaaS provider may quote as the price for the service offering, so that it may be readily accepted by the user.

5   Numerical Example

This numerical example is a hypothetical case bearing a close resemblance to a multi-national firm which had recently migrated to Cloud and had subscribed to an IaaS. Priorities of 5 executive users are captured, who had significant buy-in from the migration, so as to capture the collective preferences. The fuzzy judgments elicited from the linguistic responses were converted to crisp priorities. The individual priorities are illustrated in Figure 3.

	C1	C2	C3	C4	C5	C6	GCI
U1	0.30	0.30	0.10	0.10	0.10	0.10	0.133
U2	0.45	0.22	0.08	0.08	0.07	0.10	0.105
U3	0.15	0.48	0.13	0.07	0.08	0.08	0.098
U4	0.30	0.30	0.10	0.12	0.08	0.10	0.160
U5	0.27	0.07	0.18	0.08	0.19	0.21	0.142

Fig. 3 Individual priorities of 5 users

After these individual priorities were estimated and subsequently aggregated, the collective priority vector that was obtained was (0.303, 0.250, 0.124, 0.096, 0.106, 0.121). From the aggregated vector, it is found that $p_{max} = 0.303$, $p_{min}=0.0$. Further across all the 6 dimensions, value was quantified by considering some, assumptions, some industry specific data and firm specific data which bears resemblance to actual data. However, from case to case, such data would vary, and these numbers are only representative figures for illustration.

> *Flexibility*: Because the service was flexible, cost savings from not over-provisioning of resources was in tune of $40,000
> *Costs*: Due to a shift from major CapEx to WorkEx, the annual cost savings was estimated to be $60,000 after the migration.
> *Scope & performance*: Due to high performance, there was a reduction in manpower allocated for the maintenance of the IT. Further there was improvement in the employee productivity. These created an annual benefit of $40,000.
> *IT security & compliance*: The industry faces regular security breaches on IT infrastructure, which on an average, costs $500,000. The annual probability of a breach is 10%. The solution mitigates this risk and creates a value of $50,000.
> *Reliability & trustworthiness*: The availability of the service has improved the firm's customer orientation, which has created an improvement in the customer satisfaction. This has indirectly created a benefit of $30,000.
> *Service & cloud management*: Outsourced support created some cost-savings from employee productivity. Further individualization created benefits by improving the business workflows. This created an additional benefit of $55,000.

The total perceived value, from these six dimensions after adjusting for collective prioritization has been illustrated in Figure 4.

State	C1	C2	C3	C4	C5	C6
pi	0.303	0.250	0.124	0.096	0.106	0.121
xi	6.000	3.932	-1.077	-2.213	-1.784	-1.201
ti	0.998	0.981	0.254	0.099	0.144	0.231
Bi	$40,000	$60,000	$40,000	$50,000	$30,000	$55,000
PBi	$39,901	$58,846	$10,160	$4,931	$4,316	$12,723
Total Perceived Value TPV = $\sum$ PBi =						**$1,30,878**

Fig. 4 Total Perceived Value of IaaS

This would be further discounted by the profit margin (%) of the firm (assume 15% in this case). Hence, the annual subscription fee for the IaaS may be safely quoted up to $111,245.

6 Conclusion

This paper proposes an integrated approach for decision support for the pricing of IaaS based on total perceived value of the offering, from the multi-user perspective, for the same decision making context. Such a pricing approach would no doubt be beneficial for the service provider, who would get higher revenues from a contract. Further this would also be beneficial for the IT team of the firm that is migrating to cloud, since mapping the returns on the investment would be relatively easier and thus getting the business buy-in would be facilitated. Since business buy-in in a critical dimension of success for any technology, value based pricing would greatly facilitate the same, by providing direct insights of potential benefits from well-defined business cases with details on RoI achievement.

References

1. Aguaron, J., Escobar, M.T., Moreno-Jiménez, J.M.: Consistency stability intervals for a judgement in AHP decision support systems. European Journal of Operational Research 145(2), 382–393 (2003)
2. Aguarón, J., Moreno-Jiménez, J.M.: The geometric consistency index: Approximated thresholds. European Journal of Operational Research 147(1), 137–145 (2003)
3. Armstrong, M.: Multiproduct nonlinear pricing. Econometrica 64(1), 51–75 (1996)
4. Bolloju, N.: Aggregation of analytic hierarchy process models based on similarities in decision makers' preferences. European Journal of Operational Research 128(3), 499–508 (2001)
5. Brennan, R., Canning, L., McDowell, R.: Price-setting in business-to-business markets. The Marketing Review 7(3), 207–234 (2007)
6. Buckley, J.J.: Fuzzy Hierarchical Analysis. Fuzzy Sets and Systems 17(3), 233–247 (1985)
7. Buyya, R., Yeo, C.S., Venugopal, S., Broberg, J., Brandic, I.: Cloud computing and emerging IT platforms: Vision, hype, and reality for delivering computing as the 5th utility. Future Generation Computer Systems 25(6), 599–616 (2009)

8. Cao, D., Leung, L.C., Law, J.S.: Modifying inconsistent comparison matrix in analytic hierarchy process: A heuristic approach. Decision Support Systems 44(4), 944–953 (2008)

9. Clarke, R.: Computing Clouds on the Horizon? Benefits and Risks from the User's Perspective. In: 23rd Bled e-Conference Etrust: Implications for the Individual, Enterprises and Society, Bled, Slovenia (2010)

10. Condon, E., Golden, B., Wasil, E.: Visualizing group decisions in the Analytic Hierarchy Process. Computers and Operations Research 30(10), 1435–1445 (2003)

11. Crawford, G., Williams, C.: A note on the analysis of subjective judgement matrices. Journal of Mathematical Psychology 29, 387–405 (1985)

12. Curle, D.: There is no value if it's not relevant. Information Today 15(8), 8 (1998)

13. Escobar, M.T., Aguarón, J., Moreno-Jiménez, J.M.: A note on AHP group consistency for the row geometric mean prioritization procedure. European Journal of Operational Research 153(2), 318–322 (2004)

14. Escobar, M.T., Moreno-Jiménez, J.M.: Aggregation of individual preference structures in AHP-group decision making. Group Decision and Negotiation 6(4), 287–301 (2007)

15. Finan, J.S., Hurley, W.J.: The analytic hierarchy process: Does adjusting a pairwise comparison matrix to improve the consistency ratio help? Computers and Operations Research 24(8), 749–755 (1997)

16. Fishburn, P.C., Odlyzko, A.M., Siders, R.C.: Fixed fee versus unit pricing for information goods: competition, equilibria, and price wars. In: Conference on Internet Publishing and Beyond: Economics of Digital Information and Intellectual Property, Cambridge, MA (1997)

17. Goldman, S.L., Nagel, R.N., Preiss, K.: Agile Competitors and Virtual Organizations: Strategies for Enriching the Customer, pp. 235–266. Van Nostrand Reinhold, New York (1995)

18. Harmon, R., Demirkan, H., Hefley, B., Auseklis, N.: Pricing strategies for IT services: a value-based approach. In: 42nd Hawaii Int. Conf. on System Sciences. IEEE, USA (2009)

19. Harmon, R., Raffo, D., Faulk, S.: Value-based pricing for new software products: strategy insights for developers. In: Portland Int. Conf. on Management of Eng. & Tech. (2005)

20. Harmon, R.R., Laird, G.: Linking marketing strategy to customer value: implications for technology marketers. In: Int. Conf. on Management of Eng. & Tech. Portland, OR (1997)

21. Hinterhuber, A.: Towards value-based pricing—An integrative framework for decision making. Industrial Marketing Management 33(8), 765–778 (2004)

22. Kar, A.K., Pani, A.K.: A model for pricing emergent technology based on perceived business impact value. International Journal of Technology Marketing 6(3), 241–258 (2011)

23. Leimeister, S., Böhm, M., Riedl, C., Krcmar, H.: The Business Perspective of Cloud Computing: Actors, Roles and Value Networks. In: 18th Eur. Conf. on Information Systems (2010)

24. Marston, S.R., Bandyopadhyay, S., Ghalsasi, L.A.: Cloud Computing: The Business Perspective. Decision Support Systems 51(1), 176–189 (2011)

25. Maskin, E., Riley, J.: Monopoly with incomplete information. RAND Journal of Economics 15(2), 171–196 (1984)

26. Mikhailov, L.: A fuzzy programming method for deriving priorities in the analytic hierarchy process. Journal of the Operational Research Society 51(3), 341–349 (2000)

27. Parkhill, D.: The challenge of the computer utility. Addison- Wesley, Reading (1966)

28. Pasura, A., Ryals, L.: Pricing for value in ICT. Journal of Targeting, Measurement and Analysis for Marketing 14(1), 47–61 (2005)

29. Repschlaeger, J., Wind, S., Zarnekow, R., Turowski, K.: A Reference Guide to Cloud Computing Dimensions: Infrastructure as a Service Classification Framework. In: 45th Hawaii International Conference on System Sciences. IEEE (2012)

30. Ruoning, X., Xiaoyan, Z.: Extensions of the analytic hierarchy process in fuzzy environment. Fuzzy Sets and Systems 52(3), 251–257 (1992)

31. Saaty, T.L.: How to make a decision: The analytic hierarchy process. Interfaces 24(6), 9–26 (1994)

32. Saaty, T.L.: The Analytic Hierarchy Process. McGraw Hill, RWS Publications, New York, Pittsburgh (1980)

33. Wang, Y.M., Elhag, T., Hua, Z.: A modified fuzzy logarithmic least squares method for fuzzy analytic hierarchy process. Fuzzy Sets and Systems 157(23), 3055–3071 (2006)

34. Wei, G., Vasilakos, A.V., Zheng, Y., Xiong, N.: A game-theoretic method of fair resource allocation for cloud computing services. The Journal of Supercomputing 54(2), 252–269 (2010)

35. Weinhardt, C., Anandasivam, A., Blau, B., Borissov, N., Meinl, T., Michalk, W., Stößer, J.: Cloud-Computing – Eine Abgrenzung, Geschäftsmodelle und Forschungsgebiete. Wirtschaftsinformatik 5, 453–462 (2009)

36. Weiss, A.: Computing in the clouds. NetWorker 11(4), 16–25 (2007)

37. Wilson, R.: Nonlinear Pricing. Oxford University Press, New York (1993)

38. Wirtsch-Ing, D., Lehmann, S., Buxmann, P.: Pricing strategies of software vendors. Business & Information Systems Engineering 6(6), 452–462 (2009)

39. Wu, S., Banker, R.: Best pricing strategy for information services. Journal of the Association for Information Systems 11(6), 339–366 (2010)

40. Xu, R.: Fuzzy least-squares priority method in the analytic hierarchy process. Fuzzy Sets and Systems 112(3), 395–404 (2000)

41. Zadeh, L.A.: Fuzzy sets as a basis for a theory of possibility. Fuzzy sets and systems 1(1), 3–28 (1978)

42. Zadeh, L.A.: Fuzzy sets. Information and Control 8(3), 338–353 (1965)

43. Zimmermann, H.J.: Fuzzy set theory. Wiley Interdisciplinary Reviews: Computational Statistics 2(3), 317–332 (2010)

44. Zimmermann, H.J.: Fuzzy set theory-and its applications. Springer (2001)

Energy Efficient Congestion Control in Wireless Sensor Network

R. Annie Uthra and S.V. Kasmir Raja

Abstract. Power consumption plays crucial role in wireless sensor network (WSN). WSN is widely used for many applications in industries, military, home monitoring system etc. Data is sensed, manipulated and transmitted to the next hop nodes. Finally it reaches the destination. Certain amount of battery power is consumed by the sensor node for transmitting, receiving, listening and sleeping. In order to utilize the battery power efficiently we developed a technique which finds suitable forwarding node for transmission. The forwarding node is found based on the power level of the transmitter node. The forwarding node is chosen so that the distance to the destination is minimum compared to other neighbor nodes of the transmitter node. Moreover, energy in WSN is wasted due to packet retransmission. Network congestion is one of the primary reasons for packet drop which leads to packet retransmission. Therefore in addition to the energy efficient model, congestion control method is also proposed. Suitable outgoing rate is selected for every node in order to reduce congestion. Simulation results are compared with existing protocols and show improvement.

Keywords: Energy, Power consumption, Congestion control, Data transmission, Sensor networks.

1 Introduction

In recent years Wireless Sensor Network (WSN) established remarkable attention from both academia and industry because of its guarantee of a wide range of potential applications in both civil and military areas. A WSN consists of a large number of small sensor nodes with sensing, data processing, and communication capabilities.

Wireless sensor network is widely used in applications like military surveillance, habitat monitoring, disaster management, forest fire detection and so on. In such applications sensor nodes are deployed in the field of military, forest and the

R. Annie Uthra · S.V. Kasmir Raja
SRM University, India

S.M. Thampi et al. (eds.), *Recent Advances in Intelligent Informatics*,
Advances in Intelligent Systems and Computing 235,
DOI: 10.1007/978-3-319-01778-5_34, © Springer International Publishing Switzerland 2014

afflicted areas. Sensor nodes sense the data from these areas and may manipulate the data and transfer the data to the final destination, the sink node. Every transmission and reception of sensor node consumes some battery power. This leads to draining of battery energy which in tern causes node failure. To avoid this situation battery should be replaced. Replacing battery is not feasible for the aforementioned critical applications. Moreover the advancement in the battery technology does not yet reached to produce batteries without recharging. Therefore the battery power should be used in an efficient way so that the life time of the node can be increased.

There are several reasons that increase the power consumption of sensor nodes. One of the main reasons is packet retransmission. Packet retransmission occurs when the network or node gets congested. Congestion increases the packet drop rate, delay, and decreases throughput and the lifetime of the node. Hence congestion must be decreased in order to improve QoS [1] in a wireless sensor network in terms of link utilization, throughput, and error and delay minimization. Moreover congestion control improves the energy efficiency of the sensor network. In QoS-based routing protocols, the network has to balance between energy consumption and data quality [1]. Packet collision and channel interference are some more reasons for packet retransmission. When more numbers of nodes are present in the path for packet transmission then the battery power of all the nodes gets wasted.

Hence, efficient power management system is needed in wireless sensor environments. The problem can be concisely described as minimizing power consumption by choosing best route with minimum number of hops. To retrieve the best route in WSN, we propose a scheme to minimize power consumption by selecting best forwarding node at each step. Congestion control scheme is also employed by considering the forwarding node. Suitable outgoing rates are selected based on the incoming traffic to minimize packet drops which also minimize batter consumption.

The rest of the paper is organized as follows. Section II summarizes the related works done to improve the energy issues in WSN. Section III discuses the proposed energy efficient routing and congestion control, protocols. Section IV discuses the performance of the proposed system.

2   Related Works

Currently the main problem in WSN is how to optimize the power usage and the Quality-of-Service (QoS). QoS is usually required to meet a fixed rate on packet loss, tolerance of packet delay, or traffic throughput. Many works have developed on-demand routing protocols to the wired networks for the past decade. The routing protocols adopting in current wired networks can retrieve one suitable path according to the criterion of minimizing link cost and balancing traffic load, but those protocols may not directly apply to wireless networks due to the effects of power consumption, noise interference, or node's mobility. Related works [5,12,14] consider power management and QoS in wireless environments. Power management is done by considering connection quality of a wireless link and co-channel interference caused by the excessive power by using same frequency, radiated by nodes. Approaches [7,2] have resolved the problem by determining what amount of power to radiate and when to radiate. To increase the utilization of a system, authors in [7] addressed the advanced schemes on wireless link re-scheduling and power management. In addition, a power distribution algorithm [2] was designed to adjust transmit power with joint congestion conditions. Thus, the utilization of the wireless system can be raised by properly scheduling the link occupation and efficiently minimizing the power consumption. A power efficient [5] routing algorithm is developed along with a bandwidth-guaranteed provision in a TDMA-based CDMA system. The algorithm can calculate adequate transmission powers and distribute them to the nodes on this path by adopting current routing protocols.

Some of the ad-hoc routing protocols like GRS[4], MFR[9] find the routing path based on the node's position. But these protocols failed to consider the energy issues of WSN. SPEED[11] ensures a certain speed for each packet in the network so that each application can estimate the end-to-end delay for the packets by dividing the distance to the sink node by the speed of the packet. Moreover, SPEED can provide congestion avoidance when the network is congested. However, SPEED does not consider any energy metric in its routing protocol. Energy efficient end –to-end delay guaranteed routing protocol is performed in CED[8]. CED considered end-to-end QoS and forward error correction (FEC) to increase the probability of accurate frame transmission. Power–cost equation is formulated based on error rate and power cost is minimized with a given rate.

In this paper, we present a power minimization scheme based on the location of the forwarding node.

In [13], a priority-based rate control mechanism is proposed to adjust the source traffic rates based on current congestion in the upstream nodes and the priority of each traffic source for congestion control and service differentiation in wireless multimedia sensor networks. Congestion aware routing [6] developed by Raju Kumar et al., is an application specific, differentiated routing protocol that uses data rate to find out congestion and considers data priority to overcome congestion. This protocol is not suitable for applications that have equal priority data. Interference minimized multipath routing I2MR [10] evaluates multipath for load - balancing. Long-term congestions are determined by monitoring the size of their data transmit buffers, by using exponential weighted moving averages (EWMAs).When source node is congested, loading rate is reduced. However, the number of control packets transmitted during path discovery increases. Energy efficiency is not considered in CAR and I2MR. Congestion is detected in DPCC [15] based on buffer occupancies at the nodes along with the predicted transmitter power. Traffic-aware dynamic routing [3] algorithm routes packets around the congestion areas and scatter the excessive packets along multiple paths consisting of idle and under loaded nodes. A hybrid virtual potential field using depth and queue length forces the packets to steer clear of obstacles created by congestion and move toward the sink.

The proposed approach discovers the most suitable path with congestion control technique employed in every node, so that the power consumption at the nodes is minimized. Our model is constrained by a given power level of transmitting node. The proposed method also minimizes the number of retransmission by employing congestion control method.

We present the results of simulations conducted to test the effectiveness of our proposed approach and compare its performance with that of [4,8,9,11]. The numerical results show that our algorithm obtains the minimum energy consumption compared to other protocols.

3   Proposed Algorithm

3.1   Energy Efficient Routing:

The sensor nodes are deployed in the sensor field with the density, ρ. The nodes deployment follows Poisson distribution. The sensor nodes transmit the data packets to its neighbor nodes according to the power level of the node. Hence every node communicates with the farthest node within the communication range of that node towards the sink. The network model is shown in Fig.1.

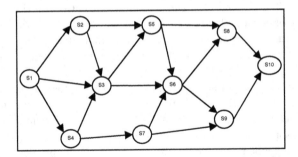

Fig. 1 Network Model

Fig.1 shows all possible paths from any node to sink node s10. Node s1 may take s2, s3 or s4 as the next hop node. In such case the battery power of s2, s3 or s4 is consumed proportional to the transmission rate of s1. Moreover s2 and s4 may also consider s3 as next hop node. In this scenario the batter power of s3 drains fast. Instead, if the power level of node s1 is high enough to communicate with s5, s6 or s7, then any one of these nodes can be selected for transmission. Hence the transmission and reception power of node s2, s3 or s4 are saved. This process is repeated for every node to find out the furthest node for packet forwarding. The new transmission path based on the power level is shown in fig.2. The packets can be transmitted from s1 to s10 by means of 4 hops in fig.1, where as it takes only 2 hops in fig 2. This reduces the power consumption of the intermediate nodes.

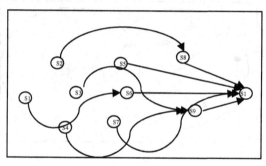

Fig. 2 New transmission path based on power level

The following section discusses the numerical analysis and algorithms to select best forwarding node.

Let P be the power level of the node and it can communicate to a maximum distance of P_d. Our aim is to find out a node within the range P_d and the distance to the sink node is minimum compared to other nodes. This is illustrated in fig 3. The best forwarding node from node i to sink node is node j, whose distance d_j to sink node is minimum.

The probability for at least one node to be present in the intersection area A_{ij} is given as,

$$p_1 = p(\text{at least one node to be present at } A_{ij})$$

$$= 1 - p(\text{no node in } A_{ij})$$

$$= 1 - \exp(-\rho A_{ij}) \tag{1}$$

(1) specifies the acceptance probability of a node at area A_{ij}.

Area A_{ij} is set to the power level of the node.

The average transmission power of a node $P_{t\text{-avg}}$ is the transmission power used for the total number of transmissions made by the node . $P_{t\text{-avg}}$ is given by,

$$P_{t\text{-avg}} = P_t \cdot N_t \tag{2}$$

where N_t is the average number of packets transmitted and P_t is the transmission power needed for single transmission.

The average power consumption of a node $P_{r\text{-avg}}$ is the reception power of a node for the number of packets received with the acceptance probability p_1 .

$$P_{r\text{-avg}} = P_r \cdot N_r \cdot p_1 \tag{3}$$

where N_r is the average number of packets received and P_r is the reception power.

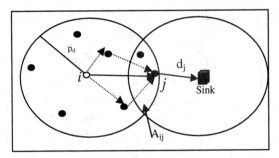

Fig. 3 Forwarding node selection

The total power consumed by a node is given by,

$$PT = P_{t\text{-avg}} + P_{r\text{-avg}} + P_{sleep}$$

where P_{sleep} is the power consumed when the node is on sleep mode.

The furthest node j from node i is identified based on the distance between node j and sink node. Node j within A_{ij} is identified using the following algorithm.

Algorithm FindDistance(node i)

```
{
1.   cost[n]=0.0   // cost of sink node is set to zero
2.   for k=n-1 to i
        {
           Choose a node j such that c[k,j]+cost[j] is minimum;
           cost[k] = c[k,j]+cost[j];
        }
}
```

cost[k] vector specifies the minimum distance from node k to sink node. Vector c[k,j] specifies the distance between node k to node j. Every time node j is selected such that c[k,j]+cost[j] is minimum (ie. distance (k→j) +distance(j→sink) is minimum). This process is repeated until it reached node i. Finally node j is the next hop node for node i which is the furthest from i and with minimum d_j distance to the sink node. The distance vector c[k,j] is assigned with weights based on the received signal strength from node k to j.

3.2 *Congestion Control:*

Packets are being transmitted in the energy efficient route that is found using the above algorithm. Network congestion occurs when either the incoming traffic exceeds the node capacity or the link bandwidth drops due to channel fading. The capacity of the node can be measured in terms of buffer occupancy of that node.

When the incoming traffic is more compared to the out going traffic, the buffer may become full. The further packets received are dropped at the buffer. In the following section we discuss a method to avoid congestion in the node. This reduces the packet drop which intern increases the energy of the node.

Let $q_i(t)$ be the buffer capacity of node i, $u_i(t)$ is the incoming traffic rate at time instant t, $v_i(t)$ is the outgoing traffic rate at time t and $d(t)$ is the disturbance. Then the buffer capacity can be calculated as,

$$q_i(t+1) = q_i(t) + u_i(t) - v_i(t) + d(t)$$

Let α_i be the threshold value for the buffer of node i. When $q_i(t)$ exceeds the threshold value α_i, then congestion is doomed in the node. Hence congestion needs to be controlled at this point. Congestion is mitigated using the following proposed method.

Split Protocol (SP):
In majority of the congestion control protocol congestion mitigation is performed by means of back propagation messages. That is the source rates of previous hop nodes are reduced using feedback messages. This reduces the fidelity of the source node. To overcome this situation we use SP. In SP the node that becomes congested forwards the packets to more than one upstream neighbor nodes as shown in fig 4. Node D gets congested as it receives from node A, B and C. Node D receives from 3 nodes and transfers to one node F. The split protocol increases the out going data rate and chooses an additional forwarding node. In fig 4 node E is selected as additional forwarding node.

When the buffer capacity $q_i(t)$ of node i exceeds the threshold value α_i, additional neighbor nodes are selected for packet forwarding towards gateway node in order to meet the incoming traffic. When node i picks additional neighbor nodes for packet transmission, it sends a request to that node. The neighboring node is the forwarding node which is selected using forwarding node selection algorithm as specified in energy efficiency routing. Upon receiving the request from node i, the upstream node checks its residual energy. If the residual energy is more than 50% then the upstream node accepts the request of node i. This way we employ load balancing in the network. If the request is rejected then node i selects another neighbor node for transmission. If none of the neighbor node accepts the request then a new route should be established.

Additionally, the outgoing traffic of the receiver node i is increased proportional to the incoming rate. That is $v_i(t)$ is set proportional to $r = u_i(t)/v_i(i)$. While increasing the outgoing rate we need to consider the channel capacity also. Let B be the channel capacity of the outgoing link of node i. If $v_i(t) > B$ then $v_i(t)$ is set to B.

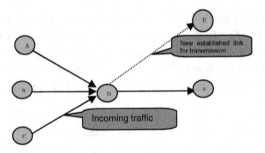

Fig. 4 Packet transmission using split protocol

4   Performance Analysis

We simulated the proposed method using the topology given in fig 1 to evaluate its performance. Simulation is executed for 30secs. We analyzed the performance in terms of power consumption. Energy consumption model uses the default parameters in the simulators. We assumed the sensors contain 2A batteries. The capacity of each battery is 1.5v and 1733mAh. Therefore the initial energy is taken as 18720J. The transmission and reception energy consumption is considered as 54.5mW and 62mW respectively. The power level of each node is set as -5dBm. The interval between packet transmissions is set as 100ms.

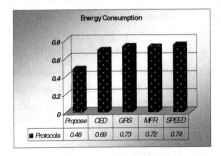 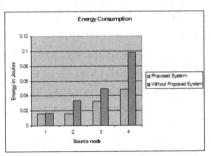

Fig. 5 Power consumption in the network   **Fig. 6** Node level power consumption

Fig.5 shows the power consumed by the nodes in the network. All the nodes are considered as the source node. Every node transmits packets at the rate of 10 packets/sec. The power consumption by the proposed method is .48J for the simulated time of 30sec. The proposed method consumes less power compared to other protocols.

Fig 6 shows the power consumed by every node using the proposed method and without using the proposed methodology. 4 intermediate nodes are selected for observing the energy consumption. Source node 1 generates data and transmits to next hop node. Source node 2 is located at 2 hop distance from the initial node which performs data generation, reception from previous hop node and transmits to next hop node. Source node 3 is placed at 2 hop distance from the initial node. The packets of previous two hop nodes are redirected to node 3. Hence the number packet received and transmitted are more in case of node 3. The energy consumption also increases with increase in packets as shown in fig 6.

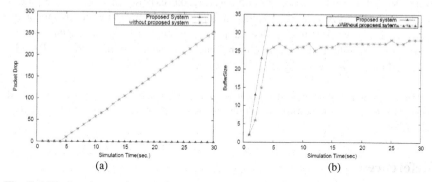

Fig. 7 (a)Packet drop at intermediate node whose in coming traffic is thrice that of out going traffic (b)buffer size of the intermediate source node

Source node 4 at 3 hops distance to the initial node. It receives data from 2 different paths whose length is 3 and 2 respectively. The increase in energy consumption is shown in fig 6. The same set up is repeated with the proposed method whose energy consumption is compared in fig 6.

Fig 7(a) and fig 7(b) show the packet drop and buffer size of the intermediate node whose in coming traffic is three times grater as much as it's out going traffic. There is no packet drop in the proposed system in the intermediate node as the out going rate is set proportional to the incoming traffic and the packets are transmitted to an additional neighbor node. Initially every node transmits data packets with a speed of 10 packets/sec. Threshold value α_i is set to 25 in our simulation. We chose an intermediate node which receives packets from 2 neighbor nodes and generates its own data. This makes the incoming traffic as 30 packets/sec in the node. Where as the outgoing traffic is the initial set value which is 10 packets/sec. Therefore the number of packets dropped raises approximately 250 packets. The proposed system increases the outgoing traffic up to 20 packets/sec when the buffer size exceeds the fixed threshold value which results in no packet loss. The buffer size of every node is set to 32 packets. The buffer size is approximately maintained in 27 in the proposed system as shown in Fig 7(b). The optimal usage of buffer is achieved by setting its rate to the initial rate if the buffer size falls below the threshold value.

5   Conclusion

In this paper we have proposed an energy efficient congestion control method. The proposed method chooses best forwarding node so that the energy of the nodes can be used in optimal way. The proposed energy efficiency is analyzed in terms of packet drops and retransmissions. Energy consumption is minimized using the proposed congestion control method by minimizing the number of retransmissions. Simulation results show that the energy consumption of the network as well as node is reduced compared to the existing protocols. The packet drop is reduced to zero in every node on applying the congestion control mechanism. The future work is to enhance the proposed method using predictive congestion control technique along with channel capacity. Secondly per node load can be balanced based on the energy level of the node.

Acknowledgments. We thank the faculties of computer science and engineering, SRM University for providing constructive comments that improved the quality of the paper in all aspects.

References

[1] Annie Uthra, A.R., Kasmir Raja, S.V.: QoS Routing in wireless sensor network – A survey. ACM Computing Surveys 45(1) (2013)

[2] Chiang, O.M.: Balancing transport and physical layers in wireless multihop networks: jointly optimal congestion control and power control. IEEE Journal on Selected Areas in Communications 23(1), 104–116 (2005)

[3] Fengyuan Ren, U., He, T., Das, S.K., Lin, C.: Traffic-Aware Dynamic Routing to Alleviate Congestion in Wireless Sensor Networks. IEEE Transactions on Parallel and Distributed Systems 22(9) (September 2011)

[4] Finn, G.G.: Routing and Addressing Problem in Large Metropolitan-Scale Internetworks. ISI res. Rep ISU/RF-87-180 (March 1987)

[5] Kim, I.D., Min, C.H., Kim, S.: On-demand SIR and bandwidth guaranteed routing with transmit power assignment in Ad Hoc mobile networks. IEEE Transactions on Vehicular Technology 53(4), 1215–1223 (2004)

[6] Kumar, S.R., Crepaldi, R., Rowaihy, H., Harris, A.F., Cao, G., Zorzi, M., Porta, T.F.L.: Mitigating Performance Degradation in Congested Sensor Networks. IEEE Trans. Mobile Computing 7(6), 682–697 (2008)

[7] Li, N.Y., Ephremides, A.: Joint scheduling, power control, and routing algorithm for ad-hoc wireless networks. In: Proceedings of the 38th Annual Hawaii International Conference on System Sciences (January 2005)

[8] Lu, B.Y.-J., Sheu, T.-L.: An efficient routing scheme with optimal power control in wireless multi-hop sensor networks. Computer Communications 30, 2735–2743 (2007)

[9] Takagi, H., Kleinrock, L.: Optimal transmission ranges for randomly distributed packet radio terminals. IEEE Transactions on Communications 32(3), 246–257 (1984)

[10] Teo, T.J., Ha, Y., Tham, C.: Interference-Minimized Multipath Routing with Congestion Control in Wireless Sensor Network for High-Rate Streaming. IEEE Trans. Mobile Computing 7(9), 1124–1137 (2008)

[11] He, T., Stankovic, J.A., Lu, C., Abdelzaher, T.: SPEED: a stateless protocol for real-time communication in sensor networks. In: Proceedings of the 23rd International Conference on Distributed Computing Systems, May19-22, pp. 46–55 (2003)

[12] Sheu, T.L., Lu, Y.J.: Power Minimization with end-to-end frame error constraints in wireless multi-hop sensor networks. In: International Wireless Communications and Mobile Computing Conference (IWCMC 2006) (July 2006)

[13] Yaghmaee, Q.M.H., Adjeroh, D.A.: Priority-Based Rate Control for Service Differentiation and Congestion Control in Wireless Multimedia Sensor Networks. Computer Networks 53(11), 1798–1811 (2009)

[14] Yuan, Y., Yang, Z., He, Z., He, J.: "An integrated energy aware wireless transmission system for QoS provisioning in wireless sensor network. Computer Communications 29(2), 162–172 (2006)

[15] Zawodniok, R.M., Jagannathan, S.: "Predictive Congestion Control Protocol for Wireless Sensor Networks. IEEE Trans. Wireless Comm. 6(11), 3955–3963 (2007)

Energy Scavenging Based HybridGSM Model for Mobile Towers

Sonal Yadav, Manoj Singh Gaur, and Vijay Laxmi

Abstract. Some plentiful renewable energy resources are available like solar, wind, biomass to generate green energy without pollution. With the growth of population, telecom companies are rapidly increasing mobile towers to extend the coverage. A GSM (Global System for Mobile communications) base station consumes $3 - 5\ KW$ of electricity and releases 5.61 million ton CO_2 annually. Price of traditional energy resources like petrol, diesel, coal etc. is increasing rapidly. This motivates researchers to search for non–conventional energy resources to produce energy at low cost. In this paper, a solar–wind based hybridGSM model is proposed to supply electricity for mobile towers. In addition, it provides battery backup of 48 hrs in all weather conditions. Set up cost and operational cost of hybridGSM is less as compared to GSM and WorldGSM model. It can efficiently address the challenges of installation and maintenance of mobile towers in rural areas. To validate the proposed hybridGSM model, $ns - 3$ simulator is extended by including solar–wind based energy scavenging and non–linear battery model. Experimental results shows effectiveness of proposed model.

1 Introduction

India is the world's second largest mobile market after China since 2010 [8]. The total number of mobile subscriber has reached 919 million in March 2012. Indian wireless tele–density (number of mobile phones per hundred population) has increased from 0.50 to 74.89 during March 1999–2012 [5] [8]. Presently, mobile towers operated with diesel based GSM technique to generate electricity. There are more than 3,37000 GSM mobile towers in India that burn an estimated 2.12 billion litres of diesel/year [8]. GSM based mobile towers have problem of limited

Sonal Yadav · Manoj Singh Gaur · Vijay Laxmi
Department of Computer Engineering, Malaviya National Institute of Technology,
Jaipur, India
e-mail: {sonaldv4,gaurms,vlaxmi}@gmail.com

S.M. Thampi et al. (eds.), *Recent Advances in Intelligent Informatics*,
Advances in Intelligent Systems and Computing 235,
DOI: 10.1007/978-3-319-01778-5_35, © Springer International Publishing Switzerland 2014

coverage in remote areas that leads to higher installation and operational cost. Thus, GSM based services are expensive for users. Department of Telecom provides subsidy per site for service providers and incentives for using environment friendly non–conventional energy resources. Non–conventional energy resources cost is less than traditional resources due to 75% subsidy provided by government. Un–electrified rural villages as of March 2012 are 40,000 and remote villages are 13,000 [3]. HybridGSM can be successfully installed and maintained in such areas. HybridGSM works as follows. It generates electricity with solar photovoltaic panel and wind turbine that is generated from solar and wind renewable resources respectively. Cost of electric power supply with diesel generator is around 3.5 *lakhs*. Capital cost per *KW* of solar power system would be around Rs 3.5 − 4 *lakhs* and for wind power system would be around *Rs.* 1 *lakh*. Therefore, Setup cost is low and affordable for hybridGSM.

2 Related Work

There are various telecom companies looking for non–conventional resources to generate electricity for mobile towers. Bergey Wind power, a US based wind power solution company has successfully implemented wind power based telecom base station projects in Greece, China, Romania, Chile [8]. Similarly, Vihaan Networks Limited (VNL) is an India based end to end GSM system manufacturer that has proposed worldGSM. It has successfully implemented numerous solar powered base stations with 3 days battery backup [4] [6] [8]. It has launched a smart solution for rural and village sites with small investment of money and little maintenance of mobile towers. It is capable to fulfill needs of remote area coverage of the telecom companies. WorldGSM may be failed to generate electricity in monsoon season due to limitation of three days battery backup for all weather conditions. But, HybridGSM works fine because it is designed to face the weather challenges and its limited battery backup reduces the set up cost. World no.1 mobile tower manufacturing company Indus and Indian telecom companies like Vodafone, Idea, VNL and BSNL wants to move towards green solution for electricity [2]. VNL is the only company that gives the successful trial for practical implementation.

3 Proposed HybridGSM Model

In hybridGSM, mobile tower is running with electric charge generated with solar photovoltaic panel and wind turbine. Solar photovoltaic does not work efficiently during monsoon season and wind turbine alone can work perfectly in monsoon season. Therefore, by combining solar and wind energy resources, a more efficient and reliable energy resource is generated. In proposed algorithm electricity is generated by solar and wind and calculation steps are mentioned in Figure 1. Left side of flow graph calculate electricity generated by solar PV module that measures voltage variation from standard temperature ($25°C$) across solar PV modules. Whereas, right side of flow graph calculate wind turbine generated electricity. This generated current passes to rechargeable non–linear lead-acid battery model. Eventually, generated energy is supplied to mobile tower and lead–acid battery. In flow graph, temperature coefficient, intermediate voltage, wind turbine swept area is calculated as follows.

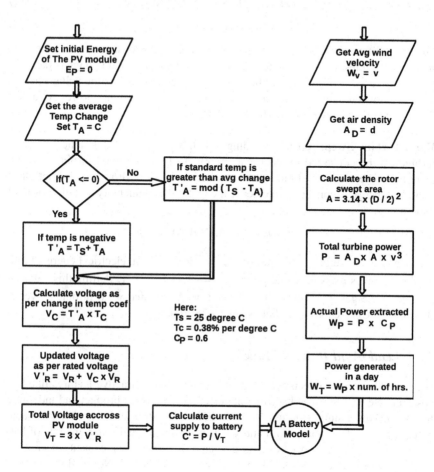

Fig. 1 Flow graph of hybridGSM

Temperature. Solar panels is classified into two categories mono–crystalline and poly–crystalline. Mono–crystalline is $3 - 4\%$ more efficient than poly–crystalline solar cells. So hybridGSM uses mono–crystalline solar cells. It has temperature coefficient between $-0.44°C$ to $-0.50°C$. Mono–crystalline performance is best with $-0.38°C$ temperature coefficient [1]. At standard temperature ($25°C$), percentage temperature coefficient is calculated as follows. Temperature coefficient at open circuit voltage (V_O) is $-0.38\%/°C$. Let average temperature variation from standard temperature is $-10°C$. Voltage change in rated voltage (V_R) is calculated with the help of eq. 1 and eq. 2. Total V_R across PV module is $18 + 18 \times 0.133 = 20.4$. In eq. 3, N is number of PV modules connected in series, here N=3. Thus, V_R at -10°C is $3 \times 20.4 = 61.2 V$.

$$\triangle T = 25°C + 10°C = 35°C \tag{1}$$

$$\triangle T_T = -0.38\%/°C \times 35°C = 13.3\% \tag{2}$$

$$V_T = N \times V_R \tag{3}$$

Voltage. Different terminologies are used to address the voltage generated across solar PV modules in intermediate steps of flow graph. Nominal voltage is the voltage that is carry over throughout the day. Open circuit voltage is the voltage across solar panel when module is not connected to any load. Maximum power voltage is the voltage of peak operating performance output.

Wind Turbine Swept Area. According to betz's law only 60% kinetic energy of turbine can be converted to electrical energy. To keep the wind turbine moving, there has to be some wind movement on outside the rotor. It results in fraction of available energy is converted to electrical energy [7]. Elementary equation of total wind power is:

$$W_P = \frac{1}{2} \times A_D \times A \times v^3 \times C_P \tag{4}$$

Wind turbine converts kinetic energy $(E_K = \frac{1}{2} \times m \times v^2)$ to electrical energy. Fluid mechanics gives mass flow rate $\rho = \frac{dm}{dt} = A_D \times A \times v$. Putting this value of ρ in eq. 4 we get $P = \frac{1}{2} \times A_D \times A \times v^3$. Generated wind turbine power is $W_P = P \times C_P$. According to betz's law $C_P = 0.6$.

3.1 Lead–Acid Battery Model.

Lead–acid battery is a deep cycle battery. It has salient attractive features like low cost, reliable, robust, overcharging tolerance, most recycled product and indefinite life that makes it suitable to use for hybridGSM. Depth of discharge for lead–acid battery is 50%. The capacity of battery is dependent on the discharging rate of battery according to Peukert's Law. As discharge rate increases, capacity decreases. Peukert's constant for lead–acid battery is between 1.1 to 1.3. Nominal voltage rating is 12 V (6 cells of 2 V each) for thoroughly charged battery and 10.5 V for fully discharge battery.

Discharging/Charging Rate (R) of Battery Model. Current withdrawal from a charged battery is called discharging of battery. During discharging, lead–dioxide (+ive plate) and lead (–ive plate) react with electrolyte of sulfuric acid to create lead sulfate, water and energy. Similarly, current supplied to a battery for charging is called charging a battery. If battery bank capacity is C and discharged in 48 hrs then we can say, Discharging rate is $\frac{C}{48}$. The formula of discharging/charging rate is

$$R = \frac{BatteryBank}{I} \tag{5}$$

Battery Capacity. Battery capacity is measured in ampere–hour. Capacity of a battery is amount of energy that can be extracted from battery under certain conditions. If T is rated discharge (hrs), C is rated capacity at the $\frac{C}{48}$ discharge rate, I is discharging current in *amperes*, k is peukert coefficient, E_C is effective capacity at the discharging rate I *(in %)*. Rated capacity $C = I^K T$ is measured in ampere–hour at 1 A discharge rate. Where I is the discharging current in amperes, T is the discharging time (*in hrs*) and K is peukert coefficient. Efficiency (E) of the battery is the ratio of discharging energy to the charging energy as shown eq. 7.

$$E_I = C \times [\frac{C}{(I \times T)}]K - 1 \tag{6}$$

$$E = \frac{E_D}{E_C} \tag{7}$$

4 $ns - 3$ **Simulation and Inferences**

Energy consumption is a key issue for mobile towers. Our proposed model is motivated by the paper [9]. $ns - 3$ simulator measures the energy of battery powered nodes. We have analyzed variation in energy consumption at different states of nodes. Ad–hoc nodes switches among these states as per variation of load (number of nodes sending packet for communication). As load increases energy consumption of node increases that will results in rapid battery discharge. There is wide scope of energy scavenging model. This approach can be used for research in wireless sensors network and other wireless areas like bluetooth, zigbee, wimax. We have used modular approach to design hybridGSM that is based on energy scavenging in ns-3 with minimum interdependence. HybridGSM is tested in $ns - 3$ simulator to ensure the performance of supporting protocols and setup cost. For experimentation nodes are arranged in cluster star topology. In experiment ad–hoc nodes are arranged in star topology. discharging/charging rates of lead–acid battery is measured in three scenario as shown in Figure 2.

- Case1, average solar insolation is 2 *hrs* and wind speed is 1 m/s, there is no electricity is generated to charge battery. Battery is discharged continuously up to 50 % depth of discharge. Eventually, battery will stop working. In such conditions tower may fail. This is a linear graph.
- Case 2, average solar insolation is 5 *hrs* and wind speed is 2 m/s, there is electricity is generated by solar photovoltaic panel in the absence of wind energy. Therefore, initially battery is discharging but when get the current it will start charging. This is non–linear graph.
- Case 3, average solar insolation is 1 hrs and wind speed is 5 m/s. there is electricity is generated by wind turbine in the absence of solar energy. Hence, initially battery is discharging but when get the current it will start charging. This is non–linear graph.

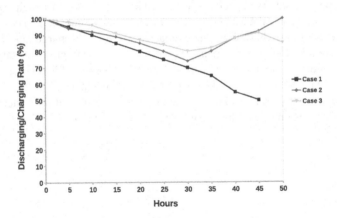

Fig. 2 Lead–acid battery discharging/charging Graph

Experiment with hybridGSM on $ns - 3$ gives the results as shown in Table 1. Installation and operational cost is less than GSM and worldGSM. Power consump-

Table 1 Comparison of HybridGSM with GSM and WorldGSM

Characteristics	GSM	WorldGSM	HybridGSM
Power Consumption	$3 - 5\ KW$	$50 - 120\ W$	$50 - 120\ W$
Cooling Requirement	Yes	No	No
Setup Cost	$5 - 10\ Lakhs$	$8\ Lakhs$	$5 - 7\ Lakhs$
Battery Back up	Fuel Availability	$3\ Days$	$2\ Days$
Running Cost	$Rs.\ 50000$	0	0
Echo Friendliness	No	Yes	Yes
Transportation	Difficult	Easy	Easy
Topology	Star	Clustered star	Clustered star

tion of mobile tower is $50 - 120\ W$ that is less than GSM. Hence, a small capacity PV module and wind turbine sufficient for generating electricity to run mobile tower. HybridGSM's estimated installation cost is $5 - 7\ lakhs$ that including around 4.5 *lakhs* cost of mono–crystalline solar PV with capacity of $50 - 120\ W$, wind turbine cost is in thousands and lead–acid battery cost is around 1 *lakh* for the capacity of $250\ W$. There is only installation cost, no maintenance cost.

5 Conclusions

HybridGSM is a generic and reliable model for electricity to run base stations. It is available with less installation and maintenance cost as compared to GSM and

WorldGSM. Estimated cost of installation and battery capacity for hybridGSM is based on average solar insolation and wind speed for indian perspective. It may vary positively and negatively from the estimated installation cost because it depends on supporting weather conditions. Hence, cost is decided by studying the weather data of particular region. A significant amount of reduction is acquired in the installation cost of hybridGSM by using two days battery backup. In addition, this particular backup is enough during monsoon as wind turbine can generate electricity to operate base stations. Experiment with extended $ns - 3$ energy model gives the favourable results for hybridGSM. It supplies current to mobile towers for 2 days without interrupting electricity. In future, hybridGSM can be extended to face environmental challenges that exists world–wide. Low availability of solar and wind energy may overcome by using biomass energy. Moreover, GPS based wind tracking can used to rotate turbine in the blowing direction of high speed wind. As a result, more wind energy is delivered. If photovoltaic panel and wind turbine does not work properly than sensors can be used to alert about the fault of the system. Eventually, these approaches will help to grab more energy with renewable resources.

References

1. Impact of temperature variation on temperature coefficient by alchemie limited inc., `http://www.solar-facts-and-advice.com/solar-panel-temperature.html` (last accessed November 2012)
2. Indian national english newspaper, `http://www.timesofindia.indiatimes.com/` (last accessed February 2012)
3. Solar photovoltaics for electrification in india, solar photovoltaics application pdf – center for study of science (last accessed November 2012)
4. Vihaan Network Limited (VNL), website homepage, `http://www.vnl.in` l(ast accessed November 2012)
5. Wireless teledensity in india, `http://www.pluggd.in/india-telecom-data-jan-2012-297/` (last accessed November 2012)
6. Hansson, A.: WHITE PAPER India's Emerging Energy and Air Pollution Crisis from GSM Site Operations. In: A Study of Critical Power and Fuel Dependencies for Indian GSM site Solutions, Awarded by USOF, DOT, India (2010)
7. Kalmikov, A., Dykes, K., Araujo, K.: Wind power fundamentals ppt. MIT Wind Energy Group & Wind Energy Projects in Action
8. Shah, B.: Online paper on cellular base stations: Infrastructure sharing, carbon emissions and CDM opportunities in india, `https://www.scribd.com/doc/58801159/Bhomik-Shah-Full-Paper-Enviriconconclave-PDF` (last accessed November 2012)
9. Wu, H., Nabar, S., Poovendran, R.: An energy framework for the network simulator 3 (ns-3). In: Proceedings of the 4th International ICST Conference on Simulation Tools and Techniques, SIMUTools 2011, pp. 222–230. ICST (Institute for Computer Sciences, Social–Informatics and Telecommunications Engineering), Brussels (2011)

Discrete Bacteria Foraging Optimization Algorithm for Vehicle Distribution Optimization in Graph Based Road Network Management

Chiranjib Sur and Anupam Shukla

Abstract. Bacteria Foraging Optimization (BFO) is a swarm intelligence optimization technique which has proven to be very effective in continuous search domain having several dimensions. In this paper a discrete and adaptive version of the Bacteria Foraging Optimization Algorithm is being introduced which will be useful in discrete search domain and all kind of multi-dimensional graph based problem. This Discrete Bacteria Foraging Optimization (DBFO) Algorithm is being analyzed and tested in the optimized route foundation phenomenon of a graph based road network and has been compared with the Ant Colony Optimization and Intelligent Water Drop with respect to global convergence. The road system is obsessed with multiple parameters which influence the management of the vehicles in the graph and needs to be analyzed and taken care of. Multiple parameters of the system demand multi-objective optimization using a weighted evaluation function which is carefully designed keeping in mind how the parameters behaves and how its variation dynamically changes the performance of the system. The new discrete version of BFO is being introduced for the first time and it readily suits all kind of graph based and combinatorial optimization problems.

Keywords: discrete bacteria foraging algorithm, vehicle routing optimization, combinatorial optimization, graph search and path planning technique.

1   Introduction

Road management has become an issue with the increase in the number of vehicles and due to the existing inefficient traffic management system. An effort has been very common on finding the least path for path planning problem, but if the

Chiranjib Sur · Anupam Shukla
ABV-Indian Institute of Information Technology & Management, Gwalior
email: {chiranjibsur, dranupamshukla}@gmail.com

S.M. Thampi et al. (eds.), *Recent Advances in Intelligent Informatics*,
Advances in Intelligent Systems and Computing 235,
DOI: 10.1007/978-3-319-01778-5_36, © Springer International Publishing Switzerland 2014

number of vehicle increases the least path criteria can become a bottleneck due to congestion. So the system must be able to analyze its data pattern and interact with vehicles by providing the relevant details. These details can be least distance path, least waiting path, fuel efficient path, least cost path, least travel time path etc. Different vehicle will follow different criteria and gradually tend to disperse the load. The load balancing criteria of the network will be taken care by the dynamic road system with various changing parameters like average waiting time, queue length etc and static parameters like distance, width of road etc. These parameters are considered because they affect the road management system and need to be considered. The work mainly focuses on the path planning for vehicles heuristically in simulation and provides them to the vehicles through smart device application. The effectiveness of this kind path planning will be reflected through the load balancing and time analysis of the system as a whole. The rest of the paper is arranged as Section 2 for basic BFO, Section 3 for DBFO, Section 4 for implementation details, Section 5 for results and Section 6 for conclusion & future works.

2   Traditional BFO Algorithm

The Bacteria Foraging Optimization (BFO) Algorithm was first introduced by Kevin M. Passino [1] in the year 2002 and its mathematical formulation implies that it more suited continuous search domains. But due to its effectiveness it dynamic system optimization it required to rethink the strategy and formulate a version for the graph based problem. For the sake of completeness the traditional BFO is revisited so that the modifications made can be clarified. BFO consists of four stages and they are as follows:

Chemotaxis: Chemotaxis is the affinity of the bacteria to move towards favorable chemicals and away from the toxic ones. The bacteria (especially E. coli) may swim or tumble following a mathematical equation [1] which acts as a variation of the positions and thus searches for global optimization. In this step the agent tumbles for turning based movement or swim for straight movement to the better positions in the search space.

Swarming: Swarming is the phenomenon in which the bacteria follow its natural behavior of being attracted to other bacteria under the influence of other bacteria and their fitness in the form of attraction (towards better ones) and repellant (towards worst ones) and this kind of combined effect of other fellow agents makes the swarm intelligent better known as social-influenced intelligence. These are a natural movement of the agents towards the regions with more food or better fitness and thus tend to form a conglomerate. This phenomenon is also represented by a mathematical equation [1]. The value of attraction and repellant parameters is proportional to some function relating to the comparative difference of the fitness between the two agents.

Reproduction: The bacteria cells possess the typical cell characteristics of periodical death and then reproducing new cells through the process of cell splitting like asexual reproduction procedure. In reproduction the healthy agents at the best positions of the search space splits into two to promote better search in that area. It compensates for the decrease in the number of bacteria agents.

Elimination/Dispersal: In this step the least fitted (with respect to optimization) bacteria either is eliminated from the solution pool or being shifted to some random place within the search space. The reproduction and elimination mutually compensate the variation in the number of agents.

The four steps are not sequential and is nested inside another loop and hence when one step occurs, it happens quite sometimes before getting out of the loop to the other.

3 Discrete BFO Algorithm

Now there are various assumptions made on the basic BFO algorithm and its implications are redefined so that the algorithm becomes ready for graph search based problem. According to the traditional BFO, it is considered that there is no rule for which step must occur first and will occur with turn and iteration. But for the graph based algorithm, the steps are opportunistic and will not take place with iteration but with the requirement and need of the bacteria and the search situation. The modified basic equations of Discrete BFO for the graph based problems are as follows:

$$\theta^{n+1}(i,j,k,l) = \theta^{n}(i,j,k,l) + C(n)\phi(n) \tag{1}$$

where
- n is the current iteration,
- i is the current position or current node of the graph,
- j is the index of Chemotaxis event (unlike the traditional BFO, in DBFO j is initiated with a maximum level of energy and it gradually decreases with investigation of the new nodes, as if a payment for service or energy lost due to work. It is to be noted that the optimum value of initialize of j depends on the graph diameter or something less than the total number of nodes present in the graph and also can be found out adaptively through experimentation. First start with a low value and if the bacteria fail to reach the destination then increase it slightly until it reaches. The fitness value of the bacteria and the number of nodes for best path determines what should be the optimal value of initialization of j)

Practically, with each movement the value of j decreases like $j = j - K_{constant} * C(n)$ where C(n) is the number of steps or capability that the bacteria can afford to investigate. Lower value of $C(n) \in \{1,2\}$ will make the system less complex and lead to better investigation of search pattern.

- k is the reproduction step, (here the reproduction step is generated only when the process of reproduction is required, say for example if 80% of the bacteria dies and there is a decrease in 80%, the rest 20% will reproduce to make up the eliminated ones. Now the k will be generated semi-randomly and according to the fitness value of the bacteria and $k \in Z$. Say if $k = 3$, then the bacteria will reproduce 3 times and hence in the 1st iteration, one will produce two, in 2nd iteration, 2 will produce 4, in 3rd iteration, 4 will produce 8. So the value of k will determine $2k$ new bacteria from one old bacterium in each iteration.
- l is the elimination/dispersal step, only used for last stage survival.

Step 1: DChemotaxis: It is considered that there is nutrition level everywhere unevenly scattered and the bacteria try to follow the best nutritious path. Initially the bacteria will follow a random path till it has the taste for good food. But when it has tested enough, it will estimate if the path of nutrition is better than α times the average of nutrition it has previously received. If it does then it will remain to the node where it has swum else it will tumble back to the previous node and try another path. Each of swim or tumble back will decrease the j factor by one.

If $n < 2$ [iteration less than 2]
　　Swim randomly
Else
　　If $[\alpha * J \text{ (average)}] > [J \text{ (new path parameter)}]$
　　　　$\Phi(n) = 0$; [signifies only swim or forward movement]
　　Else
　　　　$\Phi(n) \neq 0$; [implies change in direction or tumble movement, here $= 1$ that is tumble back]
Bacteria moves until it finds the destination. The path followed by it upon reaching the destination is considered the best path.

Step 2: DSwarming: The bacteria can exchange fluids of its own cells in the form of energy which can be utilized for searching purpose. Swarming occurs for bacteria if there is requirement of energy or may be the best bacteria have more energy and hence the other bacteria will try to move towards it. Swarming will tend to extend or increment the j factor and hence its life span as the energy is extended during the process of gathering together and energy exchange takes place. The energy fluid follows the process of diffusion or osmosis whichever applicable at that situation. This energy can be used during analysis or movement in the step Chemotaxis.

Step 3: DReproduction: After a certain span of time or number of iterations, or rather due to the death of certain percentage of bacteria it is required to regenerate the dead bacteria through the process of reproduction. However the algorithm can be formulated as (if the rate of death is low) after n^{th} iteration β percentage of the bacteria dies on their fitness value and out of the left bacteria, reproduction is performed to regenerate on the vacancy. It is to be kept in mind that the natural death of the bacteria is due to failure in the graph and there is no harm of loosing of valuable information or searched path. If out of 100 bacteria, 80% dies then the rest 20% will be responsible for the reproduction and regeneration. The number of times a bacterium will reproduce will depend randomly and the choice of bacteria may or may not depend on the fitness value of the search criteria of the bacteria. However the choice of fitness value will help in searching the best region and no choice of bacteria will help in exploration. Offspring receives the partial path of the parents and continues search from there.

Step 4: DElimination/DDispersal: The bacteria will keep on moving forward and tumbling for destination until its energy gets exhausted and it will die. But before that it will make its last move with the help of l (small L) parameter which remains one throughout until it is used at last and makes zero. So

If j = 0 && l = 1
    Use l and make last move
    Make l = 0
Else-If j = 0 && l = 0
    The bacterium dies

So if the bacterium has the last trump card of l = 1, it can actually move a considerable distance (depends on the overall graph size) without investigation and then will investigate the path using the Equation (1) and following the DBFO Chemotaxis. If it satisfies, then the bacterium survives, else this time no turning back and it dies. So it is actually a round for Elimination or there will be Dispersal to a certain new place through the l (small L) parameter movement which is arbitrary and unlike previous step where the bacteria moves carefully and investigates each C(n) paths at a time, here there is a certain forward movement of certain steps and in arbitrary direction. With DDispersal bacteria will retain the previous partial path and also the new route to the dispersed place and from here the search begins. The DDispersal is equivalent to mutation where the value of j is retained to some considerable value, depends on the graph size and progress of iterations.

4   Optimized Route Search Using DBFO

The following is the pseudo code for DBFO for graph based road network routing.

Step 1: Initiate the road graph matrix $G = (V,E)$ and its parameter matrix for each edge E as $\{EdgeMatrix\}_{ab} = [\ a,b, \{p_1, p_2, \ldots \ldots, p_k\}]$, Define BacteriaMatrix = $\{ID, Sc, Des, Reachability, PathTraversed, NodeCount, \{\sum p_1^t, \sum p_2^t, \ldots, \sum p_k^t\}$, Fitness, i, j, k, l $\}$ where Sc is starting node, Des is destination node, Reachability = 1 (if destination reached) or = 0. Define Fitness functions for each type of optimization say z and $\{fitness\}_z = \zeta\{\ \sum p_1^t, \sum p_2^t, \ldots, \sum p_k^t\ \}$, Initialize Matrix $\{BestPath\}_z = [Sc, Des, PathTraversed, NodeCount, Fitness\ ]$

Start for loop for each agent
Step 2: Perform DChemotaxis if j > 0 and update i, j, k, l for bacteria
Step 3: Perform DSwarming
Step 4: Perform DElimination/DDispersal if required
Step 5: Perform DReproduction for best agents if number of agent is less.
End for loop
Step 6: Calculate {fitness}$_z$ of bacteria for each z type of optimization according to fitness (here w.r.t. total, travelling and waiting time)
Step 7: Select the best fitness value and update {BestPath}$_z$ matrix
Step 8: Provide Data to Vehicles through smart devices.
Step 9: Update {EdgeMatrix}$_{ab}$ due to variation in dynamic parameters with time. These parameters are like waiting time, cost of travel, queue length etc.
Step 10: Update BacteriaMatrix{ } for each movement of bacteria from a to b.
Step 11: Restart for loop from step 2 for the updated network parameters.

The time complexity of the algorithm depends on the implementation or coding of the experimenter and it is very difficult to give a strong comment on the time complexity, but according to our implementation we have $O(n^3 + n^3 + n^3 + n^2)$ where each term represents the four steps of the DBFO algorithm.

5   Computational Results and Graphs

The following are the various plots of the results that are generated for a certain instance of the road network with some model parameters and with 40 agents of each of DBFO, Ant Colony Optimization & Intelligent Water Drop Algorithms are run to find the global optimization with respect to a fitness function (Fig 3) which is a linear combination of travelling time and waiting time. However the global optimization with respect to travelling and waiting time (Fitness Function = Total Time = Travelling Time + Waiting Time) individually is also plotted in Fig 1-2. The results show that the convergence rate of DBFO is better or rather considerable than the other two. The data used is modeled for an experimental road graph model that describes the busy roads and the long bypasses just outside the main city network. But the waiting time is the average for any crossing considered. The result shows that for graph based problems balance between exploration and exploitation yields better result mainly if exploitation occurs in the later part of the iteration. Exploitation tends to saturate the search opportunity in the graph network and thus affects convergence rate of the algorithm as shown in the other two. DBFO being a explorative algorithm which is not greedy to take the best path, found out of local heuristic measure, but is always eager to explore more combination of nodes/events in the network and thus promises to be an efficient discrete graph based algorithm. For declaration of the other parameters of the DBFO algorithm, knowledge of the experimental graph is important and care must be taken so that the agent do not die immaturely before reaching the destination. It is better to introduce them with adequate margin so that they have excess capability for proper and full search. Proper vehicle guidance will through optimized path planning will help in distribution scattering of the vehicles throughout the network.

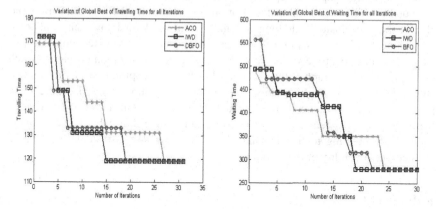

Fig. 1 Variation of Global Best of Travelling Time for all Iterations   **Fig. 2** Variation of Global Best of Waiting Time for all Iterations

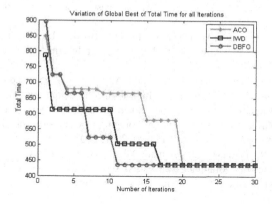

Fig. 3 Variation of Global Best of Total Time for all Iterations

6   Conclusion and Future Works

From the simulation results of the modeled system optimized with linear combination of travelling and waiting time reveal the convergence rate of new DBFO and its path searching capability as a new met-heuristics. The description of the algorithms is provided and the details for routing of the network are also given. But there is a lot of scope in the application of the DBFO algorithm and may be some hidden or more analyzed and experimented parameters or adaptive parameter can improve the performance of the DBFO. The DBFO algorithm has been a promising algorithm in the multi-objective optimization field which offers good results for both weighted multi-parameter cost function and parameter separated cost functions and a proper system oriented cost function will help in achieving better result with respect to optimization and time. DBFO algorithm can also be extended to utilize for combinatorial optimization problems for optimization and also in the coordinated searching for multi-agents in both known and unknown environments. There is also scope for scalability analysis and detection of performance for other real time variants for situation based optimization problems which are represented in graph forms. The DBFO being a explorative algorithm will always help in proper optimization of solution for dynamic network where some parameters tend to change with feedback from the system and also because of these changes we require a quick convergent algorithm which can provide solutions responding to the demand of the situation and also at the same time respecting the validity of the solution.

References

1. Passino, K.M.: Biomimicry of Bacterial Foraging. IEEE Control System Magzine 22, 52–67 (2002)
2. Datta, T., Misra, I.S., Mangraj, B.B., Imtiaj, S.: Improved Adaptive Bacteria Foraging algorithm in Optimization of Antenna Array for Faster Convergence. PIER C 1, 143–157 (2008)
3. Liu, W., Chen, H., Chen, H., Chen, M.: RFID Network Scheduling Using an Adaptive Bacteria Foraging Algorithm. Journal of Computer Information Systems (JCIS) 7(4), 1238–1245 (2011)
4. Biswas, A., Dasgupta, S., Das, S., Abraham, A.: Synergy of PSO and BFO-A Comparative Study on Numerical Benchmarks. In: International Symposium on Hybrid Artificial Intelligent Systems (HAIS), Salamanca, Spain, pp. 255–263 (November 2007)
5. Zhang, Y., Wu, L., Wang, S.: Bacterial Foraging Optimization Based Neural Network for Short term Load Forecasting. JCIS 6(7), 2099–2105 (2010)
6. Sastri, G.S.V.R., Pattnaik, S.S., Bajpai, O.P., Devi, S., Sagar, C.V., Patra, P.K., Bakwad, K.M.: Bacterial Foraging Optimization Technique to Calculate Resonant Frequency of Rectangular Microstrip Antenna. Int. J. RF Microwave Computer Aided Eng. 18, 383–388 (2008)
7. Kim, D.H., Abraham, A., Cho, J.H.: A hybrid genetic algorithm and bacterial foraging approach for global optimization. Information Sciences 177, 3918–3937 (2007)
8. Mishra, S.: A hybrid least square-fuzzy bacteria foraging strategy for harmonic estimation. IEEE Trans. Evol. Comput. 9(1), 61–73 (2005)

Survey of Influential User Identification Techniques in Online Social Networks

Roshan Rabade, Nishchol Mishra, and Sanjeev Sharma

Abstract. Online social networks became a remarkable development with wonderful social as well as economic impact within the last decade. Currently the most famous online social network, Facebook, counts more than one billion monthly active users across the globe. Therefore, online social networks attract a great deal of attention among practitioners as well as research communities. Taken together with the huge value of information that online social networks hold, numerous online social networks have been consequently valued at billions of dollars. Hence, a combination of this technical and social phenomenon has evolved worldwide with increasing socioeconomic impact. Online social networks can play important role in viral marketing techniques, due to their power in increasing the functioning of web search, recommendations in various filtering systems, scattering a technology (product) very quickly in the market. In online social networks, among all nodes, it is interesting and important to identify a node which can affect the behaviour of their neighbours; we call such node as Influential node. The main objective of this paper is to provide an overview of various techniques for Influential User identification. The paper also includes some techniques that are based on structural properties of online social networks and those techniques based on content published by the users of social network.

Keywords: Online social networks, Influential user, Content, Active user, Viral marketing.

1   Introduction

Social Network [1] is a structure of individuals or organizations tied to any of interdependencies such as friendship, fellowship, Co authorship, Collaboration

Roshan Rabade · Nishchol Mishra · Sanjeev Sharma
School of Information Technology, Rajiv Gandhi Proudyogiki Vishwavidyalaya,
Bhopal, M.P., India
e-mail: roshan.it2010@gmail.com, {nishchol,sanjeev}@rgtu.net

S.M. Thampi et al. (eds.), *Recent Advances in Intelligent Informatics*,
Advances in Intelligent Systems and Computing 235,
DOI: 10.1007/978-3-319-01778-5_37, © Springer International Publishing Switzerland 2014

etc. Online social networking provides a platform to build the communities of special interests virtually. It gives a new dimension to business models, inspires viral marketing [2], provides trend analysis and sales prediction in market, assists counterterrorism efforts [3] and acts as a foundation for information sources. In a physical world, according to [4] 83% of people prefer consulting family, friends or an expert over traditional advertising before trying a new restaurant, 71% of people do the same before buying a prescription drug or visiting a place, and even 61% of people prefer consultation with friends before watching a movie. In short, before people buy or take decisions, they talk and listen to others opinion, experience and suggestion. The latter affect the former in their decision making, and are aptly termed as the influential [4].

The Internet is the most influential invention of the 20th century. Since its commercialization in the 1990s, is steadily penetrating almost all dimensions of modern human life. With the emergence of the World Wide Web (WWW) and the evolution of information technologies, however, online social networks reached a new level. Online social networks can be thought as a number of nodes (or persons) connected by a series of links (or relationships). A relationship or link is defined as a pattern of social interaction between two or more persons that involve in meaningful communication and awareness of the probable behaviour of the other person.

Among all nodes in a given social network, it is important and interesting to discover nodes, which can affect the behaviour of their neighbours and, in turn, all other nodes in a stronger way than the remaining nodes, we call such nodes Influential nodes [4]. It is a fundamental issue to find a small subset of influential nodes in the network such that they can attract the largest number of members in a social network.

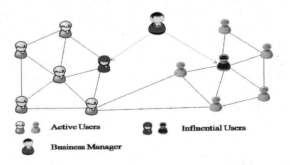

Fig. 1. Structure of Online Social Network

Figure 1 visualizes two different social groups. There is an influential node (selected by any technique) in both groups. If business manager want to advertise product via social network, then instead of interacting with each nodes, he focuses on these influential nodes only. In this way ad spread in to whole network quickly. It reduces the efforts of one to one interaction. The rest of this paper is organized as follows. Section 2 provides the basics of social network. Section 3 elaborates the related work in the area of influential user identification in the online social networks, Section 4 describes the applications of influential user identification in social network and Section 5 includes the conclusion and future work.

2   Definitions of Online Social Network

In this section of paper, definition of social network and its component discussed. The general idea of construction of society is the basis for the social network definitions. A society is not just a simple collection of individuals; it is rather the sum of the relationships that connect these individuals to one another. So the social network can be defined as the finite set of nodes (actors) and connecting edges (relationships). Every researcher defines social network in different forms, stated as follows:

In 1994, Wasserman and Faust [5] proposed a very sociological approach which defines: Actor- An actor is a discrete individual, corporate or collective social unit. Relation- A set of ties of a specific type; a tie is a linkage between a pair of actor. And Social network- The finite set or sets of actors and one or more relations defined on them.

In 2006, Yang, Dia, Cheng, and Lin [6] proposed a very formal way which defines as: Actor- A node in a graph; each node represents a customer. Relation- The undirected, unweighted edges in the graph; each edge represents the connectedness between two nodes. And Social network- An unweighted and undirected graph.

3   Related Work

Social and economic networks have been studied for decades in order to mine useful information not only to organize these networks in a way that maximum efficiency is accomplished but also to understand the role of nodes or groups in a network. By the emergence of social networks in recent years, finding in Influential nodes has absorbed a considerable amount of attention from researchers in this area. In this section, we are going to discuss the earlier methods to identify influential users in a social network, the methods to measure influential power of a user:

3.1   Techniques Based on Structural Measures of Online Social Networks

Social network research community describes a variety of structural measures for the identification of existing Influence in a network. This paper will briefly summarize the well-known centrality measures and number of link topological ranking measures.

Centrality Measures. Structural location of the node is advantageous to find the relative significance in the graph. There are various types of centrality measures of a node used to find the importance in the social network structure.

Degree Centrality. The degree centrality of any node in a graph or network is defined by the count of edges that are incident with it, or the count of nodes adjacent to it. Degree centrality of node in case of directional networks is given in two ways as in-degree and out-degree. The out-degree centrality is defined as [7]:

$$C_{DO}(i) = \sum_{j=1}^{n} a_{ij} \tag{1}$$

Where a_{ij} is 1 in the binary adjacency matrix A if a link from node i to j exists, or it is 0. Similarly, the in-degree centrality is defined as [7]:

$$C_{DI}(i) = \sum_{j=1}^{n} a_{ji} \tag{2}$$

Where i describe the node i and a_{ji} is 1 if a link from node j to i exists, or it is 0.

Closeness Centrality. Closeness centrality is a measure that identifies how fast it will take to flow information from node to all other nodes sequentially. Beauchamp [8] explains, nodes occupying a central position with respect to closeness are very productive in distributing information to the other nodes. Hakimi [9] and Sabidussi [10] developed a measure of closeness centrality as:

$$C_C(i) = \frac{1}{\sum_{j=1}^{n} d(i,j)} \tag{3}$$

Where $d(i,j)$ represent the distance between node i and j, which is the measured minimum length of any path connecting i and j.

Betweenness Centrality. Betweenness centrality counts the number of times a node founds (acts) as a bridge along the shortest route between two other nodes. It was introduced by Linton Freeman [11] as:

$$C_B(i) = \frac{\sum_{i \neq j \neq l} g_{jl}(i)}{g_{jl}} \tag{4}$$

Where $g_{jl}(i)$ is the number of shortest route linking the two nodes j and l containing node i.

Eigenvector Centrality. Eigenvector centrality measures the influence of a node within a network. It is based on the concept that links to high-scoring nodes attract more in comparison to the low-scoring nodes. A variant of Eigenvector centrality measure is Google's PageRank [12].

Let A again be the binary adjacency matrix of the network and $\vec{x}$ be the principal eigenvector corresponding to the maximum eigenvalue θ. The eigenvector centrality for a node i can be defined as a single element of the eigenvector, calculated as [7]:

$$C_E(i) = x_i = \frac{1}{\theta} \sum_{j=1}^{n} a_{ji} x_j \tag{5}$$

Edgevector Degree Centrality. Weighted digraph describes how often a user (node) called another one, or how many text messages sent by him/her. In this case, the element of an adjacency matrix a_{ij} describe the numeric weights of a connection from node i to j. Each weighted graph can be converted into a multigraph, where the same pair of nodes can be connected by multiple edges [13]. In paper [7], edge-weighted degree centrality is defined as:

$$C_{ED}(i) = \sum_{j=1}^{n} (a_{ij} + a_{ji}) \tag{6}$$

Link Topological Ranking Measures. Centrality measures (except eigenvector centrality) did not consider the type of node in the network. There exist some very influential nodes to which connection has more value than to others. With regard to social networks, a connection with a high centrality node might be more valuable than with only one neighbour node [7]. Web search engines leverage link topological ranking by means of HITS (Hyperlink-Induced Topic Search) and PageRank algorithms.

HITS. Kleinberg [14] proposed a Web search algorithm called HITS which identifies authoritative pages and a set of hub pages. An iterative algorithm is used to find the equilibrium values for the authority and hub weights of a web page or node in a network, respectively. Equilibrium is reached if the difference of the weights between two iterations is less than a threshold value. For each page i a nonnegative authority weight $C_A(i)$ and a nonnegative hub weight $C_H(i)$ are exist. The weights of each type are normalized so their squares sum to 1 and are defined as:

$$C_A(i) = \sum_{j=1}^{n} a_{ji}\, C_H(j) \tag{7}$$

$$C_H(i) = \sum_{j=1}^{n} a_{ij}\, C_A(j) \tag{8}$$

Where a_{ji} is 1 if an edge from node j to i exists otherwise 0.

PageRank Algorithm. The PageRank algorithm, which was initially developed by Brin and Page [12], the founders of the Google search engine; they maintain only a single metric for each web page. The so called PageRank is transmitted from the source page to the link target, and the value depends on the PageRank of the source page. The PageRank of the page or node i is the sum of contributions from its incoming links or edges. A constant damping factor f is the probability at each page that the "random surfer" will get bored and requests another random page. Additionally $(1-f)$ is added to each node. This is done because if a node has an out-degree of zero then his PageRank would be zero. This zero-value would be passed down to the real node. To avoid this, a constant is added to the PageRank. The PageRank can be defined as:

$$C_{PR}(i) = (1-f) + f \sum_{z \in Mi} \frac{C_{PR}(j)}{C_{Do}(j)} \tag{9}$$

Where Mi is the set of source pages that link to i and $C_{Do}(j)$ is the out-degree of page j, as described in the previous subsection. The damping factor f is often set to a value of 0.85 [12].

3.2   Techniques Based on the Diffusion Model of the Online Social Network

The diffusion model of the social networks also helps in the identifying the influential user in networks. The diffusion models [15] are classified into three categories: linear threshold model, independent cascade model and a model that combines both the features of the linear threshold model and independent cascade model.

Linear Threshold Model. The model proposed by Granovetter and Schelling [16] is based on the use of thresholds. Linear Threshold Model (LTM) is a variant of the original model. In this model a node u is influenced by its neighbour v with weight $b(u,v)$, subject to the constraint $\sum_v b(u,v) < 1$.

Each node has a predefined threshold value $\theta u \in [0,1]$. This show how difficult is to influence the node u when their neighbours are active nodes. Normally the threshold value is chosen randomly with uniform distribution, but in some cases entire network has a uniform threshold ($\theta u = 1/2$). the process starts with a set of active nodes. The diffusion process continues in discrete steps, where the nodes that become active in step t, remain active until the end. At each step, every node u whose neighbour's total weight is at least θu is activated ($\sum_v b(u,v) \geq \theta u$).

Independent Cascade Model. The Independent Cascade Model (ICM), proposed by Goldenberg, Libai and Muller [17] starts with a set of active nodes Ao, then the process triggers a cascade of activations in discrete steps. When a node u becomes active in step t, then it has only one chance to activate each inactive neighbour v with a success probability $P(u,v)$. the order of attempts to activate inactive neighbours is arbitrary. Independently of the success or not of the activation, u cannot make further attempts to activate v until the end of the process. The process runs until no more activation is possible. This process is called independent because the activation of any node v is independent of the history of activations.

3.3   A Technique Based on Community Mining in Online Social Networks

Kempel [18] state that optimization problem of influence maximization is NP-Hard. Accordingly to solve this type of problem, Greedy algorithms with provable approximation may give better outcome. But Greedy methods are expensive in computation, so as a result it is not practicable to social network. Yu Wang, Gao Cong, Guojie Song, Kunqing Xie [19] proposed novel method called "Community based Greedy algorithm for mining top-K influential nodes" which divides the network into a number of communities, and then selects individual community to find top-K influential nodes. The community structure is a main property of social network features: Individuals within a community have frequent contact; in contrast, individuals across communities has much less contact with each other and thus is less likely to influence each other. This property suggests that it might be a good approximation to identify influential nodes within communities instead of the whole network. This work gives several directions to expand research in construction of location based social network to find influential over time.

3.4   Techniques Based on Content Mining in Online Social Networks

In early studies, influential users identified based on the structural properties of a social network. In some social networking platforms (for e.g. blog network) structural properties may not exhibit the actual influence between users. In online social networks, user's content can bring on many activities that address its dynamic nature. This section includes content based influential user identification techniques.

Topic Level Influence. Lu Liu, Jie Tang, Jiawei Han, Meng Jiang, and Shiqiang Yang proposed a method [20] that brings out the idea of mining the strength of direct and indirect influence. This proposed a generative graphical model that uses heterogeneous link information as well as the textual content related to each node in the online social networks to identify topic-level direct influence. Based on this learned direct influence, proposed topic-level influence propagation and aggregation algorithms apply to derive the indirect influence between various nodes. This paper focuses on mining topic level influence and propagation method to propagate whole network. In future by basis of this method behaviour prediction model can be employed in a social network.

ExpertRank Algorithm. G. Alan Wang, Jian Jiao, Alan S. Abrahams, Weiguo Fan, Zhongju Zhang presents a novel algorithm "ExpertRank" [21], to identify a topic-aware expert for online knowledge communities. The ExpertRank algorithm evaluates expertise based on both documents-based relevance and one's authority in his or her knowledge community. For this they modify the PageRank [12] algorithm to evaluate one's authority, so that it reduces the effect of certain biasing communication behaviour in online communities. ExpertRank algorithm explores three different expert ranking strategies for combining document-based relevance and authority: Linear Combination, Cascade Ranking and Multiplicative Scaling. This method is proficient to develop expert databases or organizational memory systems for providing key knowledge exchange among employees.

Content Power Users on Blog Network. Seung-Hwan Lim, Sang-Wook Kim, Sunju Park, and Joon Ho Lee [22] proposed a new method to identify content power user in Blog network. There are special users who bring on other users to actively utilize blogs, these users called influential users. Identifying such influential users in the blogosphere is important when establishing new business policy for the blog network. This paper defines the user, whose content exhibit significant influence over other users as content power users (CPUs) and proposed a method of identifying them. They analyse the performance of the proposed method by applying to an actual blog network and compare results with pre-existing methods for identification of power users. The experimental results of their analysis, demonstrate that the definition of content power user is adequate to address the dynamic nature of the blogosphere and the main concerns of the blog industry.

3.5   A Technique for Identifying Influential Users in Micro-blog Marketing

Micro-blog marketing has become an important business model for online social networks now a day. Micro-blog marketing make possible to publish advertisements directly to customers for attracting to buy products. Fei Hao, Min Chen, Chunsheng Zhu, Mohsen Guizani proposed a method [23] to discover the influential user in micro-blogging site (for e.g. Twitter, Sina Weibo). They try to analyse the influence of nodes in a micro-blog network and proposed the "Community Scale-Sensitive Max Degree (CSSM)", an algorithm for maximizing the influence when placing advertisements. This work contributes mainly in the First, the influence analysis of the various nodes in micro-blog social networks. Second, a Community Scale-Sensitive Max Degree (CSSM) based influence maximization algorithm and third, an evaluation of the CSSM algorithm, using very popular micro-blog service Twitter dataset.

They observed that the influence of a node depends on following three matrices: first, centrality based on node degree. Second, sum of neighbour's degree and third, attributes of nodes. The attributes of a node in the micro-blog network, includes activity degree, interaction degree and social prestige.

3.6   A Technique Based on Link Polarity in Online Social Networks (Blogs)

The main objective of above discussed techniques to understand the spread patterns of influence in a social network. All these approaches used to identify the influence between individual users, but do not take into account the question of what kind of influence inclined among them. So Keke Cai, Shengha Bao, Zi Yang, Jie Tang, Rui Ma, Li Zhang, Zhong Su introduced novel approach, an "Opinion Oriented Link Analysis Model (OOLAM)", [24] to characterize user's influence personae. In particular, three kinds of influence personae which take place widely in social network includes: Positive Persona, Negative Persona and Controversy Persona. Within the OOLAM model two factors, i.e., opinion consistency and opinion creditability are defined to capture the persona information from public opinion perspective. Extensive experimental studies have been performed to make obvious the effectiveness of the proposed approach to influence personae analysis using real web datasets.

4   Applications

The significant role of influence is studied extensively in sociology, communication, marketing and political science and in understanding peer pressure, trend analysis, obedience and leadership.

4.1   Opinion Leader Finding

Finding dominating people in societies has been a key question for marketing policy makers, political managers, social researchers, security analysts, engineers and computer scientists. Since any society can be considered as a network, network analysis has provided significant insight in this area. To find out the Leader, community based influential user identification approach applied.

4.2   Viral Marketing

Influence maximization has the obvious application in viral marketing [15] through social networks; in this company try to promote their products and services through the word-of-mouth propagations among friends in the social networks. The ultimate objective of the marketers to create viral messages that strongly convey to the influential users to spread in a short period of time.

Table 1 Key features of various Influential identification techniques

Name of Technique	Year launched	Key Features
Centrality Measures	1966	Identify the relative importance of Nodes in Network.
HITS Algorithms	1998	Identify Authority Pages and Hub Pages in Network.
PageRank Algorithms	1998	Maintain a single metric for information of all Web Pages.
Linear Threshold Model	1978	Focuses on Threshold (whole) behaviour of Nodes.
Independent Cascade Model		Focuses on Individual's Interaction in Network.
Community Modelling	2010	Efficient over Greedy method and orthogonal to existing algorithms of Influential detection.
Topic Level Algorithm	2010	Consider the presence of Indirect Influence with Direct Influence in Online Social Network.
ExpertRank Algorithm	2013	Document based relevance and Authority of Individuals.
Content Power User	2011	Illustrate the dynamic nature of Online Social Networks.
CSSM Algorithm	2012	Includes Activity degree, Interaction degree and Social prestige of the user.
OOLAM Algorithm	2011	Opinion consistency and Opinion creditability are used to capture the persona of user.

5   Conclusion and Future Work

This paper, presents the brief knowledge about the previous work that has been done in the field of identifying the influential users in the online social networks. This paper includes the techniques based on the topological structure of the social network as well as content of users. The topological analysis of the social networking makes use of graph theory but, sometimes it becomes complex and typical when the size of the social networks increases, particularly in present scenario. A methodology based on usage of the diffusion history of the activities of the user is also useful for identifying the Influential. In future we are going to propose a novel approach which includes topological measures, content power and link polarity values in account to identifying influential users in online social network. Another major area for extending works by basis of temporal and location based methodologies.

References

1. Aggarwal, C.C.: Social Network Data Analytics, pp. 1–14. Springer Science Business Media, LLC (2011)
2. Richardson, M., Domingos, P.: Mining knowledge-sharing sites for viral marketing. In: ACM SIGKDD International Conference on Knowledge Discovery and Data Mining, pp. 61–70. ACM Press, New York (2002)
3. Coffman, T., Marcus, S.: Dynamic Classification of groups through social network analysis and HMMs. In: Proceedings of IEEE Aerospace Conference (2004)
4. Keller, E., Berry, J.: One American in ten tells the other nine how to vote, where to eat and, what to buy. They are The Influential. The Free Press (2003)
5. Wasserman, S., Faust, K.: Social Network Analysis: Methods and Applications. Cambridge University Press, New York (1994)
6. Yang, W.S., Dia, J.B., Cheng, H., Lin, H.T.: Mining Social Networks for Targeted Advertising. In: Proceedings of the 39th Hawaii International Conference on Systems Science, vol. 6, p. 137a. IEEE Computer Society (2006)
7. Kiss, C., Bichler, M.: Decision Support Systems 46, 233–253 (2008)
8. Beauchamp, M.: An improved index of centrality. Behavioural Science 10, 161–163 (1965)
9. Hakimi, S.: Optimum locations of switching centers and the absolute centers and medians of a graph. Operations Research 12 (1965)
10. Sabidussi, G.: The centrality index of a graph. Psychometrika 31, 581–603 (1966)
11. Freemann, L.C.: A set of measures of centrality based on betweenness. Sociometry 40, 35–41 (1977)
12. Brin, S., Page, L.: The anatomy of a large scale hyper textual web search engine. In: WWW Conference, Australia (1998)
13. Newman, M.E.J.: Analysis of weighted networks. Physical Review E 70
14. Kleinberg, J.: Auth. sources in hyperlinked environment. In: CM-SIAM Symposium on Discrete Algorithms (1998)
15. Singh, S., Mishra, N., Sharma, S.: Survey of Various Techniques for Determining Influential Users in Social Networks. In: IEEE International Conference on ETCCN, India, pp. 398–403 (2013)
16. Granovetter, M.: Threshold models of collective behaviour. American Journal of Sociology 83(6), 1420–1443 (1978)
17. Goldenberg, J., Libai, B., Muller, E.: Talk of the network: A complex systems look at the underlying process of word-of-mouth. Marketing Letters, 211–223 (August 2001)
18. Kempel, D., Kleinberg, J., Tardos, E.: Maximizing the spread of influence through a social network. In: ACM SIGKDD, pp. 137–146 (2003)
19. Wang, Y., Cong, G., Song, G., Xie, K.: Community-based Greedy Algorithm for Mining Top-K Influential Nodes in Mobile Social Networks. In: KDD 2010, Washington, July 25-28 (2010)
20. Liu, L., Tang, J., Han, J., Jiang, M., Yang, S.: Mining Topic-level Influence in Heterogeneous Networks. In: CIKM 2010, Toronto, Canada (October 2010)
21. Alan Wang, G., Jiao, J., Abrahams, A.S., Fan, W., Zhang, Z.: ExpertRank: A topic-aware expert finding algorithm for online knowledge communities. Decision Support Systems 54, 1442–1451 (2013)

22. Lim, S.-H., Kim, S.-W., Park, S., Lee, J.H.: Determining Content Power Users in a Blog Network: An Approach and Its Applications. IEEE Transactions on Systems, Man, and Cybernetics – part-A: System and Human 41(5), 853–862 (2011)
23. Hao, F., Chen, M., Zhu, C., Guizani, M.: Discovering Influential Users in Micro-blog Marketing with Influence Maximization Mechanism. In: Globecom 2012 - Ad Hoc and Sensor Networking Symposium (2012)
24. Cai, K., Bao, S., Yang, Z., Tang, J., Ma, R., Zhang, L., Su, Z.: OOLAM: an Opinion Oriented Link Analysis Model for Influence Persona Discovery. In: WSDM 2011, Hong Kong, China, February 9-12 (2011)

k-Fault Tolerant Topology Control in Wireless Sensor Network

Suman Bhowmik, Deepsikha Basu, and Chandan Giri

Abstract. In recent years research in wireless sensor network has gained immense importance. Fault tolerance is an important issue to be addressed in WSN application. Fault tolerance can be achieved at various levels of applications in WSNs. Specifically the categories of applications are: node placement, topology control, target and event detection, data gathering and aggregation and sensor monitoring. This paper proposed a distributed algorithm to control the topology of network to make it fault tolerant.

1 Introduction

In recent years Wireless Sensor Networks (WSNs) have numerous applications. The low cost and low powered sensor devices, can be deployed in various medium such as in the air, in vehicles, on bodies, under water and inside buildings. These nodes are prone to failure due to energy depletion, communication link errors, environmental disturbances, malicious attacks etc. Most wireless sensor nodes are battery operated are prone to failure. The main goal of any WSN application is to send gathered data to sink node. So, network topology should be fault tolerant as far as possible. The network is usually divided into clusters with a cluster head provides path to the sink node. Two clusters are connected via a special node called connector. One way to make the topology fault tolerant is to maintain alternative path information along with the most suitable one. In this paper we proposed a fault tolerant topology control algorithm, which maintains k number of path information along with

Suman Bhowmik
Dept. of CSE, College of Engg. & Mgmt. Kolaghat, India
e-mail: bhowmik.suman@gmail.com

Deepsikha Basu · Chandan Giri
Dept. of IT, Bengal Engg. & Sc. University, Shibpur, India
e-mail: {deepu.basu3,chandangiri}@gmail.com

S.M. Thampi et al. (eds.), *Recent Advances in Intelligent Informatics*,
Advances in Intelligent Systems and Computing 235,
DOI: 10.1007/978-3-319-01778-5_38, © Springer International Publishing Switzerland 2014

the best path, with reduced communication cost. Sensing the failure of any CH or connector node the next alternative path can be used, thus making the topology fault tolerant. This section gives the introduction and performs literature survey. Section 2 describes the system model used and defines the problem statement. In Section 3 our proposed algorithm is discussed. Simulation results are shown in Section 4. Finally the work is concluded in Section 5.

Literature Survey: In recent years a lot of work has been done on fault tolerant sensor network. Authors in [1] have presented multi state fault tolerant topology control algorithm which reduces the transmission power of nodes to the required power level and then activates and deactivates the different existing paths to maintain k number of route from sensor nodes to base station. In [2] authors proposed a Distributed Geography-based Fault Tolerant topology control algorithm (DGFT). Scale free characteristic of complex networks is introduced into the topology of large scale WSN. DGFT enables wireless nodes to define the topology by neighbour relationship, based on certain rule. In [3] authors have proposed a simple global algorithm ($GAFT_k$) to preserve k-connectivity and reduces the maximal transmission power (TP). Based on $GAFT_k$, they have proposed a simple local algorithm ($LAFT_k$) which preserves k-vertex connectivity while maintaining bi-directionality of the network. These two together provide the k-connected topologies so that the network is fault tolerant. Authors in [4] have presented a distributed topology control algorithm to calculate the per-node minimum transmission power, so that (a) reachability between any two nodes is guaranteed to be the same as in the initial topology; and (b) nodal transmission power is minimized to cover the least number of surrounding nodes. Many researchers have worked on the unreliability and asymmetry of wireless links [5]. According to [6, 7, 8] the low-power wireless links can be classified into three regions : connected, transitional and disconnected. The connected region is characterised by stable, symmetric links, in transitional region there are unreliable and asymmetric links and the disconnected region presents no link for transmission.

2   System Model and Problem Statement

System Model: In this work it is assumed that the sensor nodes are deployed randomly in a $r \times r$ area. The nodes in the network are randomly deployed and divided into several clusters. Each cluster has a cluster head (CH) which communicates with the other neighbouring cluster heads or base station. The cluster heads send data coming from member nodes either directly or indirectly to the base station. The cluster heads cannot communicate with each other directly since the distance between them is greater than τ_{fd} [9]. Here the system followed the fuzzy communication model as in [9]. So the nodes, which are present in the "far" region [9] between two cluster heads, are the connector nodes between them, as shown in Fig.1. In this way two cluster heads communicate with each other through this connector node. Running any clustering algorithm with above connectivity model, the whole network will be partitioned into a set of interconnected clusters. Thus the set of cluster heads, connector nodes and base station forms a connected graph $G = \{V, E\}$ (see Fig. 2), where $V = \{v_1, \ldots, v_n\}$, the set of cluster heads, connectors and base station and E is the set of communication links between cluster heads through connectors as well as to base station.

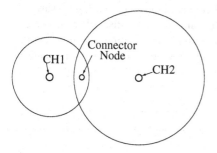

Fig. 1 Sensor Nodes   **Fig. 2** A Sensor Network

Problem Statement: A sensor network can be represented as a graph as shown in Fig.2. In the figure there are six cluster heads $CH_1 - CH_6$, five connector nodes $S_1 - S_5$ and one base station BS. The cluster heads (CHs) are sources and the base station is the sink. Our objective is to design a fault tolerant topology control algorithm which will take the similar type of graph as input and find k different paths from each cluster head to base station. If there is any failure of node (cluster head or connector) the network will still be able to find path to base station thus making the network fault-tolerant. In short algorithm will find k number of paths from every cluster head to the base station. If one route fails the alternate path can be used to move data. Therefore, the network is upto k fault tolerant.

3 Proposed Algorithm

Proposed distributed algorithm (see Algo. 1) finds k alternative paths from each cluster head to the base station. Each active node (cluster head, connector and base station) individually execute the algorithm, pass information from one another. Using those information each cluster head build the k alternative paths to base station. When a cluster head sense that the currently used path is not available use the next best path to forward data. Base station initiates the algorithm. First the algorithm creates an *"eligibility"* message with its own $id = 0$ and the path cost to the base station set to one (eligibility.costToBase = 1) and send the message to all cluster heads which are directly connected with it. The cluster heads at the beginning initializes the variables *duplicate* = FALSE and *no_of_path* = 0. A cluster head when receives a new *"eligibility"* message set the sender as its parent on the best path to the base station with path cost defined in *"eligibility"* message and increments the *no_of_path* variable by one. It then replaces the sender *id* by its own *id* and increments the path cost by two to reflect that the next cluster is at two hop distance (via the connector node), send the modified message to all of its outgoing links except the one from which it has received the message and makes $duplicate = TRUE$ to identify that it has already received an *"eligibility"* message. If a cluster head receives a duplicate *"eligibility"* message, stores the sending cluster information (sender, connector and path cost to base station) to the alternate path lists sorted by pat cost in ascending order and discard the message. It then increments the no_of_path variable by one. If the variable no_of_path reaches the value k, then it discards further messages received. This ensures that the CH stores k paths only. Connector nodes when receive an eligibility message from one cluster head simply forward the message to the other cluster head.

Algorithm 1. Topology Control Algorithm

Input : Cluster Head Set $(C) = \{C_1, C_2, ...C_i, ...C_N\}$, Number of paths (k)
Output: k-paths from each C_i to the base station
Base Station do **begin**
    Create *eligibility* Message **begin**
        eligibility.senderID = 0;
        eligibility.costToBase = 1;

    Broadcast *eligibility* Message;

Cluster Heads C_i do **begin**
    duplicate = *False* ;
    no_of_path = 0;
    receive an *eligibility* message in incoming link;
    if *no_of_path* < k **then**
        if *duplicate* == *False* **then**
            duplicate = *True*;
            parent.ID = *eligibility.senderID*;
            parent.path_cost = *eligibility.costToBase*;
            no_of_path + +;

            eligibility.senderID = C_i.*ID*;
            eligibility.costToBase + = 2;
            Send *eligibility* message all outgoing links;
        else
            alternate[*no_of_path*].*ID* = *eligibility.senderID*;
            alternate[*no_of_path*].*path_cost* = *eligibility.costToBase*;
            no_of_path + +;
            Discard the Message;

    else
        Discard the Message;

Connector Node do **begin**
    Receive *eligibility* Message from one cluster head;
    Forward *eligibility* Message to the other cluster head;

---

4 Simulation Results

The proposed algorithm is simulated with varying $k(=2,3,4)$ and have studied the results. Table 1 shows the parameters used for simulation. Fig. 3(a) shows that with less number of cluster head failure, the number of CHs working are almost same

Table 1 Simulation Parameters

Deployment field	$200m \times 200m$	Number of alternative paths (k)	2, 3, 4
Number of Clusters	20, 40	Number of Active Nodes	120, 430

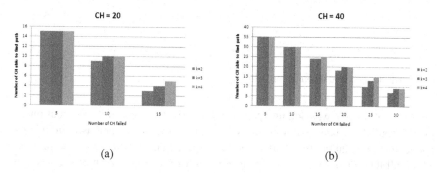

(a)   (b)

Fig. 3 Cluster Heads failure

for the three cases. But with the increase in CH failure for $k = 4$ the performance is better than the others (Fig.3(b)) where number of CHs used is 40.

In 2nd category (Fig. 4) connection between CHs fails. With 20 CH Fig. 4(a) shows that increasing k improves the performance. But with more number of CHs ($=40$), Fig. 4(b) reveals that increasing k does not always give better result ($k = 3$ performs better that $k = 4$). Thus with the increase of cluster heads more number of alternate paths may not be a good option if there are connection failures.

The 3rd category (Fig. 5) shows the combined effect of both CH and connector failure for 20 & 40 CH deployed. The performance increases with increasing k (see Fig. 5(a)). But for connection failure, increasing k does not give better result (see Fig. 5(b)).

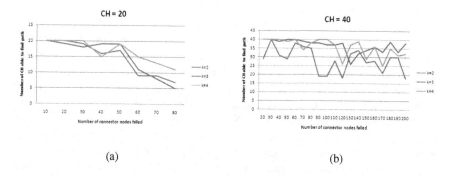

(a)   (b)

Fig. 4 Connector Nodes failure

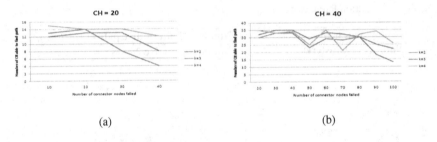

(a)   (b)

Fig. 5 Cluster heads and connector failure

Fig.6 gives the lower bounds of connection failure. There are two situations. In the first one there is no failure of cluster heads. So with the increasing number of connector nodes failure the chart shows the lower bounds of connection failure upto which all the cluster heads are able to find path for $k = 2, 3, 4$. In the second one, 5 CHs have failed. So the remaining CHs are working. The second chart gives the lower bounds of connection failure for the three values of k. The figure shows that lower bounds for the second chart is less than the first one. Moreover, in both the cases $k = 3$ is better than $k = 2$, and $k = 4$ is better than both $k = 2$ and $k = 3$.

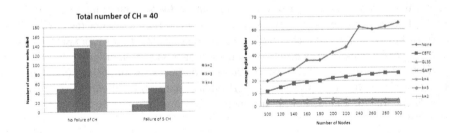

Fig. 6 The lower bound of failure   **Fig. 7** Comparison with other algorithms

Fig.7 shows the comparisons between the different topology control algorithms. **Logical neighbours** of a node is the number of nodes directly connected after the implementation of the topology control algorithm. Less number of neighbours indicates less interference between the nodes. The graph shows that CBTC $_k(\Pi/3)$ (Cone-Based Topology Control) [2] algorithm is better than None (where no topology control is implemented), and GLSS (Global Spanning Sub-graph) [2] is better than the two above [3]. The GAFT algorithm, however, is very close to GLSS but perform better. Our algorithm gives lesser average logical neighbours than all of them. The algorithm is simulated with $k = 2, 3, 4$ and for each case the proposed algorithm performs better than the existing approaches as shown in Fig. 7.

5 Conclusion

In this paper we have presented a topology control algorithm which makes the sensor network fault tolerant. Here the base station initiates the algorithm and the cluster heads store k paths. So the nodes can tolerate upto $k - 1$ number of faults. In this algorithm the cluster heads communicate with each other via the connector nodes. This work is mainly in the application layer where the topology is controlled by maintaining a set of candidate paths. Thus there are ample scopes in the near future where cross layer solutions can also be expected.

References

1. Forghani, A., Masoud Rahmani, A.: Multi State Fault Tolerant Topology Control Algorithm for Wireless Sensor Networks. In: Proc. of the IEEE 2nd Int. Conf. on Future Gen. Comm. and Net, vol. 1, pp. 433–436 (2008)
2. Wang, L., Jin, H., Dang, J., Jin, Y.: A Fault Tolerant Topology Control Algorithm for Large-scale Sensor Networks. In: Proc. of 8th Int. Conf. on Parallel and Dist. Comput., App. and Tech (PDCAT 2007), pp. 407–412 (2007)
3. Zhang, J., Chen, J., Wangt, Y., Xiao, Y., Sun, Y.: A Simple Algorithm for Fault-Tolerant Topology Control in Wireless Sensor Network. In: Proc. of IEEE 19th Int. Symp. on DOI, pp. 1–5 (2008)
4. Liu, J., Li, B.: Distributed Topology Control in Wireless Sensor Networks with Asymmetric Links. In: Proc. of IEEE Glob. Telecomm. Conf. (GLOBECOM), vol. (1), pp. 1257–1262 (December 2003)
5. Zamolla, M., Krishnamachari, B.: An Analysis of Unreliability and Asymmetry in Low-Power Wireless Links. Proc. of ACM Trans. on Sensor Net 3(2), Article 7 (June 2007)
6. Ganesan, D., Krishnamachari, B., Woo, A., Culler, D., Estrin, D., Wicker, S.: Complex behavior at scale: An experimental study of low-power wireless sensor networks. Tech. Rep. UCLA/CSD-TR, pp. 2–13, Dept. of Comp. Sc., Univ. of California at Los Angeles (2003)
7. Zhao, J., Govindan, R.: Understanding packet delivery performance in dense wireless sensor networks. In: Proc. of 1st Int. Conf. on Embedded Networked Sensor Sys., pp. 1–13 (2003)
8. Woo, A., Tong, T., Culler, D.: Taming the underlying challenges of reliable multihop routing in sensor networks. In: Proc. of 1st Int. Conf. on Embedded Networked Sensor Sys., pp. 14–27 (2003)
9. Bhowmik, S., Giri, C.: Fuzzy Communication Channel Model for Sensors in Wireless Sensor Network. In: Proc. of the IEEE Int. Conf. on Comm., Devices and Intelligent Sys (CODIS), pp. 254–257 (December 2012)

Selection of On-line Features for Peer-to-Peer Network Traffic Classification

Haitham A. Jamil, Aliyu Mohammed, A. Hamza, Sulaiman M. Nor, and Muhammad Nadzir Marsono

Abstract. The selection of features plays an important role in traffic detection and mitigation. Feature selection methods can significantly improve the computational performance of traffic classification. However, most of the selected features cannot be used online since these features can only be calculated after the completion of a flow. Another requirement is that an optimum number of features must be chosen so as to classify the P2P traffic within the minimum time possible. Out of more than ten feature selection algorithms, it was discovered that Chi-squared, Fuzzy-rough and Consistency-based feature selection algorithms were the best for P2P feature selection. The proposed algorithm gives better feature subset for online Peer-to-Peer (P2P) detection using machine learning (ML) techniques. The process of validation and evaluation were done through experimentation on real network traces. The performance is measured in terms of its effectiveness and efficiency. The experimental results indicate that J48 classifier with online subset feature selection produces a higher accuracy (99.23%) and low testing time (1.12 second).

1 Introduction

Bandwidth-intensive applications such as Peer-to-Peer (P2P) applications have changed the dynamics of the Internet. These applications consume most of the Internet bandwidth with balanced traffic in both directions [19]. This affects the performance of traditional Internet applications and may raise some security problems on the network and to the management of the organizations [5]. Therefore, traffic

Haitham A. Jamil · Aliyu Mohammed · A. Hamza · Sulaiman M. Nor ·
Muhammad Nadzir Marsono
Faculty of Electrical Engineering, UTM, 81310 Skudai, Johor, Malaysia

Haitham A. Jamil
University of Elimam Elmahdi, Kosti, Sudan
e-mail: {haitham.jamil,maliyyu,aehihamza3}@fkegraduate.utm.my,
    {sulaiman,nadzir}@fke.utm.my

S.M. Thampi et al. (eds.), *Recent Advances in Intelligent Informatics*,
Advances in Intelligent Systems and Computing 235,
DOI: 10.1007/978-3-319-01778-5_39, © Springer International Publishing Switzerland 2014

management plays a vital role in monitoring the performance fairness of the network [6]. Recently, many researchers have proposed hybrid classifiers, e.g. [20, 10]. One of the most influential parameter in traffic classification is the type of features that can discriminate P2P traffic from the background traffic.

Feature selection is a task of choosing the smallest subset of features that can identify instances of a class effectively and efficiently. By using the most relevant features, traffic classification can be improved in terms of the accuracy and computational performance [11]. An extensive amount of research has been carried out over the last two decades to obtain reliable methods for feature selection [20]. These methods differ in their evaluation of feature subsets. The evaluation measures are determined by the physiognomies of the features and the specific objectives of the classifier [14]. A feature subset for which the average class overlap is minimal is considered to be an optimal feature subset.

Presently, most of the researches focus on the effectiveness of different classifiers [21]. However, the effect of using different sets of statistical and heuristic features has not been investigated in-depth. Peter and Maarten [16] discovered that feature selection has a positive value to improve the performance than the choice of the classification algorithm. Zhao et al. [21] indicated that some types of selected features cannot be utilized in online traffic classification since the features cannot be determined before a flow is completed.

This paper is an extension of the previous work done in [7]. The purpose of the previous work is to determine the optimal feature selection algorithms for P2P traffic classification. However, the paper does not take into account on-line P2P classification. This paper describes how on-line P2P classification using J48 is carried out based on the combination of features from the optimal feature selection algorithms. Then, the feature subsets that are suitable and viable to be extracted from on-line traffic are identified. The effectiveness and efficiency of J48 classifier is evaluated using this optimal features subset. The results show that J48 classifier using the proposed optimal features has high accuracy (99.23%) and low testing time (1.12s). This feature subset can be extracted from on-line flow packets and can be easily applied for on-line traffic classification.

The remainder of this paper is as follows. Section two introduces some related works on the feature selection using Machine Learning. Section three describes the feature selection algorithms used in our approach and the methods of extracting the online features for P2P traffic classification. Moreover, the section gives an overview about J48 algorithm. Section four details the datasets and the evaluation metric. The results of the implementation are given in Section five. Section six concludes our work and examines the future work.

2   Related Works

Several research works have been done in the field of feature selection and traffic classification. Works in [4, 6] review the state-of-the-art approaches, techniques and available application for the classification of network data. Moore in [12] used Fast Correlation Based Filter (FCBF) for feature reduction and Nave Bayes algorithm to assess the feature reduction effect. The result of the overall classification accuracy based on the reduced sets is 84.06%, which is much better than using all features. On the other hand, the work in [2] also uses the Nave Bayes algorithm with only the first five packets of the flow. The resultant effect is that only two feature, the size and direction of the initial data packet with TCP connection provides the distinction for all the applications. Jun et al. [9] applied two optimal features subsets to provide a good traffic classification accuracy. The accuracy of using the flow features subsets on Support Vector Machine (SVM) classifier is 70% while the training time was reported at 40 seconds. Yang et al. [18] used random search algorithm for features reduction to identify P2P traffic by using SVM. However, this work did not include UDP traffic although P2P traffic consists of both TCP and UDP packets.

Feature selection algorithms were used to choose the best feature subsets in [13] but this process consumes much time. Moreover, most of these features are hard to be extracted from on-line traffic for online traffic classification. Auld et al. in [1] used 249 features derived from packet streams consisting of one or more packet headers. A full description of Moore features is available in [13]. One of the features selection algorithms available in WEKA tool is a correlation-based feature selection (CFS) which is a subset heuristic evaluation that takes into account the usefulness of separate features for predicting the class along with the degree of inter-correlation among them. It assigns high scores to subsets containing features that are greatly correlated with the class and have low inter-correlation with each other. The other algorithm is the Consistency-based feature selection (CON) which evaluates all of the subset of features concurrently and selects the optimal subset.

In recent time, researchers put more effort on the application of ML algorithms for the achievement of high level classification accuracy. However, very limited efforts are centered towards the selection of feature subsets for online classification. This paper is intended to study the flow features of the internet traffic for online classification. Subsequently, three feature selection algorithms are chosen to select a set of features that can be applied to detect P2P network traffic in term of accuracy and computational performance. The set of features is examined in order to choose the optimal subset of online features to discriminate P2P over non P2P. Supervised ML algorithms and controlled data are used for validation and verification.

3 Methodology

In this section, we introduce the feature selection algorithm, the methods of extracting the online features and J48 algorithm used as the traffic classifier.

3.1 Feature Selection Algorithms

Feature selection is the process of choosing an optimal subset of features that can be used efficiently and effectively to classify network traffic. In term of computational performance, feature selection is important since it reduces the modeling and testing time which speeds up the classifier performance [12].

In this subsection, initially more than ten feature selection algorithms have been evaluated (CSF, CON, Filter-Sub, Fuzzy-rough, Symmetrical-Uncert, Chi-squared, Info Gain, Relief, Principal and Latent-semantic). Then we select the top three algorithms with higher computational performance to be used as the first stage in our approach. The three algorithms are Chi-squared, Consistency and fuzzy-rough. Moreover, these algorithms used by [3, 10, 11] and information about them can be found in [17].

- Chi-Squared is a feature selection algorithm that is based on 2 statistic [11]. It is built on the top of the entropy method and evaluates features individually by measuring their chi-squared statistics with respect to their classes [10]. The 2 value of an attribution is defined as:

$$x^2 = \frac{(f_0 - f_e)^2}{f_e} \tag{1}$$

where f_0 is the observed frequency, f_e is expected frequency. After calculating the x^2 value of all considered features, the features can be ranked based on these values. The features that have higher rank are more important than the others.

- The consistency-based algorithm evaluates simultaneously feature subsets and selects the smallest subset of features that can detect instances of a class as consistently as the complete feature set [21].The consistency measure is applied to the feature selection task as follows. Given a candidate feature subset S we calculate its inconsistency rate IR(S). If IR(S) where is a user given inconsistency rate threshold, the subset S is said to be consistent [3].

- Fuzzy-rough algorithm provides a filter-based tool by which knowledge may be extracted from a domain in a concise way; retaining the information content whilst reducing the amount of knowledge involved. The main advantage that rough set analysis has is that it requires no additional parameters to operate other than the supplied data. The reduction of attributes is achieved by comparing equivalence relations generated by sets of attributes. Attributes are removed so that the reduced set provides the same predictive capability of the decision attribute as the original [8].

3.2   Extracting On-line Features

Flow features can be either behavioural or statistical calculated from numerous information of a flow. Traffic classification uses these features to assign an object to a class [21]. Moore in [12] uses 248 features derived from packet streams consisting of one or more packet headers. A full description of Moore features is available in [2]. Here, we try to select a subset of feature that can be used towards online P2P detection.

Table 1 A summary of the 248 flow features

Viability	Type	Features
Off-line	Flow(70)	duration, packet-count, total bytes,...
	Time(32)	inter-arrival time (med, var, mean, min, max),...
	Size(82)	payload size, packet size, data control size (med, var, mean, min, max) , ...
	Flag(30)	SYN, FIN, ACK,...
On-line	Phy. Entities (2)	Port number at server, port number at client
	Flow(8)	Uplink and downlink first quartile of total bytes in IP packet, control data, payload data,...
	Time(3)	First quartile inter-arrival time, uplink First quartile inter-arrival time, downlink First quartile inter-arrival time,...
	Size(9)	First quartile of bytes in IP packet, control data, payload data, ...
	Flags(10)	SYN, FIN, ACK, ...

In this paper, our method focuses on reducing the feature set to improve the classification performance in term of accuracy and cost. This is in addition to selecting a feature set that can be used for online classification. On-line feature is the feature that can be calculated on the fly, for example, first quartile inter-arrival time. However, for the rest of features in table 1, they are offline features since they cannot be calculated before the flow is ended, for example, flow duration.

In order to realize the P2P online traffic classification, a feature subset is created based on Chi-square, Consistency and fuzzy-rough algorithms as mention above. These algorithms are used because of their performance to select positive features and avoiding the heavy resource needed to use the full set of features. Then, optimal feature subset is built for on-line P2P traffic classification by examining each feature from the candidate features set, if it can be use on-line. Fig.1 illustrates these two phases Fig 2 describes the process of extracting online features. We first use the labeled dataset D as input for each feature selection algorithm as mentioned above. Then the candidate features of these algorithms are examined in such way to select the online features. At last, we evaluate the proposed feature subset.

3.3 J48 Algorithm

J48 is a Machine Learning algorithm. It makes a decision tree from a set of training data examples, with the help of information entropy idea. The training dataset consists of a wide number of training samples which are defined by various features and it also consists of the target class. J48 selects one accurate feature of the data at each node of the tree which is used to divide its set of samples into subsets improved in one or another class. It is based upon the principle of normalized information gain that is obtained from selecting a feature for splitting the data. The feature with

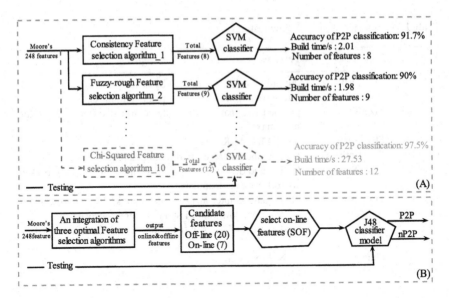

Fig. 1 Extracting on-line features flowchart

Algorithm 1. Selection of on-line feature algorithm

INPUT

Let D = {f1, f2, ..., fn, C}. D is the dataset, F is the vector of features to characterize network flow (Moores 249) features, n=number of features, C is the class.

F = Fsub0 ∪ Fsub1, Fsub0 is the offline features, Fsub1 is the online features.

M is the number of selected algorithms for feature selection.

OUPUT

SOF[] // selectedonlinefeature.

for $i = 1 \, to \, M$ **do**

    Fsub= get features(list,K); //k is the number of feature subset.

    Fsub ε F for each dataset Dp

    for $j = 1 \, to \, K$ **do**

        Let L ε (0, 1), //offline,online

        if $L = 1$ **then**

            Buildup the SOF;

        else

            Remove the feature from the list;

        end if

        SOF = get feature(list);

    end for

end for

---

the highest normalized information gain is selected, and a decision is made. After that, the J48 algorithm repeats the same action on the smaller subsets. In the present research work, J48 algorithm is used for internet traffic classification.

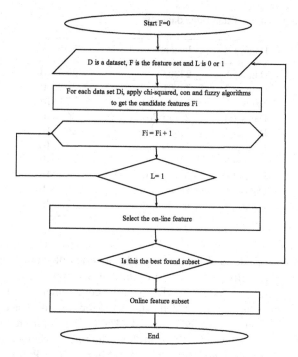

Fig. 2 Extracting on-line features flowchart

4   Evaluation Methodology

In this section, we offer the traffic traces and the evaluation metric in order to evaluate the effectiveness and efficiency of the approach.

4.1   Dataset

Datasets used in this work were downloaded from special shared resources. Cambridge datasets [23] were used to generate the training model while UNIBS datasets [24] were used to test this model. Table 2 and 3 illustrate these datasets.

- Cambridge datasets are based on the traces captured on the Genome Campus network in August 2003. They are published by the computer laboratory in the University of Cambridge. There are ten different datasets each from a different period of the 24-hour day [23]. The number of flows in each dataset is different, due to a variable density of traffic during each constant period. These datasets cover most of the statistics of absolute TCP flows; moreover, each flow example is high dimensional since it consists of 248 features that are derived from the TCP headers by using tcptrace [20].

Table 2 The samples of the Cambridge datasets

Dataset	Instances	Size
Dataset 1	24863 flows	29.7MB
Dataset 2	23801 flows	28.3MB
Dataset 3	22932 flows	27.5MB
Dataset 4	22285 flows	26.6MB
Dataset 5	21648 flows	25.8MB
Dataset 6	19384 flows	23.1MB
Dataset 7	55835 flows	66.0MB
Dataset 8	55494 flows	65.6MB
Dataset 9	66248 flows	78.3MB
Dataset 10	65036 flows	77.1MB

- UNIBS traces include packets generated by a series of workstations, located at the University of Brescia (UNIBS) in Italy in September and October 2009. These traces were captured by Tcpdump on the edge router which connects the network to the Internet through a dedicated 100 Mbps uplink. Traces were saved to particular files on a dedicated hard disk. The disk is connected to the router internals through a dedicated ATA controller. The traces occupy around 2.7 GB (78998 flows) which includes Web (61.2%), Mail (5.7%), P2P traffic (32.9%) and other protocols (0.2%).

Table 3 The Traces of UNIBS datasets

Dataset	Size
unibs20090930.anon	317 MB
unibs20091001.anon	236 MB
unibs20091002.anon	1.94 GB

4.2 Evaluation Metric

Sensitivity, Precision and F-measure are used to evaluate the effectiveness of the proposed approach. These metrics depend on true positive, false positive, true negative and false negative as shown in Table 4. TP is the number of P2P class that are correctly classified, FP is the number of nP2P class that are classified as P2P class, TN is the number of nP2P class that are correctly classified, and FN is the number of P2P class that are classified as nP2P class. Training and testing times are used to illustrate the efficiency improvement.

Table 4 The evaluation metrics

Metric	Equation	Description
Accuracy	$\frac{Tp+TN}{TP+TN+FP+FM}$	The percentage of correctly classified instances over the total number of instances
Precision	$\frac{TP}{TP+FP}$	Ratio of true positive to the sum of true positive and false positive
Sensitivity (Recall)	$\frac{TP}{TP+FN}$	The fraction of correct instances among all instances that belong to the relevant subset
F-measure	$2.\frac{Precision \times Recall}{Precision+Recall}$	A measure that combines precision and recall (harmonic mean of precision/recall)
Specificity	$\frac{TN}{TN+FP}$	True negative rate

5 Experimental Results and Discussion

5.1 Feature Selection Results

This subsection explains the feature selection result. Firstly, we apply three algorithms of features selection to generate candidate combined subset of features. The features chosen for each training dataset using Chi-squared, Fuzzy-rough and Consistency-based algorithms is demonstrated in Table 5.Then we select the best feature subset that is suitable for online P2P traffic detection and mitigation by examining the combined subset of features. As shown in Table 6, the optimal features selected by our approach are the feature 1, 4, 18, 25, 161, 182 and 203.

5.2 Classification Results

In order to show the ability of our approach to mitigate the high computational overhead difficulty when using ML P2P detection, we designed experiments to compare our algorithm with chi-squared, fuzzy-rough and consistency-based feature selection algorithms. In the experiment, feature selection algorithms and J48 were implemented using WEKA tool [25]. WEKA is a collection of open source state-of-the-art machine learning algorithms and data preprocessing tools. The classification algorithm was trained and tested by using the Cambridge and UNIBS datasets. The classification results of the datasets are shown in Table 7. As can be seen the table defines the average results of the accuracy, precision, sensitivity (recall), F-Measure, testing time and the number of features. The accuracy performance results of the J48 classifier using our proposed features subsets are improved. In addition, the computational performance and the number of features results have shown significant improvement also. These results are the best compared to the results from applying individual feature selection algorithm. Table 8 shows the comparison between the proposed approach results with the work proposed in [22]. The work in [22] used ten skewed datasets and Nave Bayes to evaluate his proposed feature

Table 5 The candidate feature subset using feature selection algorithms

Dataset	Chi-squared	Fuzzy-rough	Consistency-based
Dataset 1	**1**,95,7,96,90,93 186,180,82,179	2,9,16,**18**,24,**25** 32,43,45	1,3,7,13,93,121 184,222
Dataset 2	**1**,180,95,96,93,85 187,186,179,184	2,9,16,**18**,24,**25** 32,43,45	1,3,7,13,93,121 184,222
Dataset 3	**1**,95,187,184,180,177 93,96,90,82	2,9,16,**18**,24,**25** 32,43,45	1,2,3,5,30,119 123,224
Dataset 4	**1**,95,96,94,90,83 187,186,179,82	2,9,16,**18**,24,**25** 32,43,45	1,2,3,4,20,44 99,119
Dataset 5	**1**,187,184,95,100,44 96,48,180,177	2,9,16,**18**,24,**25** 32,43,45	1,8,16,30,115 119,184
Dataset 6	**1**,95,83,187,96,186,184 180,179,**182**,82	1,117,122 158,**203**,142	1,48,115,118,119 131,156
Dataset 7	**1**,95,96,83,90,187,180 184,94,186	2,9,16,**18**,24,**25** 32,43,45	1,2,4,9,90,117,123,131 187,222
Dataset 8	**1**,95,83,96,187,184 186,179,180,82	1,117,122 158,**203**,142	1,2,3,5,7,93,94,131 224,236
Dataset 9	**1**,96,95,93,83,187 180,184,**161**,186	1,117,122 158,238,142	1,4,30,93,110,115,127 209,216
Dataset 10	**1**,187,93,95,96,180 184,99,159,94	2,9,16,**18**,24,**25** 32,43,45	1,2,3,93,117,185 209,234

[a] Features in bold are features that can be extracted on-line.

Table 6 Proposed feature subset for online P2P traffic detection

No	Feature	Description
1	Server Port	Destination port number
2	q1 IAT	First quartile inter-arrival time
3	q1 data ip	First quartile of total bytes in IP packet
4	q1 data control	First quartile of control bytes in packet
5	q1 data ip ba	First quartile downlink of total bytes in IP packet
6	q1 IAT ba	First quartile downlink inter-arrival time
7	q1 data ip ab	First quartile uplink of total bytes in IP packet
8	class	Protocol/ application class

selection algorithm (named BFS). As compared to this work which has high accuracy (90.92%), our proposed method has 93.20% accuracy when using the same datasets and classifier whilst the accuracy is 99.23% when using decision tree classifier. In term of computational performance, the proposed on-line features have shown an improvement. This improvement is a result of reducing the number of features for our system. In addition our system is able to calculate these features on-line.

Table 7 Evaluation results using J48 detector

Evaluation using J48	Chi squared	Fuzzy rough	Consistency based	Proposed subset
Accuracy	99.22%	99.02%	99.22%	99.23%
Precision	0.978	0.964	0.977	0.977
Sensitivity	0.968	0.862	0.904	0.985
F-measure	0.973	0.910	0.939	0.981
Testing time	1.82s	22.62s	1.73s	1.12s
Number of features	12	9	8	7

Table 8 Effectiveness of the proposed approach

Dataset Generation	work in [22]	Proposed method Naive Bayes	Proposed method J48
Accuracy	90.92%	93.20%	99.23%
Sensitivity	0.60	0.64	0.985
Number of features	10	7	7

Acknowledgements. The authors would like to thank University of Cambridge, University of Brescia and CAIDA for providing their datasets.

References

1. Auld, T., Moore, A.W., Gull, S.F.: Bayesian neural networks for internet traffic classification. IEEE Transactions on Neural Networks 18(1), 223–239 (2007)
2. Bernaille, L., et al.: Traffic classification on the fly. ACM SIGCOMM Computer Communication Review 36(2), 23–26 (2006)
3. Dash, M., Liu, H.: Consistency-based search in feature selection. Artificial Intelligence 151(1), 155–176 (2003)
4. Erman, J., et al.: Semi-supervised network traffic classification. In: ACM SIGMETRICS Performance Evaluation Review. ACM (2007)
5. Gomes, J.V.P., et al.: The Nature of Peer-to-Peer Traffic. In: Handbook of Peer-to-Peer Networking 2010, pp. 1231–1252. Springer (2010)
6. Gomes, J.V.P., et al.: Detection and Classification of Peer-to-Peer Traffic: A Survey (accessed April 2011)
7. Jamil, H.A., Zarei, R., Fadlelssied, N.O., Aliyu, M., Nor, S.M., Marsono, M.N.: Analysis of Features Selection for P2P Traffic Detection Using Support Vector Machine. In: ICoICT, March 20-22, IEEE (2013)
8. Jensen, R., Shen, Q.: Fuzzyrough attribute reduction with application to web categorization. Fuzzy Sets and Systems 141(3), 469–485 (2004)
9. Jun, L., et al.: P2P traffic identification technique. In: 2007 International Conference on Computational Intelligence and Security. IEEE (2007)

10. Lei, D., Xiaochun, Y., Jun, X.: Optimizing traffic classification using hybrid feature se-lection. In: The Ninth International Conference on Web-Age Information Management Web-Age Information Management, WAIM 2008. IEEE (2008)
11. Liu, H., Setiono, R.: Chi2: Feature selection and discretization of numeric attributes. In: Proceedings of the Seventh International Conference on Tools with Artificial Intelli-gence. IEEE (1995)
12. Moore, A.W., Zuev, D.: Internet traffic classification using bayesian analysis techniques. ACM (2005)
13. Moore, A.W., Zuev, D., Crogan, M.: Discriminators for use in flow-based classification, Technical report, Intel Research, Cambridge (2005)
14. Rezaee, M.R., et al.: Fuzzy feature selection. Pattern Recognition 32(12), 2011–2019 (1999)
15. Szabó, G., Orincsay, D., Malomsoky, S., Szabó, I.: On the validation of traffic classifi-cation algorithms. In: Claypool, M., Uhlig, S. (eds.) PAM 2008. LNCS, vol. 4979, pp. 72–81. Springer, Heidelberg (2008)
16. Van Der Putten, P., Van Someren, M.: A bias-variance analysis of a real world learning problem: The CoIL challenge 2000. Machine Learning 57(1), 177–195 (2004)
17. Witten, I.H., Frank, E.: Data Mining: Practical machine learning tools and techniques. Morgan Kaufmann (2005)
18. Yang, Y., et al.: Solving P2P traffic identification problems Via optimized support vector machines. In: IEEE/ACS International Conference on Computer Systems and Applica-tions, 2007. AICCSA 2007. IEEE (2007)
19. Zarei, R., Monemi, A., Marsono, M.N.: Retraining Mechanism for On-Line Peer-to-Peer Traffic Classification. In: Abraham, A., Thampi, S.M. (eds.) Intelligent Informatics. AISC, vol. 182, pp. 373–382. Springer, Heidelberg (2013)
20. Zhang, H.L., et al.: Feature selection for optimizing traffic classification. Computer Com-munications 35(12), 1457–1471 (2012)
21. Zhao, J.-J., et al.: Real-time feature selection in traffic classification. The Journal of China Universities of Posts and Telecommunications 15(suppl.), 68–72 (2008)
22. Zhen, L., Qiong, L.: A new feature selection method for internet traffic classification using ml. Physics Procedia 33, 1338–1345 (2012)
23. Cambridge data sets, `http://www.cl.cam.ac.uk/research/srg/netos/nprobe/data/papers/sigmetrics/index.html` (cited November 18, 2012)
24. Université Brescia data sets, `http://www.ing.unibs.it/ntw/tools/traces/download/` (cited November 19, 2012)
25. WEKA. Data Mining Software in Java (2012), `http://www.cs.waikato.ac.nz/ml/weka/`

Real Time Insignificant Shadow Extraction from Natural Sceneries

Subramanyam Muthukumar, Ravi Subban, Nallaperumal Krishnan, and P. Pasupathi

Abstract. In Computer Vision, shadow free object recognition is a wide phrase covering a range of applications such as human motion capture, video surveillance, traffic monitoring, segmentation and tracking of foreground objects. Unfortunately, shadows in these applications may appear as foreground objects, when in fact they are caused by the interaction between light and objects. The inability to distinguish between foreground objects and shadows can cause malicious problems such as object merging, false segmentation, misclassified as foreground objects and identification failure, all of which significantly affect the performance of detection and tracking systems. However in most situations, it is essential to avoid shadow as it becomes undesired and unwanted part which deteriorates the outcome. Therefore, an effective shadow detection method is necessary for accurate object segmentation. One of the main challenging problems is identifying insignificant shadow from natural images by computing systems. Though many researchers try to deal with these problem using different methodologies, yet it is intriguing problem. This paper deals with the problem of identifying and extracting regions that correspond to shadow from natural scenes. Also, it aims to produce a comprehensive evaluation on the state-of-the-art methods of detecting shadows from natural images.

Keywords: Object Segmentation, Shadow Detection, Shadow Extraction, Still Image Shadow Identification, Shadow Tracking, Single Image, Object Localization and Detection.

Subramanyam Muthukumar · Nallaperumal Krishnan · P. Pasupathi
Centre for Information Tech. and Engg., M.S. University, Tirunelveli, Tamilnadu, India
e-mail: {sm.cite.msu,pp.cite.msu}@gmail.com, krishnann@ieee.org

Ravi Subban
Dept. of Computer Science, Pondicherry University, Pondicherry, India
e-mail: sravicite@gmail.com

S.M. Thampi et al. (eds.), *Recent Advances in Intelligent Informatics,*
Advances in Intelligent Systems and Computing 235,
DOI: 10.1007/978-3-319-01778-5_40, © Springer International Publishing Switzerland 2014

1   Introduction

The visual system relies on patterns of light to provide information about the layout of objects that populate our environment. Light is structured by the way it interacts with the three dimensional shape, refection, and transmittance properties of objects. The input for vision therefore, is a complex, conflated mixture of different sources. These sources of physical variation that, the brain must somehow disentangle to recover the intrinsic properties of the objects and materials that fill the world. It is very common in real world that the shadow will appear as long an object is in front of the light source. For human eyes, it is not difficult to distinguish shadows from objects. In digital scenes, shadows are often accidental and /or unwanted artifacts that in some conditions which cannot be avoided. Shadows reduce the total energy incident at the background surfaces where the light sources are partially or totally blocked by the foreground objects. Shadows are first of all, a local decrease in the amount of light that reaches a surface. Secondly, they produce a local change in the amount of light reflected by a surface toward the object.

If a scene is illuminated by two or more sources, then the shadow and non shadow region of an object differ not just in terms of their relative brightness, but also in terms of their relative color. Hence, shadow points have lower luminance values but similar chromaticity values. There are two different types of absence of light (i.e. Shadows), namely, cast shadow and self-shadow. Self shadow is the part of an object that is not illuminated (Fig 1, region A). Cast shadow, this is the area projected on the scene by the object (Fig 1, region B and C), it can be further classified into umbra which corresponds to the area where the direct light is totally blocked by the object (Fig 1, region B) always have a violent contrast to background and penumbra, which corresponds to the area where the direct light is partially blocked by the object (Fig 1, region C) and do not have clear boundaries. Shadows which are dark and the surface under that boundary are not visible are called hard shadows (dark shadows). The surfaces below the shadow are visible and glossy are called as soft shadows (hollow shadows or thin shadows.

Fig. 1 a) Shadow Classification b) Geometry of shadow c) Hard Shadow d) Soft Shadow

2   Methodologies of Shadow Detection

Shadow Detection has been an energetic field of research for several decades. The pixels in a digital photograph combine together scene components in seemingly permanent ways and provide us valuable little information about the scene's apparent shape, illumination, and reflectance properties. Most research focus on modeling the differences in pixel properties such as color, intensity, and texture of neighboring pixels or regions. Shadow properties can be used to distinguish shadow pixels from foreground object pixels. Some shadow detection methods are based on the observation that the luminance values of shadowed pixels decrease respect to the corresponding background while maintaining chromaticity values. Broadly Shadow detection classified into two categories: Model Based Techniques and Property-based techniques.

2.1   Model Based Techniques

Model based technique makes use of certain known information about illumination and geometry of the scenes. It includes the sensor/camera localization, the light source direction, and the geometry of observed objects. These models are usually used for specific situations such as in [5, 7, 8, 9, 10], where prior knowledge of scene geometry and foreground objects are incorporated into the model. Shadow pixel as being in umbra: the pixel is not affected by the light source - it is illuminated only by ambient light [1]. Shadow detection method in [2] uses advance knowledge of the illuminant vector. Several illumination invariant color spaces have been proposed and used for shadow detection by [3, 4, 5], but they all impose the constraint of requiring white illumination. One way to avoid this constraint is to white balance the camera [6]. Another method by [7] requires a calibrated camera. Shadow light environment is estimated in [10], and also presences of shadows are determined using illumination direction in [9].

2.2   Property-Based Techniques

Property-based techniques make an effort to detect shadows in a more general way by using gradient, color or texture features that discriminate shadows from foreground objects or background. Such features are used to detect distortions in luminance and chrominance [9] to exploit color differences between background and shadow in different color spaces or to statistically analyze the changes in appearance of a shadowed pixel [10,12]. These techniques are further classified into many branches based on the characteristics of shadows in luminance, chrominance, gradient density and geometry domains.

Image Based Techniques

This technique is connected with the shadow properties such as shadow structure (umbra and penumbra hypothesis), color (or intensity), boundaries, etc., [1, 21] showed that shadow regions have certain interesting properties. Specifically, illumination changes across shadow boundaries were exhibited color ratios that were different from the ratios across material boundaries. In [2], shadow region is detected with the shadow density, which is a measure of brightness of the scene. The method proposed in [4] uses photometric color invariants to extract shadow regions and subsequently classified as self (if on the object) or cast (if on the ground plane). In [1], a lookup table is used to keep track of possible illumination changes across shadow boundaries. In [1] support vector machines are used to identify probable shadow boundaries from typical images.

Color/Spectrum Based Techniques

Color is the vital visual feature for humans. Color analysis is performed to identify those pixels in the image that respect the chromatic property of the shadow. In general, a shadow causes only a change in brightness with little or no change in chromaticity (color). Fung et al. [13] detected cast shadows for monocular color image sequences. YUV model is exhibited by T. Horprasert et al [14] proposed a computational color model which separates the brightness from the chromaticity component. Pixels in the live image with similar chromaticity but lower or higher brightness levels are identified as a shadow or highlighted background region. Photometric invariant features are adopted in the HSV color space, the c1, c2, c3 color model [4], TAM [21] and the normalized-RGB color space. Levine and Bhattacharyya [8] studied the properties of color ratios across boundaries between regions in a segmented image, and used a support vector machine to identify shadow regions based on these color ratios. HSI model is used for detecting objects, shadows and ghosts in video streams by extracting color from motion information by A. Prati et al. [20]. Finlayson et al. [7] estimated an illumination invariant image based on an invariant color model, and use this invariant image together with the original image to locate the shadow region. Similarly, Salvador et al. [15] use invariant color features to segment cast shadows.

Texture Based Techniques

The texture of foreground objects is different from that of the background, while the texture of shaded the shadow points remains unchanged since shadows do not alter the background surfaces, only their illumination. Edge and texture information are used by J. Stauder et .al, [16] for detection of moving cast shadows for object segmentation. The food grains may have similarity in color but exhibit different texture patterns. Hence, texture feature acts as local features. Sanin 2012 [24] co-occurrence matrix to obtain texture features. The five local features are extracted from co-occurrence matrix known as Haralick features. The features are homogeneity (E), contrast(C), correlation (Cor), entropy (H) and local homogeneity (LH). Based on the region of interest this technique is further classified as Small Region (SR) and Large Region (LR) based methods.

Geometry Based Techniques

Geometry property of shadow such as edges, lines or corners to object models, camera location, the ground surface, orientation, mean intensity, and center position of a shadow region with the orientation and centroid position are being estimated from the properties of object moments which are the prime factor of geometry based techniques [17] used a three-level analysis of shadow intensity and shadow geometry in an environment with simple objects and a single light source to determine shadow regions. Another approach in [18] detection of object shape and orientation in scenes from shadows, shadow utilizes geometry to solve the correspondence problem. This subject has been dealt with in [9, 14] and [19] together with the color segmentation method to determine shadow regions and recover the penumbra and umbra regions of the shadow.

3 Shadow Detection with Hybrid Technique

These types of methods utilize any combination of above mentioned methods for detection and extraction of shadows. In this proposal, author uses Information theoretic approach based light reflection model (sun, sky and indirect lighting) and intensity based TAM for shadow detection [22]. The shadow extraction is done in gradient domain [23]. The shadow region refinement process uses hellinger distance similarity patches across the nearby neighbors of shadow boundary [11, 22].

Table 1 Strategy & Performance of Shadow Detection

Strategy	Categories	Characteristics	Geo	Sr	Lr	Cha	Phy	Hyb
Scene Type	Indoor, Outdoor							
Surface	Textures [Grass, Road, White, Wall, Floor, Steps, Concrete, Sand, Wood, Carpet, Reflective, Clothes]	Scene independence	M	M	H	M	M	H
	Non _textures	Object Independence	L	H	H	H	H	H
Roboustness to Noise	High, Medium, Low, Very Low, No_Noise	Shadow Independence	L	H	H	H	H	H
Object Type	Human, Trees, Ball, Boat, Cycle, Bike, Car, Van, Building, Baby, Can, Cat, Cup, Bridge, Bag, Mouse, Street Lamp, Beam, Mug, Boy	Penumbra detection	M	H	H	L	M	H
Shadow Direction	Horizontal, Vertical, Multiple, Complex	Robustness to Noise	M	H	H	L	M	H
Shadow Strength	Strong, Medium, Weak, Very Weak	DR, DT Trade-off	L	H	L	H	M	L
Shadow Size	Very High, High, Medium, Low, Very Low	Computation load	L	H	M	L	L	M

4   Experimental Results and Discussion

In Perception of the Visual Word [19], Gibson declares, "The elementary impressions of a visual world are those have clear understanding of an image". Vision based object detection systems experimented in several different ways, depending on the conditions in which they are designed to operate like, the environments: indoor and outdoor, the type and number of sensors, the objects and level of details to be traced etc. There are typically a number of challenges associated with the chosen scenario in realistic environments. Based on various test conditions different methods of shadow detection approaches are analyzed and tabulated.

In the criteria, high indicates that the method performed well in all cases. Low penumbra detection means that a method is not suitable for detecting shadow borders. Low robustness to noise means that shadow detection performance is affected by scene noise. For the last two criteria (i.e., 6 and 7) lower means better. A high detection/discrimination trade-off means that and adapt to the appearance of cast shadow in the background. Shadow detection performance can be measured by each method in every testing image are tabulated with quantitative and qualitative results. To test the shadow detection performance of these six methods the two metric proposed by Prati et.al. [20] viz., shadow detection rate (DR) η and shadow discrimination rate (DT) ξ are used, which are given as:

$$\eta = TP_F / TP_F + FN_F , \quad \xi = TP_S/(TP_S + FN_S) \qquad (1)$$

Fig. 2 (Row 1, 3) Input and (Row 2, 4) Output of Natural Sceneries

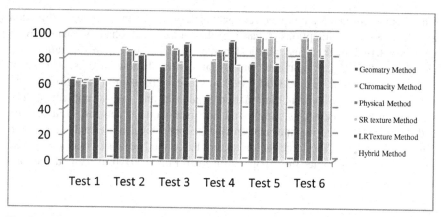

Fig. 3 Shadow Detection rate of different techniques

Table 2 Comparison of Computation Time (Milli Seconds)

Test	Geo	SR	LR	Chro	Phy	Hyb
1	6.49	120.	11.75	6.82	7.15	13.80
2	8.91	223.64	21.59	11.28	12.81	22.40
3	17.68	253.82	22.73	8.95	15.34	24.73
4	13.94	243.07	24.70	10.82	14.08	26.60
5	24.75	341.32	34.71	10.73	16.93	34.82
6	8.41	144.87	16.25	7.14	8.51	9.57
7	9.44	156.46	20.76	8.72	10.00	23.59
Avg	12.12	211.93	21.78	12.80	12.12	24.86

where, TP and FN stand for true positive and false negative pixels with respect to either shadow (S) or foreground objects (F). The shadow detection rate is concerned with labeling the maximum number of the cast shadows and shadow discrimination rate is concerned with maintaining the pixels that belong to an object as foreground. Shadow of the flower, tree, ball, bird and etc. are detected despite the very low quality of the image with a strong noise and blurred image. However, the camouflage problem sometimes appears due to the image low quality. Complex and overlapping shadows limits the performance of these methods.

5 Conclusion

From the analysis of shadow region detection, the hybrid algorithm shows better performance without losing large amount of pertinent data. Hybrid method works well independently on the quality of spectral and texture features. Physical method adapts better to the different scenes, Geometry based method offers fine result when each shadow has a unique direction which differs from the object's direction and has strong assumptions regarding the object and shadow shape. Both geometry based method and physical based methods need more operations for calculating the central moments and updating the shadow models. The small region texture based method works well in scene with textured backgrounds and this is also robust to various illumination conditions and easy to implement. The chromacity based method is simple and fast to implement and run. Both texture and chromacity based method are sensitive to pixel-level noise and scenes with low saturated color. The large region texture based method works reasonably well in most cases, but higher the computation load. Finally, Hybrid method presents the best results in most cases.

References

[1] Barnard, K., Finlayson, G.: Shadow identification using color ratios. In: IS & T/SID, 8th Color Imaging Conference: Color Science, Systems and Appl., pp. 97–101 (2000)

[2] Charit, R., et al.: Complex shadow-boundary segmentation using the entry-exit method. In: CVPR, pp. 536–541 (1998)

[3] Gevers, T.: Adaptive image segmentation by combining photometric invariant region and edge information. IEEE Trans. Pattern Anal. Machine Intell (PAMI) 24, 848–852 (2002)

[4] Gevers, T., et al.: Color-based object recognition. Pattern Recognition 32, 453–464 (1999)

[5] Salvador, et al.: Shadow identification and classification using invariant color models. In: ICASSP, pp. 1545–1548 (2001)

[6] Fieguth, P.W., Wesolkowski, S.: Highlight and shading invariant color image segmentation using simulated annealing. In: EMMCVPR, pp. 314–327 (2001)

[7] Finlayson, G.D., Hordley, S.D., Drew, M.S.: Removing shadows from images. In: Heyden, A., Sparr, G., Nielsen, M., Johansen, P. (eds.) ECCV 2002, Part IV. LNCS, vol. 2353, pp. 823–836. Springer, Heidelberg (2002)

[8] Levine, et al.: Removing shadows. Pattern Recognition Letters 26(3), 251–265 (2005)

[9] Horprasert, T., et al.: A Statistical Approach for Rreal-time Robust Background Subtraction and Shadow Detection. In: Proc. ICCV Frame-rate Workshop (1999)

[10] Benedek, C., Sziranyi, T.: Bayesian Foreground and Shadow Detection in Uncertain Frame Rate Surveillance Videos. IEEE Trans. Image Processing 17(4), 608–621 (2008)

[11] Ravi, S., Muthukumar, S., et al.: Image Inpainting Techniques – A Survey And Analysis. In: International Conference on IIT, 978-1- 4673-6203-0/13© IEEE (2013)

[12] Mikic, I., et al.: Moving Shadow and Object Detection in Traffic Scenes. In: Proc. IEEE Conf. on Pattern Recognition (ICPR), pp. 321–324 (2000)

[13] Fung, G.S.K., et al.: Effective moving cast shadows detection for monocular color image sequences. In: Proc. of 11th Int. Conf. on Image Analysis and Processing, pp. 404–409 (2001)

[14] Horprasert, T., et al.: Statistical approach for real-time robust background subtraction and shadow detection. In: Proc. ICCV Frame-rate Workshop (1999)

[15] Salvador, E., et al.: Spatio-temporal Shadow Segmentation and Tracking. In: Proc. of Visual Communications and Image Processing (2003)

[16] Stauder, J., et al.: Detection of moving cast shadows for object segmentation. IEEE Transactions on MM 1(1), 65–77 (1999)

[17] Jiang, C., et al.: Shadow identification. In: IEEE CV and Pattern Recognition, pp. 606–612 (1992)

[18] Funka-Lea, G., et al.: Combining color and geometry for the active, visual recognition of shadows. In: Proceedings of IEEE Int. Conference on Computer Vision, pp. 203–209 (June 1995)

[19] Funka-Lea, G., Bajcsy, R.: Combining color and geometry for the active, visual recognition of shadows. In: Proc. of IEEE Int. Conf. on Computer Vision, pp. 203–209 (1995)

[20] Prati, A., et al.: Detecting moving shadows: Algorithms and evaluation. IEEE Transactions on Pattern Analysis and Machine Intelligence, 918–923 (2003)

[21] Muthukumar, S., et al.: Fuzzy Information based on Image Segmentation by using Shadow Detection. In: IEEE Internal Conference on ICCIC 2010, pp. 1–6 (2010)

[22] Muthukumar, S., et al.: Hybrid shadow detection and compensation for plausible visual scene Reconstruction. IJISE 1, 141–146 (2011)

[23] Arbel, E., Hel-Or, H.: Shadow removal uing intenity surface and texture anchor point. PAMI 33(6), 1202–1216

[24] Sanin, A., Sanderson, C., Lovell, B.C.: Shadow detection: A survey and comparative evaluation of recent methods. Pattern Recognition 45, 1684–1695 (2012)

Burning Ship and Its Quasi Julia Images Using Mann Iteration

Shafali Agarwal and Ashish Negi

Abstract. An invented form of Mandelbrot set came into existence in 1992 when Michelitsch and Rossler have applied seemingly small changes in complex analytic Mandelbrot set function and got an image resembled to a ship going into flame. He named it burning ship. Our goal in this paper is to apply Mann Iteration method to burning ship function and produce a collection of stunning images. We have also calculated the fixed points of such images to measure the convergence rate and those fixed points can be further useful in various fractal applications such as fractal cryptography. Hence we are in position to examining numerically the stability of the fractals.

Keywords: Mann Iteration, M-Burning Ship, Quasi Julia Set, Fixed Point.

1   Introduction

Mandelbrot set is a well known fractal for many mathematicians. A lot of researches about the various format of Mandelbrot set have been done by researchers [1, 2, 4, 8]. In 1992, Michelitsch and Rossler have applied a small change in standard Mandelbrot set function in terms of its imaginary part. Initially Michelitsch had done experiment in 1992 with the absolute part of the function and got a completely new fractal images see Michelitsch [5, 6]. Appearances of obtained images are incredibly beautiful and completely different from previous fractal images known as Burning ship. The Burning ship fractal is generated by iterating the function:

Shafali Agarwal
JSS Academy of Technical Education, Noida, India
e-mail: shafali.agarwal@gmail.com

Ashish Negi
Dept. of Computer Science, G.B. Pant Engg. College, Pauri Garwal, Uttarakhand, India
e-mail: ashish.ne@gmail.com

S.M. Thampi et al. (eds.), *Recent Advances in Intelligent Informatics*,
Advances in Intelligent Systems and Computing 235,
DOI: 10.1007/978-3-319-01778-5_41, © Springer International Publishing Switzerland 2014

$$z_{n+1} = (|\operatorname{Re}(z_n)| + i |\operatorname{Im}(z_n)|)^2 + c$$

Here the real and imaginary components of the complex quadratic equation are:

$$x_{n+1} = x_n^2 - y_n^2 - c_x$$

And

$$y_{n+1} = 2x_n y_n - c_y$$

For $x_0 = 0$ and $y_0 = 0$, the above equations yield the fractal images in the c-plane (parameter space) which will either escape or remain bounded. Now we are considering absolute value of x_n and y_n which helps to shown a lot of curviness and turn it into angles and lines. It makes the difference between Mandelbrot set and burning ship fractal as in case of Burning ship the values to be considered in its absolute form of its real and imaginary components before squaring, whereas in Mandelbrot set no absolute part is considered.

Based on these differences, the Mandelbrot set contains images of classical beauty, organic forms and ornate scrollwork and the burning ship contains cartoonist forms and patterns that look like war print, paw print, tokens and towers.

It is a kind of escape time fractal. Escape time measures time of escaping to infinity and time is measured in steps (Iterations) needed to escape from the covered area. This paper contributes towards the generation of various astonishing images of iterated burning ship & its corresponding quasi Julia sets. Basically Julia set is obtained for a particular complex point in the Mandelbrot set. Later we will calculate that after how much iterations, this point will get converge. This convergence or stable point is known as fixed point. The knowledge of iterated or escape time fractal and the quantitative study of the fixed point that naturally related to the iteration theory and dynamic system is possible through the computer system [3].

2   Structure of Burning Ship

Burning ship is a region of chaos that contains wild and noisy images. Most fractals look like oil or watercolours as they were spray painted on a brick wall. The intensity of colour assigned to each pixel is determined by how fast a pixel tends to infinity. e.g. A pixel shown in black colour remains in the centre of fractal and never diverges to infinity. It will either oscillate or converges to single point. Change the colour of pixel from black to light colour, it shown the divergence of pixel very quickly to infinity. Even the Mandelbrot set creation used the same technique.

The main figure shows diversified images like on top border you will be surprised to see a lady as passenger on the ship. This image shows a lady is waving her hands, probably calling for help. She is standing on the deck where the lady structures take a deep dip down. On the upper side of burning ship, there is a large dusty pattern in contrast to other corresponding parts. Within the dust dumb-bell like voids of all sizes are randomly distributed with some mini burning ships.

A tower like structure is strangely ordered and as well magnify that we are astounded to find endlessly varying kaleidoscopic images. These patterns are weird, wonderful, and undeniably beautiful and still show that twisted edginess we have come to expect from burning ship. This structure is somewhat related to image reported in Rossler [12] for the simplest non-analytic Julia set.

On zooming the sail of burning ship, there are infinite incredibly intricate small mini ships are shown the presence of midgets [7] on the external ray. Burning ship image can also be extended with different power values.

3   M-Burning Ship

Initially we have described the basic geometrical structure of burning ship function, which by default used Picard iteration method [9]. *i. e.*

$$x_{n+1} = f(x_n)$$

Here $f(x_n)$ is a function, after applying Picard iteration we will get output as x_{n+1}. Now we define burning ship with respect to Mann iterates. We named it M-Burning ship.

Mann Iteration method
The method is given as below. See Negi A & Rani M. and Rani M. & Kumar V. [10,11]

$$z_d = sf(z_{d-1}) + (1-s)z_{d-1},$$

where $f(z)$ is the function on which we will apply the given iteration method, z is a complex number and $0<s<1$ and s is convergent to a non-zero number. Mann iteration is based on two step feedback machine. Here function $f(z)$ is given as follow:

$$f(z_{d+1}) = (real(z) + imag(z))^d + c$$

After applying Mann iteration, we got some strange but self explanatory images which can be useful for further research work.

4   Quasi Julia Set

Michelitsch and Rossler entitled to Julia set as Quasi Julia Set because it refers to non analytic images which don't obey Cauchy-Riemann conditions: See Michelitsch M. and Rossler O.E. [6]

$$\partial f_1(x, y) / \partial x = \partial f_2(x, y) / \partial y$$
$$\partial f_1(x, y) / \partial y = -\partial f_2(x, y) / \partial x$$

The above given equation represents Cauchy-Riemann conditions. The creation of Quasi Julia Set along real axis is identical to Julia set taken from Mandelbrot set because at $y=0$ implies $z^2 = x^2$. Otherwise for $y!=0$ not a single Julia set point will be common between Mandelbrot set and burning ship.

To consider different Julia set from Mandelbrot set we have to consider the orbit of all points that will take a point somewhere in second or fourth quadrants i.e. $(-x,y)$ or $(x,-y)$. The y value of next iteration from within one of these quadrants will be:

$$z = (-x)^2 + (y)^2 + i * 2 * (-x) * (y) + c$$

Or

$$z = (x)^2 + (-y)^2 + i * 2 * (x) * (-y) + c$$

If a point is already in one of these quadrants, condition is satisfied. Otherwise we have three conditions:

(1) If $x^2 = y^2$, then the series will be

$$(x^2 - y^2, 2*|x|*|y|) = (0, 2*|x|*|y|)$$

Next iteration will be

$$(0 - (2*|x|*|y|)^2, 2*0*2*|x|*|y|) = (-(2*|x|*|y|)^2, 0) = (-x, 0)$$

(2) If $x^2 < y^2$, then the series will be

$$(x^2 - y^2, 2*|x|*|y|) = (-x, y)$$

(3) If $x^2 > y^2$, then the series will be

$$(x^4 - 6*x^2*y^2 + y^4, 4*|x|^3*|y| - 4*|y|^3*|x|)$$
$$(x^2 - y^2, 2*|x|*|y|) = (x, y)$$
$$((x^2 - y^2)^2 - 4*(x^2 - y^2, 2*|x|*|y|) =$$
$$(-x, y)*x^2*y^2, 2*2*|x|*|y|*(x^2 - y^2))$$

After solving it gives

$$(x^4 - 6*x^2*y^2 + y^4, 4*|x|^3*|y| - 4*|y|^3*|x|)$$

If we keep iterating it and ignore the smaller chunk of fractal, we will get different Julia set than Mandelbrot set's [12].

5 Result and Analysis

We have used standard burning ship fractal function and obtained Julia set images for different seed values using ultrafractal software which maps different points in burning ship.

5.1   Description of Julia Set Fractals

If we divide main image into regions then each region exhibit different properties. Julia set behaves differently for lower power values like $d=2$ and 3 as compared to higher power values like $d=10$.

For less power values and $s=1$,

- The lower side of burning ship produced Julia set with one bulb. See Fig. 9.
- We move towards upper side of burning ship, two bulbs Julia set is seen. See Fig. 10.
- A disconnected Julia set is seen for the same values also known as cantor set.
- At the lower middle and upper middle position of burning ship, images having spiritual feeling are seen which shows the south-western paw print motif. See fig. 8.
- A frame of light seems coming out from it gives an extraordinary appearance of image.
- Move to top layer of main image where no Julia set is available.

After changing the value of s from 1 to 0.5, the existence of Julia set is restricted to the lower side with maximally single and double bulbs fractals.

For higher power value such as $d=10$ and $s=1$, only single bulb Julia set is obtained from centre to upper core of main image. Now change the value of constant s from 1 to 0.5 then the range of produced Julia set is increased such as

- The upper part of main image erects only Julia sets with single bulb.
- There is no Julia set available corresponding to lower part of burning ship.
- A collection of astonishing Julia sets are obtainable in the middle section of main image.

Like an image is visible like an ornament for $s=0.3$ whereas one image gives an impression of insect for $s=0.5$.

5.2   Iterated Images of Burning Ship and Its Quasi Julia Set

Fig. 1 $f(z)= (|Re(z_n)| + i|Im(z_n)|)^2 + c$

Fig. 2 Mini burning ships on external ray with Eiffel Tower structure

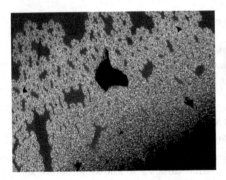

Fig. 3 Dusty area showing dumbbell like voids with mini-burning ships

Fig. 4 Lady on deck

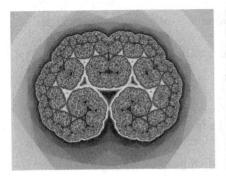

Fig. 5 Resemble to Alzheimer's brain image

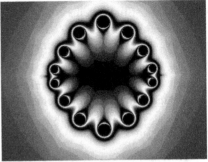

Fig. 6 Light coming out from the source

Fig. 7 Julia Set in Paw Print

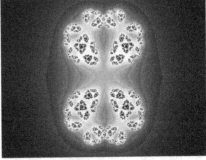

Fig. 8 Julia set in disconnected form (Cantor set)

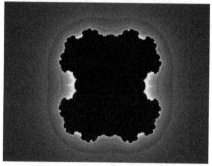

Fig. 9 Julia set with single bulb

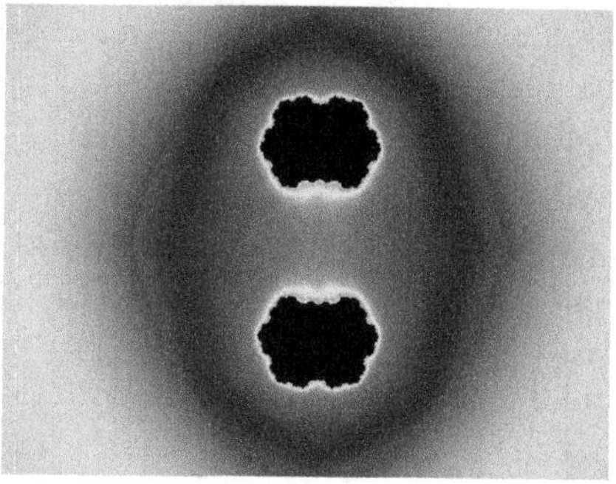

Fig. 10 Julia set with two bulbs

Fig. 11 Julia set in ornament form

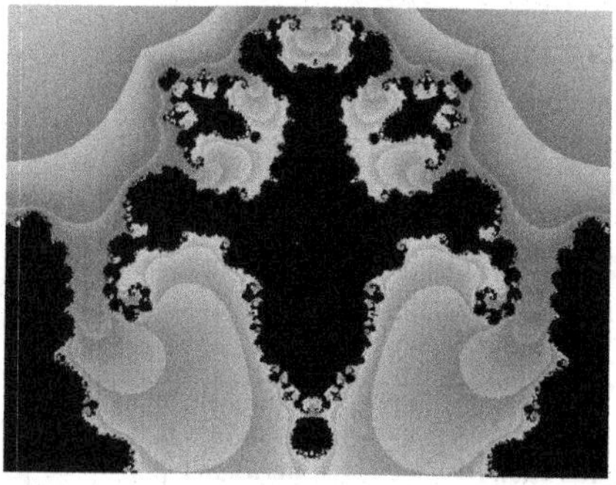

Fig. 12 Image resemble to an Insect

5.3 Fixed Points

The calculation of fixed points gives the convergence rate of the function for a particular seed value. This estimation gives the number of iterations required to the function to be stable or diverge to infinity. We have calculated it using Matlab software.

Table 1 Orbit of F(z) at s=1, (C=-0.4625,0.41875i)

Number of Iteration i	F(z)	Number of Iteration i	F(z)
1	-0.4625	4	-0.4607
2	-0.4606	5	-0.4607
3	-0.4607	6	-0.4607

Table 2 Orbit of F(z) at s=0.5, (C=-1.141459,0.578863i)

Number of Iteration i	F(z)	Number of Iteration i	F(z)
16	-0.4662	21	-0.4666
17	-0.4667	22	-0.4665
18	-0.4664	23	-0.4665
19	-0.4666	24	-0.4665
20	-0.4664	25	-0.4665

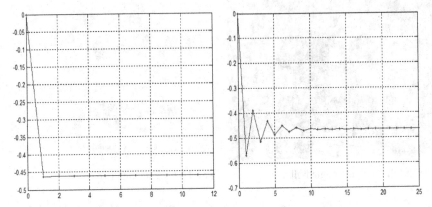

Fig. 13 Orbit of F(z) at s=1 for (C=-0.4625,0.41875i)

Fig. 14 Orbit of F(z) at s=0.5 for (C=-1.141459,0.578863i)

Table 3 Orbit of F(z) at s=1, (C=-0.4064297,-0.228333i)

Number of Iteration i	F(z)	Number of Iteration i	F(z)
1	-0.4064	4	-0.4013
2	-0.4008	5	-0.4013
3	-0.4013	6	-0.4013

Table 4 Orbit of F(z) at s=1, (C= 0.4194764,-0.7014249i)

Number of Iteration i	F(z)	Number of Iteration i	F(z)
1	0.4195	5	0.3892
2	0.3971	6	0.3890
3	0.3913	7	0.3890
4	0.3896	8	0.3890

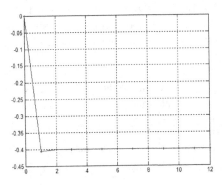

Fig. 15 Orbit of F(z) at s=1 for (C=-0.4064297,-0.228333i)

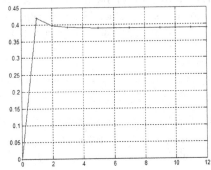

Fig. 16 Orbit of F(z) at s=1 for (C=0.4194764,-0.7014249i)

Table 5 Orbit of F(z) at s=0.3, (C=-1.0716078,1.692406i)

Table 6 Orbit of F(z) at s=0.5, (C=-0.6414225,0.4615118i)

Number of Iteration i	F(z)	Number of Iteration i	F(z)
16	-0.3985	21	-0.3981
17	-0.3979	22	-0.3982
18	-0.3983	23	-0.3982
19	-0.3980	24	-0.3982
20	-0.3983	25	-0.3982

Number of Iteration i	F(z)	Number of Iteration i	F(z)
6	-0.2898	11	-0.2908
7	-0.2912	12	-0.2907
8	-0.2905	13	-0.2907
9	-0.2908	14	-0.2907
10	-0.2907	15	-0.2907

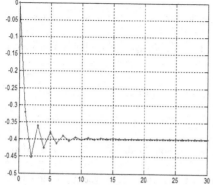

Fig. 17 Orbit of F(z) at s=0.3 for (C=-0.0716078,1.692406i)

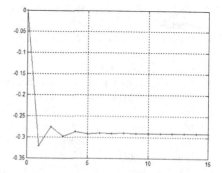

Fig. 18 Orbit of F(z) at s=0.5 for (C=-0.6414225,0.4615118i)

6   Conclusion

Burning Ship is a kind of escape time fractal. It can be analyzed on the basis of escape points set and prisoner points set after applying various Iteration methods. Burning ship has its own unique images like lady on deck, dumbbell like shape, bird of pray etc. In burning ship fractal midgets are also found on the external ray in terms of mini burning ship. An analysis of Julia sets is carried out for Mann Iterated burning ship. Here we observed main image on the basis of different regions. Accordingly we found that there are some places where no Julia set subsist. Another portion of image produced an exciting collection of Julia set images such as Julia set with single bulb, Julia set with two bulbs, Cantor set, resemblance to an ornament, an insect etc. (For ref. see fig. above). A completely different and religious Julia set images are also shown in the form of paw print which mapped to different values in burning ship see fig. 7. All the fractals form a link with realistic images which can be further useful for approaching researchers.

References

1. Ewing, J.H.: can we see the Mandelbrot Set. The College Mathematics Journal 26(2), 90–99 (1995)
2. Ewing, J.H., Schober, G.: The area of Mandelbrot Set. Numerische Mathematik 61, 59–72 (1992)
3. Garmendia, A., Salvador, A.: Fractal dimension of birds population size time series. Science Direct, Mathematical Bioscience 206, 155–171 (2007)
4. Mandelbrot, B.B.: Fractal Aspects of the Iteration of $z \to \lambda z\,(\lambda\text{-}z)$ for Complex 2. Z. Ann. NY Acad. Sci. 357, 249 (1980)
5. Michelitsch, M., Rossler, O.E.: Spiral structures in Julia sets and related sets. In: Hargittai, I., Pickover, C.A. (eds.) Spiral Symmetry, pp. 123–134. World Scientific, Singapore (1992)
6. Michelitsch, M., Rossler, O.E.: The Burning Ship and Its Quasi Julia Sets. Chaos and Fractals: Computer & Graphics 16(4), 435–438 (1992)
7. Negi, A., Rani, M.: Midgets of Superior Mandelbrot set. Chaos, Solitons, and Fractals 36, 237–245 (2008)
8. Peitgen, H.O., Richter, P.: The Beauty of Fractals: Images of Complex Dynamical Systems. Springer, Heidelberg (1986)
9. Rana, R., Dimri, R.C., Tomar, A.: Remarks on Convergence among Picard, Mann and Ishikawa iteration for Complex Space. International Journal of Computer Applications (0975-8887) 21(9), 20–29 (2011)
10. Rani, M., Kumar, V.: Superior Mandelbrot sets. J. Korea Soc. Math. Educ. Ser. D, Res. Math. Educations, 279–291 (2004)
11. Rani, M., Kumar, V.: Superior Julia set. J. Korea Soc. Math. Educ. Ser D Res. Math. Educations 8(4), 261–277 (2004)
12. Rossler, O.E., Kahlert, C., Parisi, J., Peinke, J., Rohricht, B.: Hyperchaos and Julia sets. Zeitschrift Naturforschung Teil A 41, 819–822 (1986)
13. http://theory.org/fracdyn/burningship/julias.html

A Computationally Faster Randomized Algorithm for NP-Hard Controller Design Problem

M. Jerome Moses and Ayyagari Ramakalyan

Abstract. In Finite Dimensional Linear Time-Invariant (FDLTI) control systems the following problem is important from a computational perspective. Given $A \in \Re^{n \times n}$, $B \in \Re^{n \times m}$ such that rank of the composite matrix $\begin{bmatrix} B : AB : \cdots : A^{n-1}B \end{bmatrix} \in \Re^{n \times mn}$ is full, and a nth order polynomial χ with constant coefficients, compute a matrix $K = [k_{ij}] \in \Re^{m \times n}$ such that characteristic polynomial of $A + BK = \chi$. It is proven that when $m > 1$ and when the elements of the matrix K are constrained such that $\underline{k}_{ij} \leq k_{ij} \leq \overline{k}_{ij}$, the problem belongs to the class NP-hard. In this paper, we provide a computationally efficient polynomial time algorithm to this problem using randomization. We show that the number of matrices K satisfying the given specification follows an interesting distribution w.r.t the matrix norm $\|K\|$. We give several examples wherein the algorithm outputs the desired K matrices in polynomial time.

Keywords: Randomized Algorithms, Computational Complexity.

1 Introduction

Engineers aspire to design and control larger systems. Examples vary from biology, VLSI design, to networks. In these systems, complexity is enormous. To tame complexity randomization is a promising key. And, the idea of randomization has been successfully applied in computer science, optimization, signal processing and in many diverse fields. During recent years it has been shown that a number of problems in matrix theory do not have polynomial time algorithms. From a computational perspective these problem are proven to be NP-hard. One of the problems

M. Jerome Moses · Ayyagari Ramakalyan
National Institute of Technology, Tiruchirappalli
e-mail: jeromemoses@gmail.com, rkalyn@nitt.edu

S.M. Thampi et al. (eds.), *Recent Advances in Intelligent Informatics*,
Advances in Intelligent Systems and Computing 235,
DOI: 10.1007/978-3-319-01778-5_42, © Springer International Publishing Switzerland 2014

of interest to the control engineering community is defined as follows. Given a system matrix $A \in \Re^{n \times n}$, an input matrix $B \in \Re^{n \times m}$, that the rank of the composite matrix $\left[B : AB : \cdots : A^{n-1}B \right] \in \Re^{n \times mn}$ is full, and a nth order polynomial χ with constant coefficients, compute a matrix $K = [k_{ij}] \in \Re^{m \times n}$ such that characteristic polynomial of $A + BK = \chi$. The roots of χ are p_i, $i = 1, 2, ..., n$. There is a combinatorial explosion of the gain matrix K when $m > 1$ (Multi input- Multi output system). If K is unbounded any one of the K matrices generated will work for us. But in real world applications of state feedback control involves control actuators having rate limitation and bandwidth problem. In particular any electrical or mechanical device can provide limited voltage, current, force or torque. So, searching for a suitably bounded K matrix takes an unpredictable amount of time, and hence the problem belongs to class NP-hard. It was shown [2] that when $m > 1$ and when the elements of the matrix K are constrained such that $\underline{k}_{ij} \leq k_{ij} \leq \bar{k}_{ij}$, the problem belongs to the class NP-hard. Problems belonging to this class may be solved using a technique called Randomized Algorithms which solve the problem efficiently, i.e., in polynomial time, most of the times if not at all times. In our work, we employ randomization in a way different from what has been reported in literature [5]. We have a polynomial time algorithm [1] & [3] to compute a K matrix given the matrices A and B (satisfying the rank condition mentioned earlier), and the roots of χ. We use this algorithm to generate a large pool of K matrices. Consequently, we show that this pool has an interesting distribution w.r.t the norm of K. This distribution allows us to pick up the right K matrix, i,e. $K : \underline{k}_{ij} \leq k_{ij} \leq \bar{k}_{ij}$ in polynomial time.

Rest of the paper is organized as follows. Section II describes the proposed randomized algorithm. Section III presents examples demonstrating the efficacy of the proposed algorithm. The paper is concluded in section IV.

2 Algorithm

In this section we present our algorithm. The algorithm consists of 3 steps and we use *randomization* in all the steps.

Inputs : $A \in \Re^{n \times n}$, $B \in \Re^{n \times m}$, a polynomial χ, with constant coefficients, whose roots are p_i, $i = 1, 2, ..., n$.

Step 1

Following [1] & [3] we generate $K \in \Re^{n \times m}$ as follows.

1.1 For each p_i, form the composite matrix

$$M_c = \left(p_i I - A : -B \right)$$

1.2 Solve for the null space of M_c and partition the dependent & independent components as shown below

$$\zeta_i = \begin{pmatrix} \mu_i \\ \cdots \\ v_i \end{pmatrix}, \quad \text{where } \mu_i \in \mathfrak{R}^{n \times 1}, \quad v_i \in \mathfrak{R}^{r \times 1}$$

and we need to ensure that v_i is dependent on μ_i.

1.3 Stack the vectors column-wise using the dependency such that

$$K \begin{pmatrix} \mu_1 & \mu_2 & \cdots & \mu_n \end{pmatrix} = \begin{pmatrix} v_1 & v_2 & \cdots & v_n \end{pmatrix}$$

from which we solve for $K \in \mathfrak{R}^{m \times n}$. The independent matrix $\begin{pmatrix} \mu_1 & \mu_2 & \cdots & \mu_n \end{pmatrix}$ takes *random* values within any given interval, and hence we will have infinitely many K matrices to choose from, and thus the problem is NP-hard [2].

Step 2

Next, we *randomly* generate a large sample of $\mathcal{K} \in \mathfrak{R}^{m \times n}$ matrices within the specified limits $\underline{k_{ij}} \leq k_{ij} \leq \overline{k_{ij}}$. For each of the above matrices, we determine the norm of each matrices and take the mean N_{avg} over the sample. Thus, the constraint on the K matrix – $\underline{k_{ij}} \leq k_{ij} \leq \overline{k_{ij}}$ – is converted into mean 2-norm measure N_{avg}.

Step 3

3.1 From the large pool of generated K in step 1, We filter out the matrices with norm N such that
$$N_{avg} - t \leq N \leq N_{avg} + t$$
where $t \in 1, 2, .., n$ is chosen according to the order of the system. And these are stored in a cell array Ce.

3.2 From Ce, we *randomly* choose a matrix and check element-wise, if it satisfies the constraint; if yes, we return the matrix.

It is easy to see that this algorithm has polynomial time (that depends only on the pool size of the generated matrices) complexity.

3 Numerical Examples

For all the examples below, the number of samples generated was 10000 in step 1. We executed the algorithm for 1000 trials.

Example 1: Let

$$A1 = \begin{bmatrix} 1 & 29 \\ 6 & 8.5 & 9 \\ 7.5 & 4.3 & 9 \end{bmatrix}, \quad B1 = \begin{bmatrix} 4 & 2.5 & 1 \\ 5 & 1 & 0 \\ 4 & 3 & 1 \end{bmatrix}$$

The characteristic polynomial be $\chi_1 = \lambda^3 + 14\lambda^2 + 55\lambda + 42$: the roots are $p_{1i} = \{-6, -7, -1\}$.

We chose $k_{ij} \in [-5, 10]$, and $t = 2$

The mean size of the cell-array Ce was around 1120. The N_{avg} was found to be 17.40, and the mean of iterations in which we obtained K was around 4.01. One of the K matrices picked up at random is

$$K = \begin{bmatrix} -1.0249 & -4.2521 & -1.8609 \\ -0.8051 & 4.9519 & -0.5281 \\ -1.1818 & 6.8933 & 0.1308 \end{bmatrix}$$

so that

$$A + BK = \begin{bmatrix} -6.2941 & 4.2646 & 0.3670 \\ 0.0704 & -7.8086 & -0.8326 \\ -0.1967 & 9.0406 & 0.1029 \end{bmatrix}$$

and $\chi_{A+BK} = \lambda^3 + 14\lambda^2 + 55\lambda + 42$. One of the K matrices picked up at random with large norm is ($\|K_1\| = 250.18$).

$$K_1 = \begin{bmatrix} -0.5931 & \mathbf{-44.8966} & \mathbf{-11.4426} \\ -2.3939 & \mathbf{155.9465} & \mathbf{34.8055} \\ -0.8382 & \mathbf{-49.3471} & \mathbf{-31.4219} \end{bmatrix}$$

Thus we typically have several matrices that would give us the needed χ_{A+BK} but, not within the bounds of K. However, the distribution of these matrices w.r.t $\|K\|$ is rather surprising, and we have a larger pool candidate matrices with low $\|K\|$. In this example(ref Fig. 1), we could see that the no. of matrices with smaller norm (around 75) is larger compared to matrices with larger norm (around 250).

Example 2: Let

$$A2 = \begin{bmatrix} -6 & -6 & -2 & -5 \\ 4 & 5 & 4 & -7 \\ -3 & -6 & -9 & 2 \\ 8 & 6 & 3 & -6 \end{bmatrix}, \quad B2 = \begin{bmatrix} -1 & -8 & 9 & -9 \\ -6 & -8 & 3 & 3 \\ -1 & 8 & 7 & -7 \\ 4 & -5 & -3 & 2 \end{bmatrix}$$

$\chi_2 = \lambda^4 + 16\lambda^3 + 91\lambda^2 + 216\lambda + 180$: the roots are $p_{2i} = \{-2, -3, -5, -6\}$

We chose $k_{ij} \in [-10, 10]$, and $t = 2$

The mean size of the cell-array Ce was around 152. The N_{avg} was found to be 27.18, and the mean of iterations in which we obtained K was around 3.28. One of the K matrices picked up at random is

$$K = \begin{bmatrix} -6.7459 & 4.7312 & 1.0120 & -0.5857 \\ -1.9001 & 2.3136 & 0.6755 & -0.8154 \\ -9.4758 & 6.3920 & 1.4253 & -0.2014 \\ -7.5628 & 3.3752 & 0.5336 & -0.0500 \end{bmatrix}$$

Example 3: Let

$$A3 = \begin{bmatrix} 1\,2\,4 & 7\,8 \\ 3.5\,5\,6 & 7\,1 \\ 2\,0\,4 & 3\,1 \\ 2\,4\,0 & 1\,5 \\ 4\,0\,1 & 3.2\,5 \end{bmatrix}, \quad B3 = \begin{bmatrix} 1\,2\,0\,0\,1 \\ 5\,2\,4\,3\,1 \\ 5\,6\,2\,1\,8 \\ 1\,0\,0\,3\,1 \\ 0\,4\,1\,2\,5 \end{bmatrix}$$

$\chi_3 = \lambda^5 + 21\lambda^4 + 160\lambda^3 + 540\lambda^2 + 784\lambda + 384$: the roots are $p_{3i} = \{-1, -2, -4, -6, -8\}$

We chose $k_{ij} \in [-20, 20]$, and $t = 2$

The mean size of the cell-array Ce was around 80. The N_{avg} was found to be 66.57, and the mean of iterations in which we obtained K was around 7.70. One of the K matrices picked up at random is

$$K = \begin{bmatrix} -3.9639 & -4.9019 & 3.0024 & -2.6815 & -7.9946 \\ -6.9545 & -5.9529 & 4.8355 & -4.9006 & -14.7509 \\ 7.4833 & 8.5779 & -8.0223 & 3.8651 & 17.3521 \\ -4.3739 & -5.7437 & 3.8507 & -2.8762 & -9.6886 \\ 7.0247 & 7.6802 & -6.6250 & 4.1426 & 14.9744 \end{bmatrix}$$

Example 4: Let

$$A4 = \begin{bmatrix} 12 & 5 & -9 & 2 & 2 & 0 \\ 3 & 15 & -7 & 3 & 14 & -11 \\ 10 & -3 & -9 & 2 & 8 & 1 \\ 2 & 13 & 6 & 11 & 11 & -2 \\ -8 & -1 & 13 & -3 & -10 & -2 \\ 3 & 0 & 7 & 7 & -7 & 9 \end{bmatrix}, \quad B4 = \begin{bmatrix} 2 & 14 & -5 & -3 & -7 & -9 \\ -11 & 3 & 11 & -5 & -15 & -5 \\ 2 & 13 & -9 & 7 & 14 & -12 \\ -4 & -9 & 11 & -3 & -2 & 13 \\ 4 & 5 & -3 & 7 & 9 & 1 \\ -2 & -2 & 4 & -7 & 3 & 1 \end{bmatrix}$$

$$\chi_4 = \lambda^6 + 30\lambda^5 + 352\lambda^4 + 2046\lambda^3 + 6127\lambda^2 + 8724\lambda + 4320$$

The roots are $p_{4i} = \{-1, -3, -4, -5, -8, -9\}$
We chose $k_{ij} \in [-20, 20]$, and $t = 5$.

The mean size of the cell-array Ce was around 88. The N_{avg} was found to be 80.04, and the mean of iterations in which we obtained K was around 39.14. One of the K matrices picked up at random is ($\|K\| = 78.07$)

$$K = \begin{bmatrix} -9.8092 & 4.2271 & 8.3806 & -0.3087 & 5.4256 & -8.6294 \\ -9.4148 & 4.8890 & 8.8510 & 1.3834 & 6.8169 & -7.9784 \\ -9.8498 & 2.2236 & 7.9708 & -1.4176 & 3.4295 & -7.5158 \\ 16.7665 & -9.7601 & -16.1202 & -2.0096 & -12.3905 & 15.7564 \\ -12.6874 & 7.4932 & 12.1943 & 1.3270 & 8.8585 & -11.9253 \\ 19.5513 & -10.0200 & -19.7173 & -2.2365 & -11.8084 & 16.5656 \end{bmatrix}$$

Clearly, randomization was effective for a successful search. The results are summarized in the table below for a quick reference.

SUMMARY OF RESULTS

A B	roots of χ	k limits	$\overline{Ce}$	Mean iterations
A1 B1	p_{1i}	$[-05, 10]$	1120	4.01
A2 B2	p_{2i}	$[-10, 10]$	152	3.28
A3 B3	p_{3i}	$[-20, 20]$	80	7.70
A4 B4	p_{4i}	$[-20, 20]$	88	39.14

There are many more examples we have generated; for brevity we have shown a few significant ones.

4 Conclusions

The distributions of the candidate K matrices with respect to the norm $\|K\|$ for examples 1 and 4 are shown in figure 1 & figure 2 respectively.

It is noteworthy that, as the interval $[\underline{k}_{ij}, \overline{k}_{ij}]$ becomes larger the number of mean iterations increases significantly; simultaneously the pool size comes down. But, given practical problems such as actuator constraints in process control, bandwidth problem[4], the very motivation for imposing the limits on k_{ij}, it is rather surprising to note that obtaining a matrix with smaller $\|K\|$ can be pretty easy using randomization. We are currently investigating this phenomenon further with higher order systems. We are also investigating a similar problem called partial pole-placement, whose complexity is not known till now.

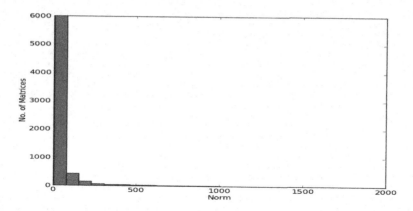

Fig. 1 Distribution of Norm Vs No. of Matrices(Ex:1)

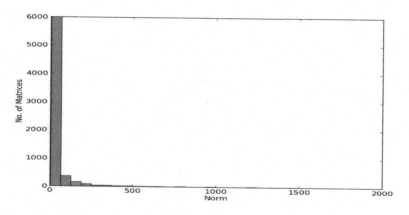

Fig. 2 Distribution of Norm Vs No. of Matrices(Ex:4)

References

[1] Ayyagari, R.: Control Engineering:A Comprehensive Foundation, 1st edn. Vikas Publication (2003)
[2] Blondel, V.D., Tstiskilis, J.N.: Np-hardness of some linear control design problems. SIAM J. Control Optim. 35, 2118–2127 (1997)
[3] Brogan, W.L.: Modern Control Theory, 3rd edn. Prentice Hall (1991)
[4] Callier, F.M., Desoer, C.A.: Linear System Theory, 1st edn. Springer, NY (1991)
[5] Vidyasagar, M., Blondel, V.D.: Probabilistic solutions to some np-hard matrix problems. Automatica 37, 1397–1405 (2001)

An Exploration of Wavelet Transform and Level Set Method for Text Detection in Images and Video Frames

V.N. Manjunath Aradhya, M.S. Pavithra, and S.K. Niranjan

Abstract. In texture-based text detection method, text regions are detected by obtaining textural properties of an image. In order to obtain textural properties of an input image, the proposed system performs single-level 2D DWT. The resultant detail coefficients are averaged to get a better texture properties and to localize for further processing. Then, 2D DWT is explored with a level set method to address the problem of text detection especially curving portions of text present in images and video frames. Thus, the proposed system implements the level set method to detect the true text regions effectively based on contours in images and video frames. Experimental results prove that the proposed level set based method is competitive when compared with other existing methods in reducing false positive rate and mis detection rate. Hence, the proposed system is encouraging and useful to carry out further research on text extraction in images and video.

Keywords: Single-Level 2D DWT, Level Set Method, Text Detection.

1 Introduction

With the advanced development of television, internet and wireless network, the demand of video indexing and retrieval is increasing. In this developed multimedia technology, an information is very well exhibited with the text and other media. The automatic detection, extraction and recognition of such a text is of great need to determine the topics of interest in digital video libraries. In this regard, detection of text in images and video frames plays a vital role.

V.N. Manjunath Aradhya and S.K. Niranjan
Department of MCA, Sri Jayachamarajendra College of Engineering, Mysore, India
e-mail: aradhya.mysore@gmail.com, sriniranjan@gmail.com

M.S. Pavithra
Department of MCA, Dayananda Sagar College of Engineering, Bangalore, India
e-mail: ngspavithra@gmail.com

S.M. Thampi et al. (eds.), *Recent Advances in Intelligent Informatics*,
Advances in Intelligent Systems and Computing 235,
DOI: 10.1007/978-3-319-01778-5_43, © Springer International Publishing Switzerland 2014

1.1   Related Research Results

Many methods on text detection in images and video frames have been proposed. TV commercial detection based on shot change and text extraction is proposed in [1]. In this, text detection algorithm is based on maximum gradient difference. A method [2], is proposed based on wavelet and color features for text detection in video. Recently, a method for arbitrarily-oriented text detection in video is presented in [3]. In this dominant text pixel selection, text representatives and region growing procedures are performed. Multioriented video scene text is detected in [4] using Bayesian classification and boundary growing methods. In this approach, product of Laplacian and Sobel operations is performed to enhance text pixels and a Bayesian classifier is used to classify true text pixels. Finally, multioriented scene text lines are detected using boundary growing method. Automatic caption extraction of news video and its implementation is presented in [5]. This algorithm uses adaptive method and projection method respectively. Fuzzy cmeans method is applied to obtain the character area.

From the literature survey, it is clear that the text orientation information is needed to detect arbitrarily-oriented and multioriented texts. In this regard, methods such as boundary growing methods, trained classifier, etc. are used. Towards this, we explored an idea "wavelet transforms together with the level set method "to detect a arbitrarily-oriented and multioriented text of English and South Indian languages in images and video frames. Level set framework is mainly meant for curve evolution and is being used in many of the computer vision and medical image applications. The remaining structure of the paper is organized as: first proposed method is described. Second experimental results and performance evaluation is presented and finally conclusion and future work is presented.

2   Proposed Method

In this section two stages of the proposed system are explained: 1) Single-level 2D DWT 2) Level set method. Initially, in the pre-processing stage an input color image/video image is converted into gray image then the resulted image is processed with the image size 500x500.

2.1   Single-Level 2D DWT

A study of discrete wavelet transform (DWT) in [6] describes series expansion of signals in terms of wavelets and scaling functions. With DWT, the signal is analyzed at different frequency bands with different resolutions by decomposing the signal into a coarse approximation and detail information. In the present work, we applied a single level 2D DWT, with that a signal is passed through a High Pass Filter(HPF)

(a) Input image of 101 video images dataset.

(b) Result of single level 2D wavelet decomposition.

(c) Result of averaged detail coefficients.

(d) Resulted text detection image of level set method.

Fig. 1 Step-by-step results of the proposed method

and a Low Pass Filter(LPF), with coefficients associated with each wavelet system. We experimented on Daubechies family's first order Daubechies wavelet function i.e., db1 because of its compact supportness property. Though the other wavelets may yield similar results, in this experiment we used db1 wavelets. As a result of single-level 2D wavelet decomposition, approximate and detail coefficients are obtained with texture features and edge information which is shown in Figure 1(b) for an input image of 101 video images dataset Figure 1(a). Further, detail coefficients (LH, HL, HH) of the resultant image are averaged and is shown in Figure 1(c).

2.2 Level Set Method

Level set method is a thriving technique in the field of image science [7]. The method was introduced by Osher and Sethian [8] in 1987, for capturing moving fronts. Level set method for an image is described as a method which analyzes and manipulates the level sets of a given continuous function i.e., an image. The method handles the topological changes in the curves with no emotional involvement [8]. In the proposed system, we applied a level set method of [9]. This method is general and robust. In this, the level set method uses a Gaussian filter to regularize the selective binary level set function after each iteration. The standard deviation of the Gaussian filter controls the regularization strength. The procedure of the level set method is summarized as follows:

$$\frac{\partial \phi}{\partial t} = spf(F(x)).\alpha |\nabla \phi| \tag{1}$$

Where $x \in \Omega$ and $spf(F(x)) = \dfrac{\left(F(x) - \frac{a_1 + a_2}{2}\right)}{max\left(\left|F(x) - \frac{a_1 + a_2}{2}\right|\right)}, x \in \Omega$

where a_1 and a_2 are as follows:

$$a_1(\phi) = \frac{\int_\Omega F(x).H(\phi)dx}{\int_\Omega H(\phi)dx} \quad \text{and} \quad a_2(\phi) = \frac{\int_\Omega F(x).(1-H(\phi))dx}{\int_\Omega (1-H(\phi))dx}$$

where F is the average image of detail co-efficients obtained by performing single level 2D wavelet decomposition in the proposed system and $H(\phi)$ is the Heaviside function.

1. Initialize the level set function ϕ as in [9].
2. Compute $a_1(\phi)$ and $a_2(\phi)$.
3. Evolve the level set function according to Equation 1.
4. Let $\phi = 1$ if $\phi > 0$; otherwise $\phi = -1$.
5. Regularize the level set function with a Gaussian filter i.e., $\phi = \phi * G_\sigma$.
6. Check whether the evolution of the level set function has converged. If not, return to step 2.

In step 5, We used Gaussian filter to smoothen the level set function. The standard deviation σ of the Gaussian filter G_σ is a critical parameter. If σ is too small, the proposed method detects text blocks with non-texts present in the image. On the other hand, if σ is too large, edge leakage may occur and the detected boundary of the text may be inaccurate. Hence, we iteratively experimented and set $\sigma = 7$. As a result the level set method detects true text regions effectively. The resulted image of level set method is shown in Figure 1(d). In step 6, we empirically set the number of iterations, so that the evolution process is stopped till the completion of the specified iterations. As a result, closed boundary around the text blocks got evolved.

3 Experimental Results and Performance Evaluation

The proposed system is experimented on intel CORE Duo 2.0 GHz machine with MATLAB R2008a. In the experiment, we used three challenging datasets: (1) A dataset of 101 video images of [10]. (2) Our own collected multilingual dataset. (3) MSRA Text detection 500 database [11].

3.1 Experiment on 101 Video Images

A dataset of 101 video images [10] is a collection of images, which are extracted from news programmes, sports, videos and movie clips. The dataset comprises both graphic and scene texts. The performance of the proposed system is evaluated at the block level. The blocks are determined as the categories described in [10]. The performance of the system is evaluated by manually counting the Actual Text Blocks(ATB) in an each image of the dataset. We compared the proposed method with the existing text detection methods of [12], [10], [13] & [14]. To evaluate the

Table 1 Results obtained for the dataset of 101 video images

Method	ATB	TDB	FDB	MDB
Edge-based [12]	491	393	86	79
Laplacian [10]	491	458	39	55
Transforms and Gabor based [13]	491	481	78	53
Gabor and k-means based [14]	491	486	78	15
Proposed	491	465	64	3

Table 2 Performance results obtained on the dataset of 101 video images

Method	DR	FPR	MDR
Edge-based [12]	80.0	18.0	20.1
Laplacian [10]	93.3	7.9	12.0
Transforms and Gabor based [13]	97.9	13.9	11.0
Gabor and k-means based [14]	98.9	13.7	3.0
Proposed	94.7	12.09	0.64

performance of the proposed method, we used the performance measures defined in [10].

From Table 1, it is observed that considerable number of TDBs are obtained, FDBs are sustained as in [13], MDBs are considerably reduced. From Table 2, it can be seen that the proposed method has reached lesser FPR, lowest MDR compared to existing methods and DR is sustained as of [10]. This quantitative evaluation proves that the proposed system detects most of the accurate text blocks with a very few missing information and few false alarms. Figure 2 shows the sample results of the four existing methods and the proposed method. From the figure, it is observed that the proposed method detects the texts of low and high contrast with the presence of false detected block. From this, we can observe that the low contrast text block is also detected though the boundary is not exactly around the text, but the missing of text block is no where observed in the resultant image of the proposed method. Hence the proposed system detects text by reducing much of the miss detection rate.

3.2 Experiment on Multilingual and MSRA Text Detection 500 Database

We experimented on our own collected South Indian language dataset comprising 130 images of Kannada, Telugu, Tamil and Malayalam languages. The dataset comprises texts in an image with complex and varying background. The experimental study of this dataset shows that the texts present in images were detected including its modifiers and compound bases in simple background, where as in complex

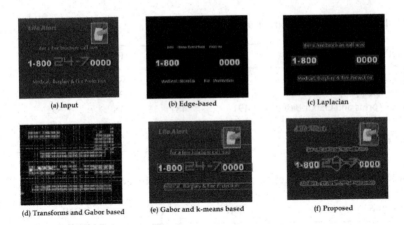

(a) Input   (b) Edge-based   (c) Laplacian

(d) Transforms and Gabor based   (e) Gabor and k-means based   (f) Proposed

Fig. 2 The detected text blocks of the four existing methods and the proposed method for an input image of 101 video images

background texts were detected with the presence of few false alarms. Few of the results of multilingual text detection images are shown in Figure 3(a). We also experimented the proposed system on MSRA Text detection 500 database [11]. The database is a collected and publicly available database, especially for detecting texts of arbitrary orientations. The database contains 500 natural images, which are taken from indoor and outdoor scenes using a packet camera. As a result, we obtained accurate text detection in simple background images with the presence of arbitrary oriented texts and results are shown in Figure 3(b). In the complexed images, texts were detected with the presence of few false alarms.

(a)

(b)

Fig. 3 Successful text detection results of: (a) our own collected multilingual dataset.(b) MSRA-TD500 database

4 Conclusion and Future Work

An effective approach of text detection in images and video frames comprising English and South Indian languages is introduced. The main contribution in the present system is that, wavelet transforms and a level set method is implemented to detect true text region with a small period of time. The most beneficiary thing in applying 2D DWT with level set method is that, the boundary of the text region are immediately get detected by obtaining the texture features of an input image. The main purpose of developing this algorithm is to detect a text of English and especialy South Indian languages text that are framed with curvings, compound bases and extra modifiers. Hence, the proposed system better suitable for detecting horizontal, vertical and multi-oriented texts of English and South Indian languages presented in images and video frames. In near future, we work for a metric to measure the textural properties inorder to reduce false alarms atmost.

Acknowledgements. We would like to thank the authors of [10] for providing a dataset of 101 video images for the purpose of experiment.

References

1. Meng, L., Cai, Y., Wang, M., Li, Y.: TV Commercial Detection Based on Shot Change and Text Extraction, pp. 10–13. IEEE (2009)
2. Shivakumara, P., Phan, T.Q., Tan, C.L.: New Wavelet and Color Features for Text Detection in Video. In: Proceedings of International Conference on Pattern Recognition, pp. 3996–3999 (2010)
3. Sharma, N., Shivakumara, P., Pal, U., Blumenstein, M., Tan, C.L.: A New Method for Arbitrarily-Oriented Text Detection in Video. In: Proceedings of 10th IAPR International Workshop on Document Analysis Systems, pp. 74–78 (2012)
4. Shivakumara, P., Sreedhar, R.P., Phan, T.Q., Lu, S., Tan, C.L.: Multioriented Video Scene Text Detection Through Bayesian Classification and Boundary Growing. IEEE Transactions on Circuits and Systems for Video Technology 22, 1227–1235 (2012)
5. Yi, X.G.: Automatic Caption Extraction of News Video and its Implementation, pp. 122–125 (2012)
6. Polikar, R.: The Wavelet Tutorial part IV, Multiresolution Analysis:The Discrete Wavelet Transform. Rowan University (2004), Available via DIALOG
   `http://users.rowan.edu/~polikar/WAVELETS/WTpart4.html`
7. Tsai, R., Osher, S.: Level Set Methods and Their Applications in Image Science. Communications in Mathematical Sciences 1(4), 1–20 (2003)
8. Osher, S., Sethian, J.A.: Fronts propagating with curvature-dependent speed: algorithms based on Hamilton-Jacobi formulations. J. Comput. Phys. 79(1), 12–49 (1988)
9. Zhang, K., Zhang, L., Song, H., Zhou, W.: Active contours with selective local or global segmentation: A new formulation and level set method. Image and Vision Computing 28, 668–676 (2010)
10. Phan, T., Shivakumara, P., Tan, C.: A Laplacian method for video text detection. In: Proceedings of 10th International Conference on Document Analysis and Recognition, pp. 66–70 (2009)

11. Yao, C., Bai, X., Liu, W., Ma, Y., Tu, Z.: Detecting Texts of Arbitrary Orientations in Natural Images. In: Proceedings of IEEE Conference on Computer Vision and Pattern Recognition (CVPR), pp. 1083–1090 (2012)
12. Liu, C., Wang, C., Dai, R.: Text detection in images based on unsupervised classification of edge-based features. In: Proceedings of ICDAR, pp. 610–614 (2005)
13. Aradhya, V.N.M., Pavithra, M.S., Naveena, C.: A robust multilingual text detection approach based on transforms and wavelet entropy. In: Proceedings of 2nd International Conference on Computer, Communication, Control and Information Technology(C3IT 2012), Procedia Technology, pp. 232–237. Elsevier (2012)
14. Manjunath Aradhya, V.N., Pavithra, M.S.: An application of K-means clustering for improving video text detection. In: Abraham, A., Thampi, S.M. (eds.) Intelligent Informatics. AISC, vol. 182, pp. 41–47. Springer, Heidelberg (2013)

Alamouti Space Time Coded Design for OFDMA Systems Based Layered FFT Structure

T. Deepa and R. Kumar

Abstract. Orthogonal Frequency-Division Multiple Access (OFDMA) is a multi-user orthogonal frequency-division multiplexing (OFDM) digital modulation scheme. Multiple accesses can be achieved in OFDMA by assigning subsets of subcarriers to individual users. This allows simultaneous low data rate transmission from several users. However, the OFDMA system exhibits high peak to average power ratio (PAPR) and introduce larger complexity by applying discrete Fourier transform (DFT) at the transmitter and inverse DFT (IDFT) at the receiver section. In this paper, we propose a space time block coding (STBC) for OFDMA based layered FFT structure, which provides better PAPR reduction property, low complexity in user terminals. The STBC scheme is also provide full diversity from both OFDMA signaling and STBC coding transmission.

Keywords: D&C FFT algorithm, OFDM, PAPR, Space Time Coding.

1   Introduction

Future broadband wireless mobile communication systems require high data rate transmissions through severe multipath channels [1].

As an effective anti multipath multicarrier modulation scheme, orthogonal frequency division multiplexing (OFDM) is adopted by leading standards such as HIPERLAN/2, IEEE802.11, IEEE 802.16 and 3GPP long term evolution (LTE). Wireless channel exhibits a number of severe impairments such as fading, which is one of the most severe. Transmit diversity is an effective technique to improve transmission over fading channels. One of the most popular transmit diversity schemes has been proposed by Alamouti[2],which uses two transmit antennas and one receive antenna. It is applied over two transmit antennas and two multiple

T. Deepa · R. Kumar
SRM University
e-mail: {deepa.t,kumar.r}@ktr.srmuniv.ac.in

S.M. Thampi et al. (eds.), *Recent Advances in Intelligent Informatics*,
Advances in Intelligent Systems and Computing 235,
DOI: 10.1007/978-3-319-01778-5_44, © Springer International Publishing Switzerland 2014

access OFDM (OFDMA) symbols [3]. The STBC is a diversity scheme, which can achieve full diversity from both coded OFDMA signaling and STBC coding transmission. However, the number of subcarriers, N, in OFDMA system is usually very high, which provides high spectral efficiency, at the expense of high peak to average ratio (PAPR), high complexity in the transceiver, and sensitivity to carrier frequency offset(CFO). Hence, OFDM system must be combined with forward error correcting codes to improve the BER performance of the overall system for high data rate applications. Recently, LDPC coding is considered the most powerful coding solution known in many digital communication systems [4]. Complexity reductions for Fast Fourier Transform (FFT) computation are useful in OFDMA systems where individual transceivers do not need to modulate or demodulate every sub channel. Divide & Conquer FFT algorithm with a larger L point FFT transform can be used for low cost transceiver structures.

In multi user communication systems, transmitters and receivers may have different cost and capability requirements. For instance, in an uplink scenario, multiple transmitters may simultaneously send separate signals to one receiver. The receiver may have high cost, provided the transmitters have low cost. Alternatively, in a downlink scenario, one transmitter sends the same composite signal to multiple receivers [5]. Each receiver may only be interested in a small fraction of the transmitted data. The transmitter may have high cost, provided the receivers have low cost. A FFT is an algorithm to convert a sampled function from its original domain to the frequency domain. A layered FFT structure concept involves the division of a single complex problem into two or more simple sub-problems using the, Divide and Conquer inverse FFT (IFFT)/FFT algorithm (D&C IFFT/FFT) works by recursively breaking down a problem into two or more smaller IFFT/FFT of the same or related type, until these become simple enough to be solved directly [6]. In this paper, we present Alamouti STC-OFDMA system with D&C IFFT/FFT algorithm which can overcome the problems such as high PAPR and high complexity in user terminals on AWGN and Rayleigh channel [7,8] .

The remainder of the paper is organized as follows. In this Section, the divide and conquer algorithm is described .In section 2, the space time coded OFDMA transceiver is described. The performance metrics of the system is detailed in section 3. Section 4 is devoted to simulation results. In Section 5, conclusion is presented.

1.1   Divide and Conquer FFT Algorithm

In this section, a computationally efficient DFT algorithm is described. It is a D&C approach in which a DFT of size N, where N is a composite number is reduced to the computation of smaller DFTs from which the larger DFT is computed [9],[10] as in Fig.1. We recall the process of the three Layered based FFT structure (Frequency domain, Intermediate domain, time domain) using D&C approach to calculate an N-point DFT of the signal $x = \{x_n\}, n \in [0, N-1]$.

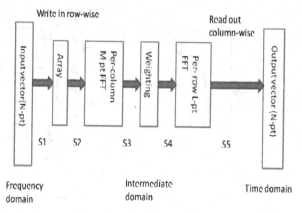

Fig. 1 Divide& Conquer FFT algorithm

The algorithm as follows:

Step1: store the N point sequence x(n) column wise into a $L \times M$ matrix
Where $l \in [0, L-1]$, $q \in [0, M-1]$ and N= $L \times M$

Step 2: Compute M point DFT for each X rows and this leads to the following

L*M new array $F(l,q) = \sum_{m=0}^{M-1} x(l,m) W_M^{mq}, \qquad 0 \le q \le M-1$ (1)

For each of the rows $l = 0,1,...L-1$

Step 3: Multiply the resulting array F(l,q) by the phase factors W_N^{lq} where,
$W_N = \exp(-j2\pi/N)$ and generate new matrix G(l,q) defined as

$$G(l,q) = W_N^{lq} F(l,q) \qquad \begin{array}{l} 0 \le l \le L-1 \\ 0 \le q \le M-1 \end{array}$$ (2)

Step 4: Finally, compute the L point DFTs for each column of G(l,q) matrix

$$X(p,q) = \sum_{l=0}^{L-1} G(l,q) W_L^{lp},$$ (3)

For each column q=0,1,...M-1 of the array G(l,q)

Step 5: Read the resulting array $L \times M$ matrix row wise.

2   System Model of Space Time Coded OFDMA

In this paper, we consider single user system with NT 2 transmits antennas and NR 1 receives antennas for OFDMA system based layered FFT structure.

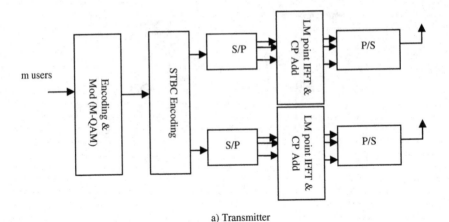

a) Transmitter

Fig. 1

The block diagram depicting the system is given in Fig. 2. In STC-OFDMA transmitter, the information data is modulated, LM point IFFTs (N=$L \times M$) are performed. From Fig.2, the information signal of proposed system is organized into an array of two dimensions, and their LM point IFFT is implemented. Unlike the general OFDMA system with N subcarriers, OFDMA system with $L \times M$ subcarriers is defined over an array of two dimensions. Firstly, N is factored as N=$L \times M$. The outputs of IFFTs are converted into parallel by using interleaved. Finally, the result is appended with cyclic prefix (CP). The length of CP is the same to that the conventional OFDMA systems.

Assume after IFFT transform of the output signal at the time k is X_k, the received signal Y corresponding to the transmitted signal can be expressed as,

$$Y = H_k X_k + n \tag{4}$$

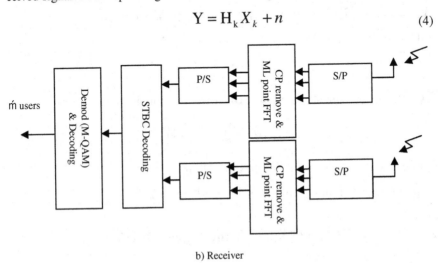

b) Receiver

Fig. 2 Block diagram of Space time coded OFDMA system

Where, H_k is the attenuation factor, with $H_k = 1$ in AWGN channel. At the receiver of the STC-OFDMA system, the received symbols are transformed to frequency domain using ML point FFT transform. Finally, the symbols are decoded by Space time decoding.

3   Performance Metrics

3.1   Complexity Analysis

The complexity of IFFT/FFT transforms is compared between the conventional and proposed OFDMA receivers. With M ary QAM modulation, a general OFDMA receiver includes an N point FFT and the complexity is $N/2 \log_2 N$. For a proposed OFDMA system, the receiver provides M point FFTs, L point weighting factors and the complexity is $N/2 \log_2 M + L\log_2 L$. From table 1, the OFDMA receiver based on ML point FFT approach requires less complexity, compared to the general OFDMA. This table indicates that complexity strongly depends on the value of L point IFFT/FFT transform.

Table 1 Layered based FFT structure system receiver complexity in terms of number of complexity multiplications

Systems	IFFT/FFT size	Complexity	Multiplication complexity
OFDMA with D&C	96 (L=24,M=4)	$N/2\log_2 M + L\log_2 L$	206
	96 (L=4,M=24)	$N/2\log_2 M + L\log_2 L$	228
OFDMA without D&C	96	$N/2\log_2 N$	316

4   Simulation Results

We consider the OFDMA system with N point DFT (N=96 is taken for an example) using 16-QAM modulation (M ary=64, where, k=4 bits/symbol), simulated by randomly generated data. Since N=LM (where L=24, M=4). In this paper, we consider the irregular LDPC 192×384 codes and the corresponding coding rate R is ½. The maximum iteration number of decoding equals 10.

Fig.3. shows the BER performance of SISO and 2×1 STBC OFDMA systems with 16 –QAM modulation. From Fig.3, it can be noticed that 2×1 STBC system achieves SNR about 13 dB at 10^{-2} BER level which is about 12 dB lower than system without space time coding. Fig.4. shows the BER performance of the Alamouti STBC OFDMA with and without D & C algorithm in AWGN channel. The STBC achieves spatial diversity gain comparing with the SISO system. A simulation result compares the BER performance of 2×1 STBC OFDMA system with different L point FFT.

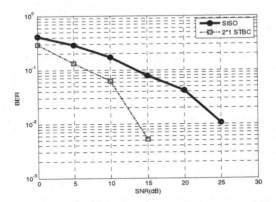

Fig. 3 BER performance of SISO and 2×1 STBC for OFDM system with L= 24 on an AWGN channel

In Fig.4, it has been shown for 2×1 STBC OFDMA systems, which the BER depends on the value of L. It can be seen that the system with L=24 point FFT has same and better BER performance than the system with L=12 and L=8 point FFT. Fig.5 compares BER performance for L=24, 12, 8 with N=96 in 2×1 STBC OFDMA system for both AWGN and Rayleigh fading channel.

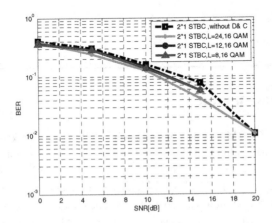

Fig. 4 BER performance of 2×1 STBC for OFDMA system with and   without D&C approach in AWGN channel

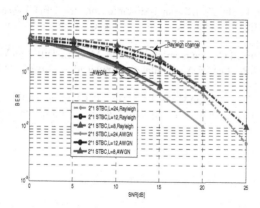

Fig. 5 BER performance of 2×1 STBC for OFDM system with L=8, 12, 24 on a Rayleigh fading channel

5   Conclusions

Complexity reduction for FFT computation is useful in OFDMA systems where individual transceivers do not need to modulated or demodulate every sub channel. In this paper, we extend MISO spatial diversity for OFDMA system based on D & C IFFT/FFT structure systems. The proposed system is also takes advantage of full diversity introduced by STBC. Also, we have investigated the system based layered FFT structure in order to realize the significant low complexity, high throughput.

References

1. Goldsmith, A.: Wireless Communications, 1st edn. Cambridge University Press, Stanford University (2005)
2. Alamouti, S.M.: A simple transmit diversity technique for wireless communications. IEEE Journal on Selected Areas in Communications 16(8), 1451–1458 (1998), doi:10.1109/49.730453.
3. Nee, R.V., Prasad, R.: OFDM wireless multimedia communications. Artech House, London (2000)
4. Zheng, X., Lau, F.C.M., Tse, C.K.: Performance evaluation of irregular low-density parity-check codes at high signal-to-noise ratio. IET Communications 5, 1587–1596 (2011)
5. Murphy, C.D.: Low-complexity FFT structures for OFDM transceivers. IEEE Transactions on Communication 50, 1878–1881 (2002)
6. Myung, H.G., Lim, J., Goodman, D.J.: Single carrier FDMA for uplink wireless transmission. IEEE Vehicular Technology Magazine 1(3), 30–38 (2006)
7. Zhang, J., Luo, L., Shi, Z.: Quadrature OFDMA systems based on layered FFT structure. IEEE Transactions on Communications, 850–860 (2009)

8. Luo, L., Zhang, J., Shi, Z.: BER analysis for asymmetric OFDM systems. In: Proceed-
    ings of GLOBECOM 2008, New Orleans, La, USA, pp. 1–6 (November-December
    2008)
9. Proakis, J.G.: Digital Communications, 4th edn. McGraw-Hill, NewYork (2001)
10. Barry, D.G.M.J.R., Lee, E.A.: Digital Communication, 3rd edn. Kluwer Academic
    Publishers (2003)
11. Costa, E., Pupolin, S.: M-QAM-OFDM System Performance in thePresence of a Non-
    linear Amplifier and Phase Noise. IEEE Transactions on Communications, 462–472
    (March 2002)

Novel SVD Based Character Recognition Approach for Malayalam Language Script

S. Sachin Kumar, K. Manjusha, and K.P. Soman

Abstract. The research on character recognition for Malayalam script dates back to 1990's. Compared to other Indian languages the research and developments on OCR reported for Malayalam script is very less. The character level and word level accuracy of the existing OCR tools for Indian languages can be improved by implementing robust character recognition and post-processing algorithms. In this paper, we are proposing a character recognition procedure based on Singular Value Decomposition (SVD) and k- Nearest Neighbor classifier (k-NN). The proposed character recognition scheme tested with the dataset created from Malayalam literature books and it could classify 94% of character images accurately.

1 Introduction

Optical Character Recognition (OCR) is the process of transforming scanned document images to machine understandable or editable form. From 1950 onwards, Optical Character Recognition (OCR) is been an active area of research in pattern recognition [8]. OCR technology successfully applied with different applications like, invoice processing, postal address readers, etc. It reduces the overhead in data entry applications and enhances the interface between human and computer. The idea of OCR was originated during making a reading aid for visually challenged. In 1955, the first commercial system was installed at the Reader's Digest for transfering sales reports into computer. Later the technology progressed like anything and now the OCR applications are existing even in mobile platforms. Commercial OCR systems available today report accuracy ranging from 71% to 98% in good quality documents. Besides that, open source OCR products also exist. OCRopus and SimpleOCR are opensource OCRs available today.

S. Sachin Kumar · K. Manjusha · K.P. Soman
Centre for Excellence in Computational Engineering and Networking, Amrita School
of Engineering, Coimbatore, India
e-mail: {sachinnme,manjushagecpkd}@gmail.com, kp_soman@amrita.edu

S.M. Thampi et al. (eds.), *Recent Advances in Intelligent Informatics,*
Advances in Intelligent Systems and Computing 235,
DOI: 10.1007/978-3-319-01778-5_45, © Springer International Publishing Switzerland 2014

With OCR, the massive data digitization of books and manuscripts is possible. Before the internet age, books were used as the information transfer medium. Digital Library of India (DLI) [4] started with the mission to provide free-to-read, searchable collection of one million books, over the internet. As per the report on 4th December 2012, DLI have scanned totally 3,71,110 books (in more than 30 different languages). But the content inside those digital archives cannot be edited to the required format or is inaccessible through normal search. Once all those digital archives are made searchable, it opens the door to the vast world of knowledge and information which can make revolutions and revelations in academic, literature, research communities etc.Even though good quality OCR systems are existing for Latin, Arabic, Chinese scripts, very few works on OCR are reported in the case of Indian language scripts and a complete OCR system for Indian language scripts is still in its nascent stage.

Unlike Roman system, Indian language scripts have alphabetic—syllabic nature [2]. The recognition of Indian language scripts is a challenging task because of the large character set, structural similarity among character classes, irregular positioning of characters in a running text and existence of both old and new version of language scripts. [8]. The visual similarity among Indian language scripts may help in developing a general OCR engine by integrating all scripts, at the same time it may create a great difficulty in case of script identification tasks [2].

Usually OCR tools use a feed forward architecture. It includes the Pre-processing, Segmentation, Recognition and Post-Processing processes. The accuracy or the efficiency of each stage is highly dependent on the accuracy of preceding stages [7]. Pre-processing include the noise removal, skew correction and binarization of scanned document. Segmentation process is a bit difficult in Indian language scripts because of the irregular positioning of character in running text. Besides that, Bengali, Devanagari and Gurumukhi scripts contain a horizontal headline called 'Sirorekha' which makes the segmentation process more challenging. Levelset based active contour segmentation can be applied for Indian language scripts [13] [6][1]. As mentioned earlier, the similar shapes among characters and large character classes are making multi stage recognition scheme more appropriate for Indian language scripts [11] [10].

In this paper, we are focusing on character recognition in Malayalam script. Section 2 describes about the specialties and the different types of characters in Malayalam script. Section 3 is discussing about the proposed character recognition scheme and the algorithms utilized. Section 4 and 5 provide details about implementation of proposed scheme.

2 Malayalam Script

Malayalam is an Indian language spoken by 40 million people mostly in the southwestern state of Kerala [8]. Malayalam is closely related to the languages of Tamil and Sanskrit. The language started as a variant of Tamil that was spoken in regions of Kerala, and evolved its own form, grammar, and vocabulary by 500 CE [8]. The word Malayalam probably originated from the Malayalam/Tamil words mala meaning hill, and elam meaning region [5].

In Malayalam script, the vowel symbols are termed as 'Swaraksharangal' and consonants as 'Vyanjanaksharangal'. Totally 13 vowels, 36 consonants and 5 pure consonants are used in Malayalam script. Compound characters are formed by combining the basic consonants either vertically or horizontally. Besides this, dependent vowels, crescent mark ('Chandrakala') and Vakar symbols are used in Malayalam language script[refer] The existence of both old and new lipi in document images makes the character recognition more complex [8].

3 Character Recognition

It is one of the crucial phases in OCR. Character recognition is the process of mapping the character image with its Unicode representation. Feature extraction algorithms [15] can be used to extract feature vector from the character image. Classifier finds the best match of this feature vector with trained feature to identify the corresponding Unicode character. In our approach, we are using SVD for extracting features from character images and k-NN for classifying this feature vector.

3.1 Singular Value Decomposition (SVD)

Singular value decomposition is a dimensionality reduction technique for matrices. It orders the information contained in the matrix so that, the *dominating parts* become visible [14]. SVD represents or transforms the data in the matrix to those axes in which variation among the data is maximum while keeping the total variation among data same.

Let A be an m×n matrix and it can be represented as the product of two orthogonal matrices (U and V) and a diagonal matrix (Σ).

$$A = U\Sigma V^T \tag{1}$$

U and V are two orthogonal matrices with dimensions m×m and n×n respectively. Σ is called the singular matrix, is an m×n diagonal matrix whose diagonal entries are non-negative real numbers called *Singular values*. If the rank of matrix A is r, then there will be only r nonzero diagonal entries inside Σ. The diagonal entries of Σ have the property that they will be in decreasing order. SVD provides the best low-rank approximation of matrix A [3].

3.2   k-Nearest Neighbor (k-NN) Classifier

It is one of the simplest non-parametric machine learning algorithm. The idea in k-NN is to dynamically identify k observations in the training data set that are similar to the new observation and to use those observations to classify the new observation to target class. Finding neighbors can also be termed as finding the dissimilarity measure and can be calculated by using methods for finding distance such as Euclidean distance. It needs no explicit training, it keeps all training data which makes it time consuming when number of training data increases. When a character image comes for classification, the algorithm finds the k closest neighbors of the character image and classifies it. Accuracy wise, k-NN performs well among large number of character classes. The performance of the algorithm is greatly depended up on the selection of k. When k becomes 1, the algorithm finds the nearest neighbor to the test sample.

3.3   Proposed Methodology

In this paper, we are proposing a character recognition approach, which combines the strength of SVD and the k-NN. SVD can closely approximate the given training matrix such that even in lower dimensions it captures the latent relations existing among data present. k-NN can accurately classify among large number of classes by compromising time. The proposed character recognition procedure can be summarized as follows. For training, all the character images are resized to 32×32.

1. Read the training images, vectorize it and append it vertically to matrix A. Finally the columns of matrix A represent the training character images. The dimension of A will be 1024×number of training images.
2. Find SVD of A

$$A = U\Sigma V^T \tag{2}$$

3. Diagonal of Σ matrix contains the singular values in the decreasing order. Take only the first few values and build Σ^* and V^*. Find H such that,

$$H = \Sigma^* V^* \tag{3}$$

4. Now an approximation to A can be written as

$$A = U^* H \tag{4}$$

As U is the orthogonal bases, we can say that each column in H represents the projection co-efficient of each training image to U. H can act as the features extracted from training images.

5. During testing, the vectorized character image Img will be projected to U^* to obtain the feature vector.

$$\text{Img}_{feature} = U^{*T} \text{Img} \tag{5}$$

6. Find the close k neighbor classes of $\text{Img}_{feature}$ in H. Among them, take those classes whose distance measure is less than the fixed tolerance and take the mostly occurring character class as the target class of test image.

4 Experimental Setup

The proposed algorithm is evaluated on both scanned and synthetic images. Scanned database includes document images from various Malayalam literature books and newspapers at different resolutions. Synthetic database includes Malayalam characters with various styles and sizes. For the experiment purpose we considered 75 classes of Malayalam Unicode characters with all the consonants, vowels, half consonants and some commonly used compound characters. Each class have more than 200 character images from which 75% taken for training and rest for testing. The experiments are conducted in MATLAB environment.

5 Results and Discussion

The performance of the algorithm is evaluated using misclassification rate, which is defined as:

$$MisclassificationRate = \frac{WronglyClassifiedTestImages}{TotalNo.ofTestImages} \quad (6)$$

The algorithm tested with different number of singular values. The result shows that misclassification rate is very high for small number of singular values. Misclassification rate decreases with the increased number of singular values. Fig. 1 shows the relation between number of singular values considered and the number of misclassifications obtained. Initially total number of misclassifications is high for small number of singular values. The number of misclassification remains almost same (varies within a range) after it reaches the minimum, then it again rises with the number of singular values. But the rate of increase in number of misclassification is at very low pace. The minimum misclassification rate of 5.44% is obtained with first 58 singular values.

Optimal value for the number of singular values (the dimension of Σ^*) is the parameter which will be used for testing process.

Fig. 2 shows the zoomed minimum misclassification region of fig.1. From the graph, we can observe that the misclassification rate is abruptly varying with the number of singular values. The result shows that the computation of number of singular value at which optimal misclassification attained is an iterative process.

Confusion matrix is created to find the mostly misclassifying character classes. The most misclassified character classes are shown in Table:1. Most of the misclassification occurred in between similarly shaped characters. So the proposed method

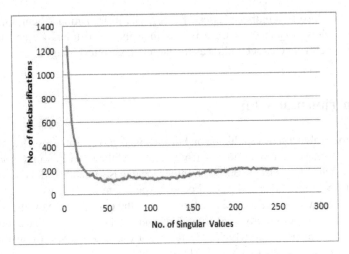

Fig. 1 Variation of number of misclassifications with number of singular values

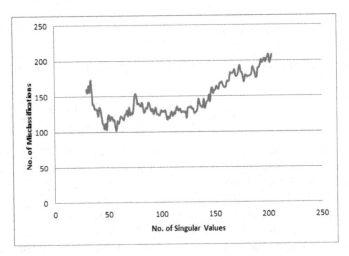

Fig. 2 Zoomed version of fig.1

is suitable for the first level of multi-stage classification scheme. Similarly shaped character classes can be grouped together in the first level of classification for reducing the error rate.

The drawback of the proposed technique is the time taken for classification increases with the number of target classes, number of training images taken and the number of singular values considered. Fig.3 shows the variation in execution time with the number of singular values considered. The drawback can be rectified by using parallel programming architecture for classification. The process of finding character class of each test image is independent and can be concurrently executed.

Table 1 Misclassified classes

Character Class	Misclassified as
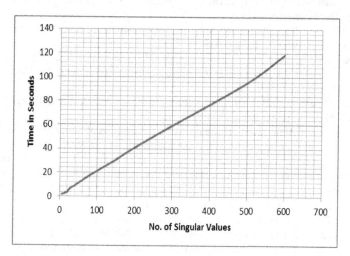	

Fig. 3 Relation between the numbers of singular values considered during each experiment and execution time

6 Conclusion

In this paper, we presented how Singular Value Decomposition (SVD) and k-Nearest Neighbor (k-NN) can be applied for character recognition. The result shows that the proposed method is suitable for large number of target classes. The classification accuracy of the proposed algorithm is 94%. For real time applications, parallel processing architecture can be used to reduce the execution time. Most of the misclassifications occurred in similarly shaped character classes. The accuracy can be improved by grouping those classes together and applying a second level of recognition among the grouped classes.

References

1. Manjusha, K., Sachin Kumar, S., Rajendran, J., Soman, K.P.: Hindi Character Segmentation in Document Images using Level set Methods and Non-linear Diffusion. International Journal of Computer Applications (0975 - 8887) 44(16) (2012)
2. Chaudhuri, B.B.: On OCR of a Printed Indian Script. In: Advances in Pattern Recognition Digital Document Processing. Springer, London (2007)
3. Kalman, D.: A Singularly Valuable Decomposition: The SVD of a Matrix. The American University, Washington (2002)
4. Digital Library of India, http://www.dli.ernet.in (cited December 10, 2012)
5. Malayalam Language, http://en.wikipedia.org/wiki/Malayalam (cited March 7, 2013)
6. Cherian, M., Radhika, G., Shajeesh, K.U., Soman, K.P., Sabarimalai Manikandan, M.: A Levelset Based Binarization and Segmentation for Scanned Malayalam Document Image Analysis. In: IEEE International Conference on Computational Intelligence and Computing Research (2011)
7. Meshesha, M.: Recognition and Retrieval from Document Image Collections. Ph. D. Thesis, IIIT Hyderabad, India (2008)
8. Neeba, N.V., Namboodiri, A., Jawahar, C.V., Narayanan, P.J.: Recognition of Malayalam Documents. In: Advances in Pattern Recognition, Guide to OCR for Indic Scripts. Springer, London (2009)
9. Pal, U., Chaudhuri, B.B.: Indian script character recognition: A survey. Pattern Recognition 37(9), 1887–1899 (2004)
10. Rahman, A.F.R., Rahman, R., Fairhurst, M.C.: Recognition of handwritten Bengali characters: a novel multistage approach. Pattern Recognition (2002)
11. Shivsubramani, K., Loganathan, R., Srinivasan, C.J., Ajay, V., Soman, K.P.: Multiclass Hierarchical SVM for Recognition of Printed Tamil Characters. In: Proc. of IJCAI (2007)
12. Mori, S., Suen, C.Y., Yamamoto, K.: Historical Review of OCR Research and Development. Proceedings of the IEEE (1992), doi:10.1109/5.156468
13. Soman, K.P., Ramanathan, R.: Level Set Theory for Image Segmentation. In: Digital Signal and Image Processing- The Sparse Way. Isa Publishers (2012)
14. Soman, S.T., Soumya, V.J., Soman, K.P.: Singular Value Decomposition A Classroom Approach. International Journal of Recent Trends in Engineering 1(2) (2009)
15. Trier, D., Jain, A.K., Taxt, T.: Feature Extraction Methods for Character Recognition-A Survey. Pattern Recognition 29, 641–662 (1996)

Experimentation and Analysis of Time Series Data for Rescue Robotics

Radhakrishnan Gopalapillai, Deepa Gupta, and T.S.B. Sudarshan

Abstract. In today's world, rescue robots are used in various life threatening situations where human help or support is not possible. These robots transfer real time data about the environment continuously. Research is focussed on techniques to analyse real time data to enable Decision Support Systems (DSS) to take timely actions to save lives. This paper discusses preliminary experiments that have been carried out to simulate a set of simple robotic environments. A robot attached with four sensors is used to collect information about the environments as the robot moves in a straight line path. Time series data collected from these experiments are clustered using data mining techniques. Experimental results show recall and precision between 73% to 98%.

Keywords: rescue robots, clustering, data mining, dynamic time warping, time series.

1 Introduction

Intelligent mobile robots are being more and more deployed in various fields. These robots navigate through their environment either autonomously or with different levels of human assistance. If a robot has to navigate through its environment effectively, it has to first understand its environment. Hence techniques for analyzing the robotic environment play an important role in the deployment of robots.

One of the areas which have received increased attention over the years is urban search and rescue situations. The advantage of using robots in search and rescue situations is that human life need not be put at risk for saving another

Radhakrishnan Gopalapillai · Deepa Gupta · T.S.B. Sudarshan
School of Engineering, Amrita Vishwa Vidyapeetham,
Bangalore Campus, Bangalore, India
e-mail: {g_radhakrishnan,g_deepa,tsb_sudarshan}@blr.amrita.edu

S.M. Thampi et al. (eds.), *Recent Advances in Intelligent Informatics*,
Advances in Intelligent Systems and Computing 235,
DOI: 10.1007/978-3-319-01778-5_46, © Springer International Publishing Switzerland 2014

human life. It has been reported that rescue robots have been deployed during the rescue operation after the collapse of World Trade Centre in New York [1]. Since then rescue robotics has got the attention of rescue teams as well as researchers. Robots are becoming more sophisticated with the use of advanced sensing devices and data processing techniques. The data captured by the sensors can be used to get a clear picture of the disastrous environment and an analyst can be assigned the job of analyzing this data to find out what actions to be performed in the respective situations. The challenge is to analyze the data faster to draw conclusions about the environment. This is where the importance of exploration of data mining techniques on robotic data comes into picture. When data from multiple robots are analyzed, it is important to identify whether a particular scenario seen by one robot is same or similar as a scenario seen by other robots. Since the robots move in unknown environments, prior information about the scenarios encountered by them will not be available Clustering techniques can be used to group together scenarios which are similar. In this paper, we design a simple indoor environment and then capture information about the environment through the sensors attached to a robot. A set of different environments have been simulated. The environments are designed in a way that the accuracy of clustering techniques employed can be measured and conclusions drawn on their effectiveness.

Remaining sections of this paper are organized as follows. Section 2 discusses some of the earlier work related to autonomous robots. The design of robotic environments used for the purpose of collecting data is described in Section 3. Section 4 discusses various data mining techniques used for clustering of the data collected. Analysis of the clusters and the accuracy of clustering are discussed in Section 5. Section 6 discusses the conclusions from this work.

2   Related Work

Mobile autonomous robots equipped with advanced sensors are increasingly being relied upon by rescue workers operating in a disaster scenario. When multiple robots move autonomously in an unknown environment, it is important to determine their locations. There have been many studies carried out on the problem of simultaneous localization and mapping [2]. There have been studies on human robot interaction in a rescue scenario [3]. The Pioneer 1 mobile robot data available in the machine leaning repository of University of California, Irvine is the result of experiments conducted to simulate robot experiences [4]. Tim Oates et al. have presented an unsupervised method for learning models of environmental dynamics based on clustering multivariate time series [5].

Dynamic time warping (DTW) is an algorithm for measuring similarity between two sequences which may vary in time or speed. It has been widely applied in the field of speech recognition. DTW has been also used in finding predefined patterns in a continuous time series stream data [6]. Application of Dynamic Time Warping method to calculate the similarity between time series data in the context of robotic environment is explained in detail in [7]. Methods combining DTW and uniform scaling (US), a technique that allows global scaling of time series, have been useful in certain kind problems where it is important to account for the natural variability of human actions [8].

This work focuses on the design of a robotic scenario to analyze the effectiveness of clustering techniques to indentify similar scenarios.

3   Simulation of Robotic Scenarios

There are many challenges in analyzing robotic data. The key challenge in search and rescue robotics is to recognize the presence of humans trapped in debris. In a multi robot scenario, it is equally important to correlate the scenarios seen by different robots. Since these robots move autonomously, it possible that robots at different locations will be inspecting the same scenario from different angles. In order to understand the scenarios which are similar, clustering techniques can be employed. In this paper, a set of simple scenarios are simulated in an indoor environment to collect time series data and then cluster these scenarios based on their similarity. Each of these scenarios contains objects that are placed at pre determined locations. A robot experiences the presence of these objects when it moves from one side of the environment to the other. The design of scenarios and objects used in the experiment has been carefully done to gain insights in to the data mining techniques used to cluster the scenarios.

3.1   Design of Objects

Objects used in this experiment are stationary objects and all of them have same size. The width of the object is designed in such a way that it is more than the distance between the light and ultrasonic sensors mounted on the robot. The height of the object is 21.5 centimeters and width of the object is 13.5 centimeters.

3.2   Design of Scenarios

Scenarios are designed in such a way that each scenario differed from another on three aspects: distance between objects, distance of objects from the robot, and color of objects. Some of the scenarios are designed to be similar to others whereas others are completely different. There are five objects in each of these scenarios and the scenarios differ by the location where these objects are placed. There are 15 different scenarios. These scenarios are numbered sequentially from S1 to S15. Fig. 1 shows scenario S2. Small colored circles seen in the figure indicate the location and color of the objects placed in the environment. Ladder like element shows the path taken by the robot. A summary of various aspects of the scenario design is given in Table 1.

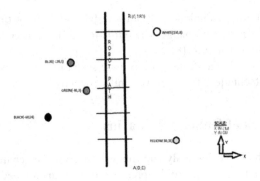

Fig. 1 Scenario S2

Table 1 Characteristics of Scenarios

Number of Scenarios	15	
Width of robot path	12 cm	
Length of robot path	150cm or 180 cm	
Number of objects	3 or 5	
Color of objects	Black, Blue, White, Green, Yellow	
Similar Scenarios	i)	S1,S3,S8,S12
	ii)	S7,S11,S13,S15
	iii)	S10, S14
Dissimilar Scenarios	S2,S4,S5,S6,S9	

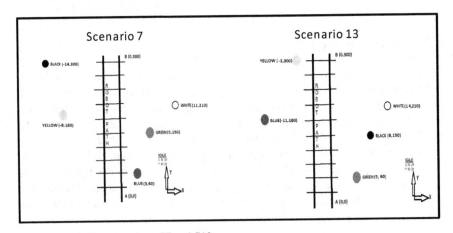

Fig. 2 Two similar scenarios - S7 and S13

Some of the scenarios have a similar pattern of placement of objects with respect to the path taken by robot. For example, scenario S1 has first object on left side of the robot path, second object on right side, third object again on right side, fourth on left side and fifth on right side. We call this an LRRLR pattern. Other scenarios which have an LRRLR pattern of object placement are S3, S8 and S12. Similarly S7, S11, S13, and S15 have similar patterns. Fig. 2 shows two scenarios S7 and S13 which are similar. Scenarios S2, S4, S5, S6 and S9 are completely different from each other.

3.3   Robot Configuration

For the purpose of data collection, LEGO Mindstorms NXT robot kit has been used. The path to be taken by the robot is programmed using NXT 2.0 programming software. At the end of each trial where the robot travels in a designated path, data collected is transferred to a desktop computer for future processing.

The experimental set up used has two ultrasonic and two light sensors on either side of the robot. Fig. 3 shows a picture of the robot with sensors mounted. Ultrasonic sensors have been used to measure the distance to the objects in the environment. The maximum distance they can measure is 233 cm with a precision of 3 centimeters. Light sensors can distinguish between light and dark environments on a scale of 0 to 50, with 50 being very bright and 0 being dark. They can also determine the light intensity of different colors.

Fig. 3 Robot with sensors mounted

3.4   Data Collection Procedure

The environment used for data collection is static. Robot is made to run in a straight line path at different speeds of 50 power and 100 power. Robot is run for 24 seconds or 14 seconds or 12 seconds or 7 seconds depending on the scenario selected and speed of the robot. Sensor readings are recorded every 100 milliseconds. Each robotic run collected 70 to 240 snapshots of the environment depending on the length of the trial run. Each snapshot contains time stamp of that snapshot in milliseconds and four sensor readings capturing light intensity values ranging from 0 to 50 and sonar values showing object distance in centimeters. Data collected from a trail run is stored as time series data. Fig. 4 shows first seven lines of one such data file.

As mentioned earlier, 15 different scenarios or environments have been used to collect data. Robot explored each of these scenarios 10 times at speed 100 and another 10 times at speed 50 thus collecting a total of 300 time series data. As the objects in the environment are stationary, time series data collected are expected to be similar at least among the trial runs with same speed. However, there are variations due to the deviation in the path taken by the robot as well as due to variation in the sensor values returned.

Time	Light1	Light2	Sonar1	Sonar2
2501	26	22	113	220
2600	28	23	113	219
2701	27	22	113	220
2803	26	22	22	220
2904	26	21	12	220
3001	30	23	9	220

Fig. 4 Sample time series data

4   Clustering of Scenarios

Time series data collected from different trials have been subjected to a set of data mining tasks to form clusters. The sequence of data mining tasks performed is shown in Fig. 5. These tasks are explained in the next few sections.

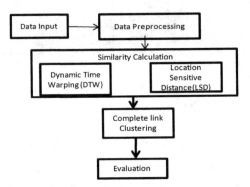

Fig. 5 Sequence of data mining tasks

4.1 Data Preprocessing

The first step in the overall clustering process is data preprocessing. It is observed that there are minor variations in values returned by sensors when no object is present in the scenario. In all 15 scenarios, all objects have been placed within 50cms from the robot path. In order to eliminate any clustering errors due to noise, sensor readings are truncated to a constant value of 50 when no objects are detected.

4.2 Distance Measures

Distance between any two points in time series data can be measured using either Euclidean or Manhattan distance. Manhattan distance has been used in this study. The time series data collected have different lengths due to two reasons - different robotic runs are carried out at two different speeds and the duration of trails are varied form 7 seconds to 24 seconds. Hence a direct measure for calculating the distance between time series data is not feasible. This problem is addressed by using a technique like Dynamic Time Warping (DTW) or by compressing the longer time series by a constant factor.

Dynamic Time Warping (DTW) Distance Measure
DTW warps time series data nonlinearly to match each other. Given two time series T1 and T2, DTW finds the warping of the time dimension in T1 that minimizes the difference between the two. DTW has been successfully used in the past to find the similarity between certain kinds of time series data [5-7].

DTW method is capable of finding scenarios which have similar pattern of object placement where as it is not capable of detecting dissimilarity between scenarios based on the distance between objects. For example, for scenarios S1 and S3 which have similar object placement pattern DTW method is not appropriate to identify them as two separate scenarios. One way to differentiate these scenarios is to use the current position of the robot to obtain position of the object.

Location Sensitive Distance (LSD) Measure
LSD measure makes an assumption about the location of the robot to calculate the distance between time series data. Location of the robot in terms of the distance travelled by it (Y co-ordinate) and the distance at which the object detected (X co-ordinate) is calculated with an assumption that the robot travels at constant speed without deviating from the path specified. Once the position of the robot for a particular snapshot of sensor readings is evaluated, dissimilarity between two time series data can be computed as the cumulative sum of the Manhattan distance between the snapshots taken from same locations.

4.3 Clustering Techniques

Two widely used clustering techniques are partitioning based clustering and hierarchical clustering [9]. Hierarchical clustering used in this study employs agglomerative clustering based on complete linkage method to calculate inter cluster distance.

5 Experimental Results and Analysis

Data mining technique explained in Section 4 have been used to cluster the time series data of 300 trials in 15 different scenarios. Some of these scenarios have similar object placement pattern. Time series data collected have different lengths based on the speed of the robot and length of the path as explained in Section 3. Cluster analysis is carried using Recall (R) and Precision (P). Here recall is calculated as the percentage of correctly obtained trial to the expected number of trails. Precision is calculated as the percentage of correctly obtained trials to the numbers of trials obtained in the cluster. The experimental results and analysis based on this data are explained in the following sub sections.

5.1 DTW Distance Based Complete Linkage Clustering

The result of hierarchical clustering with complete linkage criteria based on DTW distance measure to form 15 clusters is tabulated in Table 2. While 6 of the 15

Table 2 Results of DTW distance based complete linkage clustering for 15 clusters

Cluster No.	Expected no. of trials	No. of trials obtained	Correctly obtained trials	Recall (%)	Precision (%)
1	20	14	11	78.57	55.00
2	20	18	18	100.0	90.00
3	20	16	16	100.0	80.00
4	20	20	20	100.0	100.0
5	20	20	20	100.0	100.0
6	20	20	20	100.0	100.0
7	20	20	20	100.0	100.0
8	20	40	20	50.00	100.0
9	20	17	17	100.0	85.00
10	20	32	20	66.67	100.0
11	20	40	20	50.00	100.0
12	20	9	0	00.00	00.00
13	20	20	10	50.00	50.00
14	20	3	0	00.00	00.00
15	20	10	8	80.00	40.00
Total	**300**	**300**	**220**	**73.33**	**73.33**

clusters have 100% precision, 2 of them i.e., cluster numbers 12 and 14 have zero precision. The clustering algorithm have grouped trials belonging to scenario S12 along with trials belonging to scenario S8 in cluster 8. Similarly, 12 trials belonging to scenario 14 have been grouped together with trials belonging to cluster 10. This has happened as DTW distance measure considered S8 and S12 as similar due their similarity in object placement pattern. The 3 trials which have been grouped together to form cluster 14 should have been part of cluster 9. The relatively low recall and precision value of 73.3 is due to the fact DTW method looks for similarity in the pattern of object placement and not the match for exact object positions.

Since there are 8 distinct patterns of scenarios as shown in Table 1, experiment has been repeated to form 8 clusters. Results of this experiment are tabulated in Table. 3. It is observed that the precision and recall improved to 86%. This is to be expected DTW is able to accurately differentiate object patterns that are dissimilar. However, due to noise data, trails belonging scenario S7 have formed cluster 4 separately though they have similar patterns to scenarios clustered in cluster 8.

5.2   Location Sensitive Distance Based Complete Linkage Clustering

As discussed in Section 4.2, clustering based on Location Sensitive Distance measure is expected to give better results if the robot moves at reasonably constant speed. This has been validated in the clustering experiment whose results are tabulated in Table. 4. The recall and precision have improved to 98%. The 2% deviation is due to the inherent variations in sensor readings recorded.

Table 3 Results of DTW distance based complete linkage clustering for 8 clusters

Cluster No.	Expected no. of trials	No. of trials obtained	Correctly obtained trials	Recall (%)	Precision (%)
1	20	18	18	90.00	100.0
2	20	42	20	100.0	47.62
3	20	20	20	100.0	100.0
4	20	20	0	00.00	00.00
5	20	20	20	100.0	100.0
6	40	40	40	100.0	100.0
7	80	80	80	100.0	100.0
8	80	60	60	75.00	100.0
Total	**300**	**300**	**258**	**86.00**	**86.00**

The experiments conducted confirm the ability of on Dynamic Time Warping to find similarity in patterns. However, if robotic scenarios are to be clustered together based on exact object locations rather object placement patterns, a distance measure which considers robot location is important.

Table 4 Results of LSD based complete linkage clustering for 15 clusters

Cluster No.	Expected no. of trials	No. of trials obtained	Correctly obtained trials	Recall (%)	Precision (%)
1	20	22	20	100.0	90.91
2	20	20	20	100.0	100.0
3	20	18	18	90.00	100.0
4	20	20	20	100.0	100.0
5	20	20	20	100.0	100.0
6	20	20	20	100.0	100.0
7	20	22	20	100.0	90.91
8	20	20	20	100.0	100.0
9	20	20	20	100.0	100.0
10	20	22	20	100.0	90.91
11	20	20	20	100.0	100.0
12	20	20	20	100.0	100.0
13	20	18	18	90.00	100.0
14	20	18	18	90.00	100.0
15	20	20	20	100.0	100.0
Total	**300**	**300**	**294**	**98.00**	**98.00**

6   Conclusion

Search and rescue robots deployed in a disaster scenario have to explore disaster scenarios and gather information. Preliminary experiments have been carried out to simulate a set of simple robotic environments. A robot attached with four sensors is used to collect information about the environments while the robot moves in a straight line path. Effectiveness of data mining techniques, particularly, clustering has been investigated on collected time series data. When 15 clusters are formed based on Dynamic Time warping distance, precision and recall figures obtained are about 73%. However when scenarios which have similar object patterns have been considered together to form eight clusters, precision and recall improved to 86%. These figures have been significantly improved to 98% by using location sensitive distance measure.

This investigation can be extended to study data mining techniques to explore robotic environment from multiple paths.

References

1. Davids, A.: Urban search and rescue robots: from tragedy to technology. IEEE Intell. Syst. 17, 81–83 (2002)
2. Ko, A., Lau, H.Y.K.: Robot Assisted Emergency Search and Rescue System With a Wireless Sensor Network. International Journal of Advanced Science and Technology 3, 69–78 (2009)

3. Murphy, R.R.: Human-robot interaction in rescue robotics. IEEE Trans. Syst. Man Cybern. C, Appl. Rev. 34(2), 138–153 (2004)
4. Schmill, M.D., Cohen, P.R.: UCI Machine Learning Repository. University of California, School of Information and Computer Science, Irvine (1999), http://archive.ics.uci.edu/ml
5. Oates, T., Schmill, M.D., Cohen, P.R.: A Method for Clustering the Experiences of a Mobile Robot that accords with human Judgments. In: Proceedings of the seventeenth National Conference on Artificial Intelligence, pp. 846–851 (2000)
6. Li, G., Wuhan, Y.W., Li, C.M., Wu, Z.: Similarity Match in Time Series Streams under Dynamic Time Warping Distance. In: nternational Conference on Computer Science and Software Engineering, csse, vol. 4, pp. 399–422 (2008)
7. Radhakrishnan, G., Gupta, D., Abhishek, R., Ajith, A., Sudarshan, T.S.B.: Analysis of multimodal time series data of robotic environment. In: 2012 12th International Conf. on Intelligent Systems Design and Applications (ISDA), November 27-29, pp. 734–739 (2012)
8. Fu, A., Keogh, E., Lau, L., Ratanamahatana, C., Wong, R.: Scaling and time warping in time series querying. The VLDB J. Int. J. Very Large Data Bases 17(4), 921 (2008)
9. Han, J., Kamber, M.: Data Mining: Concepts and Techniques. Morgan Kaufmann Publishers, San Francisco (2001)

Human Face Recognition Biometric Techniques: Analysis and Review

Ravi Subban and Dattatreya P. Mankame

Abstract. Biometrics is the study of methods for identifying a person based on one or more intrinsic physical or behavioral traits. After decades of research activities, biometrics has advanced to a great extent both in practical technology and theoretical discovery to meet the increasing need of biometric deployments. This paper depicts a detailed coverage on the research problems with their solutions in face recognition biometrics. The contributions also present the pioneering efforts and state-of-the-art results, with special concentration on practical issues concerning system development.

Keywords: Biometric, Face Recognition, Principal component analysis (PCA), Linear Discriminant Analysis (LDA), 3D Face recognition, Convolutional Neural Network.

1   Introduction to Face Recognition Biometrics

A facial recognition system is a computer application for automatically identifying or verifying a person from a digital image or a video frame from a video source. Human face detection plays an important role in applications like video surveillance, border control, to identify card counters or other undesirables in casinos, shoplifters in stores, criminals and terrorists in urban areas, eliminating voter fraud, computer security, ATM and like. Face recognition has distinct advantages due to its non-contact process i.e. Face images can be captured from a distance without touching the person being identified. The face biometric system involves image capturing, extracting features, comparing templates and matching phases.

Ravi Subban
Department of Computer Science, Pondicherry University, Puducherry, India
e-mail: sravicite@gmail.com

Dattatreya P. Mankame
Department of Information Science and Engg., K.L.E. Institute of Technology, Hubli, India
e-mail: dpmankame@gmail.com

S.M. Thampi et al. (eds.), *Recent Advances in Intelligent Informatics*,
Advances in Intelligent Systems and Computing 235,
DOI: 10.1007/978-3-319-01778-5_47, © Springer International Publishing Switzerland 2014

The recognition algorithms can be divided into two main approaches; First method, geometric uses the location, shape and spatial relationships between facial attributes like eyes, eyebrows, nose, lips, chin (as shown in Fig.1). Second approach photometric treats the face image as a whole in terms of weighted combination of a number of canonical faces [1].

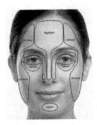

Fig. 1. Face Diagram

The outline of the paper would be as follows. Section II reviews related previous research works carried out in the areas of face recognition. Section III depicts discussion on the state-of-the-art of existing research work and their comparative analysis. Section IV reveals conclusion and future work envisaged.

2   Related Research Work on Face Recognition Biometrics

Many face recognition biometric methods developed fail in handling situation like uncontrolled background, subject's non-cooperation like-subject not looking at the camera, moving target, uncontrolled environmental conditions like lighting (shadow, glare), camera angle, image resolution, misalignment of image, occlusion, liveliness check, finding a face in a picture with variable position, the orientation, the background and the size of a face, face detection when faces are partially covered, like face with beards, glasses, long hair or hats, where most of the relevant information is hidden. The related work for face recognition are surveyed and compared with different parameters such as grayscale/multispectral images, images/video frames, using single/multiple camera, stationary/motion objects, and objects at varying distances, security and the like.

The related works surveyed are classified into categories like Face recognition in unconstrained condition, 2D Face recognition techniques and 3D Face recognition.

2.1   Related Work on Face Recognition in Unconstrained Condition

Chun-Wei Tan et al. [2] devise an approach to identify a person from the face images captured from a distance. The test results on UBIRIS.v2, FRGC and CASIA.v4 show error rate of 32.6% to 47.5%.

Chun-Nian Fan et al. [3] exemplify a scheme to recognize face in varying light intensity. The scheme use homomorphic filtering-based illumination normalization method. The test results on the Yale B and the CMU PIE face database show recognition rate of 99%.

Wonjun Hwang et al. [4] reveal face recognition system under uncontrolled variations in the illumination. The system uses a hybrid Fourier-based facial feature extraction and a score fusion scheme. The test results reports 81.49% verification rate on the FRGC v2.0 database with 2-D face images.

Amirhosein Nabatchian et al. [5] present a face recognition technique with variation in light intensity based on local matching. The experimental result on Yale B and CMU-PIE databases report recognition rate of 99.82% and 99.81% respectively.

Xiao Yang Tan et al. [6] present face recognition method under varying light conditions, which utilizes illumination normalization, local texture-based face representations, distance transform based matching, kernel-based feature extraction and multiple feature fusion. The experimental result with Extended Yale-B, CAS-PEAL-R1 and FRGC ver2 experiment 4(FRGC-204) report face verification rate of 88.1%

Carlos D.Castillo et al. [7] introduce a method to identify face in variations of face using stereo matching. The proposed method use epipolar geometry and work on 2D face images. The outcomes on CMU PIE database report accuracy of 86.8% with 34 faces and 82.4% with 68 faces.

Hyung-Soo Lee et al. [8] devise a method to recognize face in highly varying poses, expressions, and light intensity. Using the tensor-based AAM assessment results show that the tensor-based AAM with continuous variation estimation outperforms that with discrete variation estimation.

Daniel González-Jiménez et al. [9] illustrate face recognition for pose variation based on point distribution model and facial symmetry. The result on CMU PIE database RR of 97.06% is reported.

Richa Singh et al. [10] introduce mosaicing technique for face recognition in varying pose. The experimental outcome on CMU PIE, WVU visible-light and WVU SWIR face databases report 97.34%, 96.81% and 97.22% accuracy respectively.

Xiujuan Chai et al. [11] develop solution for overcoming the problem of degradation of facial image due to variation of the facial expression due to view point. The proposed method uses locally linear regression (LLR) method, which converts the non-frontal facial image to frontal facial image. The experimental result on CMU PIE database outperforms over eigen light field method.

2.2   Related Work on Face Recognition Techniques

John Wright et al. [12] comprehend an approach to identify face using sparse representation. The test result on Yale B database report recognition rate (RR) on classifiers NN=90.7%, NS=94.1%, SRC=98.1%, SVM=97.7% and on AR database 94.7 respectively. The RR of 98.3% on partial face features, 100% for random pixel corruption, and 100% for occlusion is reported on SRC classifiers.

Yanwei Pang et al. [13] devise face recognition scheme by utilizing Gabor-based region covariance matrices. Recognition accuracy of 84.24% on AR database and 85.90% on FERET database respectively reported on RCM methods.

Xudong Xie et al. [14] present elastic local reconstruction based on a single face image for face recognition. The test result with different distance RR=96%, varying lighting condition RR=100%, different facial expressions RR=99.2%, occlusion RR=94.4% is reported on various databases.

Haitao Zhao et al. [15] use a concept linear discriminant analysis (LDA) for recognising face. The experimental result shows better result of recognition rate.

Dao-Qing Dai et al. [16] explain Regularized Discriminant Analysis(RDA) for face recognition. Using ORL database RR of 98.55% and with FERET database RR of 93.47% is displayed.

Juwei Lu et al. [17] propose Kernel Direct Discriminant analysis for face recognition. The test result on UMIST database provide better error rate (approximately 34%) than other FR methods.

Juwei Lu et al. [18] present LDA-Based concepts for face recognition. The average error rate of 80.03% on ORL and 79.6% on UMIST database individually and when both combined average error rate of 79.82% is reported.

Aleix M. Martõ Ânez et al. [19] illustrate comparative analysis of PCA (Principal Components Analysis) and LDA (Linear Discriminant Analysis). The test results show that PCA performs better whenever data set is small.

Xiaofei He et al. [20] illustrate a face recognition concept using Laplacian face approach. The test result on CMU, YALE and MSRA database report recognition error rate of 8.17%, which is very less compared to eigenfaces and fisherfaces.

Chengjun Liu [21] analyses face recognition using Gabor-based kernel Principal Component Analysis (PCA) method. The method utilizes Gabor wavelet representation of face images and the kernel PCA method for face recognition. The assessment outcome on FERET and CMU PIE database outcomes other methods.

Jian Yang et al. [22] design a method for face image representation and recognition using two-dimensional principal component analysis (2DPCA). The test result on ORL, AR and Yale face databases report recognition accuracy of 98.3%, 89.8%, 96.1%.

Chengjun Liu et al. [23] explain Independent Component Analysis of Gabor Features for face recognition. The experimental result show excellent performance of 100% accuracy on ORL database.

W.ZHAO et al. [24] review face recognition related work concentrating on various issues like facial change in expression, varying light intensity. Also, listed face recognition techniques and respective work.

Keun-Chang Kwak et al. [25] illustrate Fuzzy Integral and Wavelet Decomposition Method for face recognition, which includes wavelet decomposition, fisher face and fuzzy integral methods. The RR of 100% is reported on Chungbuk National University(CNU) and Yale University face databases.

Ralph Gross et al. [26] presents Appearance-Based Face Recognition concept, which uses pixel intensity. The test result on PIE, FERET database report 66.3%, 75% recognition accuracy respectively.

Marian Stewart Bartlett et al. [27] illustrate face recognition using Independent Component Analysis (ICA). The experimental results on FERET face database under two architecture report 100% reliability and 99.8% classification performance.

Peter Mc Guire et al. [28] develop a method for face recognition using eigen-paxels as set of basic function to represent images. The test result on databases report the excellent performance.

Stan Z. Li and Juwei Lu [29] propose Nearest Feature Line (NFL) Method for face recognition (FR). The experimental data on ORL database show NFL error rate is 43.7% to 65.4% as that of eigenface method.

Laurenz Wiskott et al. [30] introduce elastic bunch graph matching technique for FR. In this method facial features are represented as wavelet components. Comparison of image graph is performed to recognize face. The method is tested on FERET, Bochum database with 11 degree rotated face and variation in expression report 92% and 94% accuracy.

2.3   Related Work on 3D Face Recognition

Di Huang et al. [31] propose a technique for 3D face recognition. The technique uses geometric recognition approach and local hybrid matching method. The technique overcomes the problems of partial occlusions, variation in facial expressions, changes in the pose. But it is limited to frontal face images. The experimental results on FRGC v2.0, Gavab and Bosphorus database report recognition rate of 97.6% and verification rate of 98.4% with 0.001 FAR.

Tolga İnan et al. [32] designs a scheme to identify a human using 3D face image. The scheme uses local shape descriptors. The experimental results on FRGC v2.0 database exhibit 98.35% detection rate with 0.001 FAR.

Georgios Passalis et al. [33] devise a method to overcome the problem of change in pose in 3D face recognition. The anticipated method uses facial symmetry to handle pose variations. The evaluation result on University of Notre Dame and the University of Houston databases show the recognition rate of 83.7%.

Ajmal S. Mian et al. [34] design automatic face recognition system. The proposed system works on both 2D and 3D face images as well as on both feature based and holistic matching schemes. The trial result on FRGC report accuracy of 99.74% and verification rate of 98.31%.

Le Zou et al. [35] introduce 3D face recognition scheme. The proposed method identifies face from range of images, with varying expressions and hair. The test result show the proposed method outperforms the other existing methods.

Kyong I. Chang et al. [36] present a solution for 3D face recognition under changing facial expression using multiple nose region matching. The experimental results report matching accuracy of 96.6%.

Volker Blanz et al. [37] present 3D Morphable model to recognize face. The success rate of 95.0% and 95.9% is reported on CMU-PIE and FERET database respectively. Chongzhen Zhang et al. [38] enlighten 3D face recognition from multiple face images using 3D morphing. The test results report recognition rate of 97%.

3   Discussion

The pursuits of knowledge on the diverse work on face biometric systems envisage that the facial recognition is not perfect and struggles to perform under certain conditions. i.e. poor light conditions, varying face expressions, sunglasses, hair, partially covered face, big smile, low resolution image and like. Facial recognition is not restricted just to identify an individual, but also to unearth other personal data associated with an individual - such as other photos featuring the

Table 1 Performance Analysis of Face Recognition Biometric Methods in Unconstrained Conditions Based on Recognition Rate (RR)

Authors	Methods	Database Used For Testing	RR (%)
Chun-Nian Fan [3]	Homomorphic filtering based illumination normalization	Yale-B CMU PIE	99%
Wonjun Hwang [4]	Hybrid fourier based facial feature extraction	FRGC V2	81.49%
Richa Singh [10]	Mosaicing Technique	CMU PIE	97.34%
		WVU-VL	96.81%
		WVU-SWIR	97.22%

Table 2 Performance Analysis of Face Recognition Biometric Methods Based on RR

Methods	RR (%)	Methods	RR (%)	Methods	RR (%)
PCA	96.0	SURF-M	80.5	LFAb	93.0
LDAg	97.0	SIFT-L&R-M	87.8	ICA	94.0
LDAt	98.0	KFLD	96.3	PCA	95.0
LFAe	93.0	LBP	97.3	KPCA	82.7
Wavelet	93.5	NFS	95.18	WLD-EU-10	98.1
GJD-BC	91.2	MS-eLBP-DFS	98.4	LBP with NUC	93.3
--	--	UNI-RF-CF5D	97.5	SIFT-L&R-M	87.8

Table 3 Performance Analysis of 3D Face Recognition Biometric Methods Based on RR

Authors	Methods	Database Used For Testing	RR (%)
Di Huang [31]	Geometric recognition + Local hybrid matching	FRGC V2	97.6%
		Gavab	98.4%
		Bosphorus	98.4%
Tolga Inan [32]	Local Shape Discirptor	FRGC V2	98.35%
Kyong I. Chang [36]	multiple nose region matching	---	96.6%
Chongzhen Zhang [38]	3D Morphing	---	97%

individual, blog posts, social networking profiles, Internet behavior, travel patterns. Facial recognition offers several advantages like non-intrusive contact free process, uses legacy databases and integrates with existing surveillance systems. Table 1-3 depicts the performance analysis of various related work in unconstrained condition, Recognition Techniques and 3D recognition methods respectively on various databases.

4   Conclusion and Future Work

This paper presented the survey on the face recognition techniques used in the literature and the performance evaluation with different parameters on existing methods. Various issues related to face recognition technique are discussed. The enumerated review plays a pivotal role for establishing identity for the industry like law enforcement, forensic science community and with general public. This leads to the invention of next generation faster and higher quality acquisition devices designed using sophisticated algorithms. Summarizing it can be ensured that better accuracy, security and performance can be achieved by adopting face recognition biometric systems.

Acknowledgements. This work is supported and funded by the University Grant Commission (UGC), India under Major Research Project to the department of Computer Science of Pondicherry University, Puducherry, India.

References

1. Bhattacharyya, D., Ranjan, R., Farkhod Alisherov, A., Choi, M.: Biometric Authentication: A Review. International Journal of u- and e- Service, Science and Technology 2(3), 13–28 (2009)
2. Tan, C.-W., Kumar, A.: Unified Framework for Automated Iris Segmentation Using Distantly Acquired Face Images. IEEE Transactions On Image Processing 21(9), 4068–4079 (2012)

3. Fan, C.-N., Zhang, F.-Y.: Homomorphic filtering based illumination normalization method for face recognition. Pattern Recognition Letters 32, 1468–1479 (2011)
4. Hwang, W., Wang, H., Kim, H., Kee, S.-C., Kim, J.: Face Recognition System Using Multiple Face Model of Hybrid Fourier Feature Under Uncontrolled Illumination Variation. IEEE Transactions on Image Processing 20(4), 1152–1165 (2011)
5. Nabatchian, A., Abdel-Raheem, E., Ahmadi, M.: Pattern Recognition 44, 2576–2587 (2011)
6. Tan, X.Y., Triggs, B.: Enhanced Local Texture Feature Sets for Face Recognition Under Difficult Lighting Conditions. IEEE Transactions on Image Processing 19(6), 1635–1650 (2010)
7. Castillo, C.D., Jacobs, D.W.: Using Stereo Matching with General Epipolar Geometry for 2D Face Recognition across Pose. IEEE Transactions on Pattern Analysis and Machine Intelligence 31(12), 2298–2304 (2009)
8. Lee, H.-S., Kim, D.: Tensor-Based AAM with Continuous Variation Estimation: Application to Variation-Robust Face Recognition. IEEE Transactions on Pattern Analysis and Machine Intelligence 31(6), 1102–1116 (2009)
9. González-Jiménez, D., Alba-Castro, J.L.: Toward Pose-Invariant 2-D Face Recognition Through Point Distribution Models and Facial Symmetry. IEEE Transactions on Information Forensics and Security 2(3), 413–429 (2007)
10. Singh, R., Vatsa, M., Ross, A., Noore, A.: A Mosaicing Scheme for Pose-Invariant Face Recognition. IEEE Transactions on Systems, Man and Cybernetics —Part B: Cybernetics 37(5), 1212–1225 (2007)
11. Chai, X., Shan, S., Chen, X., Gao, W.: Locally Linear Regression for Pose-Invariant Face Recognition. IEEE Transactions on Image Processing 16(7), 1716–1725 (2007)
12. Wright, J., Yang, A.Y., Arvind Ganesh, S., Sastry, S., Ma, Y.: Robust Face Recognition via Sparse Representation. IEEE Transactions on Pattern Analysis and Machine Intelligence 31(2), 210–227 (2009)
13. Pang, Y., Yuan, Y., Li, X.: Gabor-Based Region Covariance Matrices for Face Recognition. IEEE Transactions on Circuits and Systems For Video Technology 18(7), 989–993 (2008)
14. Xie, X., Lam, K.-M.: Face recognition using elastic local reconstruction based on a single face image. Pattern Recognition 41, 406–417 (2008)
15. Zhao, H., Yuen, P.C.: Incremental Linear Discriminant Analysis for Face Recognition. IEEE Transactions on Systems, Man and Cybernetics — Part B: Cybernetics 38(1), 210–221 (2008)
16. Dai, D.-Q., Yuen, P.C.: Face Recognition by Regularized Discriminant Analysis. IEEE Transactions on Systems, Man and Cybernetics — Part B: Cybernetics 37(4), 1080–1085 (2007)
17. Lu, J., Plataniotis, K.N., Venetsanopoulos, A.N.: Face Recognition Using Kernel Direct Discriminant Analysis Algorithms. IEEE Transactions on Neural Networks (August 2002) (accepted)
18. Lu, J., Plataniotis, K.N., Venetsanopoulos, A.N.: Face Recognition Using LDA Based Algorithms. IEEE Transactions on Neural Networks (May 2002) (accepted)
19. Ânez, A.M.M., Kak, A.C.: PCA versus LDA. IEEE Transactions on Pattern Analysis and Machine Intelligence 23(2), 228–233 (2001)
20. He, X., Yan, S., Hu, Y., Niyogi, P., Zhang, H.-J.: Face Recognition Using Laplacian faces. IEEE Transactions on Pattern Analysis and Machine Intelligence 27(3), 328–340 (2005)

21. Liu, C., Wechsler, H.: Evolutionary Pursuit and Its Application to Face Recognition. IEEE Transactions on Pattern Analysis and Machine Intelligence 22(6), 570–582 (2000)

22. Yang, J., Zhang, D., Frangi, A.F., Yang, J.-Y.: Two-Dimensional PCA: A New Approach to Appearance - Based Face Representation and Recognition. IEEE Transactions on Pattern Analysis and Machine Intelligence 26(1), 131–137 (2004)

23. Liu, C., Wechsler, H.: A Shape-and Texture-Based Enhanced Fisher Classifier for Face Recognition. IEEE Transactions on Image Processing 10(4), 598–608 (2001)

24. Zhao, W., Chellappa, R., Phillips, P.J., Rosenfeld, A.: Face Recognition: A Literature Survey. ACM Computing Surveys 35(4), 399–458 (2003)

25. Kwak, K.-C., Pedrycz, W.: Face Recognition Using Fuzzy Integral and Wavelet Decomposition Method. IEEE Transactions on Systems, Man, and Cybernetics — Part B: Cybernetics 34(4), 1666–1675 (2004)

26. Gross, R., Matthews, I., Baker, S.: Appearance-Based Face Recognition and Light-Fields. IEEE Transactions on Pattern Analysis and Machine Intelligence 26(4), 449–465 (2004)

27. Bartlett, M.S., Movellan, J.R., Sejnowski, T.J.: Face Recognition by Independent Component Analysis. IEEE Transactions on Neural Networks 13(6), 1450–1464 (2002)

28. McGuire, P., D'Eleuterio, G.M.T.: Eigen paxels and a Neural-Network Approach to Image Classification. IEEE Transactions on Neural Networks 12(3), 625–635 (2001)

29. Li, S.Z., Lu, J.: Face Recognition Using the Nearest Feature Line Method. IEEE Transactions on Neural Networks 10(2), 439–443 (1999)

30. Wiskott, L., Fellous, J.-M., Krüger, N., von der Malsburg, C.: Face Recognition by Elastic Bunch Graph Matching. IEEE Transactions on Pattern Analysis and Machine Intelligence 19(7), 775–779 (1997)

31. Huang, D., Ardabilian, M., Wang, Y., Chen, L.: 3D Face Recognition Using eLBP-Based Facial Description and Local Feature Hybrid Matching. IEEE Transactions on Information Forensics and Security 7(5), 1551–1565 (2012)

32. İnan, T., Halici, U.: 3-D Face Recognition With Local Shape Descriptors. IEEE Transactions on Information Forensics and Security 7(2), 577–587 (2012)

33. Passalis, G., Perakis, P., Theoharis, T., Kakadiaris, I.A.: Using Facial Symmetry to Handle Pose Variations in Real-World 3D Face Recognition. IEEE Transactions on Pattern Analysis and Machine Intelligence 33(10), 1938–1951 (2011)

34. Mian, A.S., Bennamoun, M., Owens, R.: An Efficient Multimodal 2D-3D Hybrid Approach to Automatic Face Recognition. IEEE Transactions on Pattern Analysis and Machine Intelligence 29(11), 1927–1943 (2007)

35. Zou, L., Cheng, S., Xiong, Z., Lu, M., Castleman, K.R.: 3-D Face Recognition Based on Warped Example Faces. IEEE Transactions on Information Forensics and Security 2(3), 513–528 (2007)

36. Chang, K., Bowyer, K.W., Sarkar, S., Victor, B.: Comparison and Combination of Ear and Face Images in Appearance-Based Biometrics. IEEE Transactions on Pattern Analysis and Machine Intelligence 25(9), 1160–1163 (2003)

37. Blanz, V., Vetter, T.: Face Recognition Based on Fitting a 3D Morphable Model. IEEE Transactions on Pattern Analysis and Machine Intelligence 25(9), 1063–1074 (2003)

38. Zhang, C., Cohen, F.S.: 3-D Face Structure Extraction and Recognition From Images Using 3-D Morphing and Distance Mapping. IEEE Transactions on Image Processing 11(11), 1249–1259 (2002)

Author Index